W9-DCY-962

WITHDRAWN

St. Louis Community College

Library

5801 Wilson Avenue
St. Louis, Missouri 63110

THE SEVENTH CONTINENT

THE SEVENTH CONTINENT

Antarctica
in a
Resource
Age

Deborah Shapley

RESOURCES FOR THE FUTURE, INC., WASHINGTON, D.C.

Published by Resources for the Future, Inc.,
1616 P Street, N.W. Washington, D.C. 20036

Resources for the Future books are distributed worldwide by
The Johns Hopkins University Press.

Library of Congress Cataloging in Publication Data

Shapley, Deborah, 1945–
   The seventh continent.

   Bibliography: p.
   Includes index.
   1. Antarctic regions—International status.
2. Marine resources conservation—Law and legislation—
Antarctic regions.   I. Title.
JX4084.A5S53   1985   341.2′9′09989   85-5581
ISBN 0-915707-17-9

This book is the product of RFF's Energy and Materials Division, Joel Darmstadter, director. Deborah Shapley, a journalist who writes on science, public affairs, and national defense topics, is a visiting scholar at Georgetown University's Center for Strategic and International Studies.

This book was edited by Jo Hinkel and designed by Elsa Williams. Maps were drawn by the James A. Smith Studio, photographs were obtained by Picture Research, and the index was prepared by The Information Bank.

Contents

List of Tables

Foreword

Resources for the Future is pleased to publish this book for at least three reasons. First, any broad-ranging study of Antarctica's history, politics, and development prospects calls for a command of issues in geography, science policy, technology, and international law, and Deborah Shapley addresses these with authority and flair.

Second, at this time, the nations of the world are struggling to fashion a legal framework to govern Antarctic resources, which some regard as the "common heritage of mankind." This debate, described vividly here, represents an ongoing application of the common-property resource concept, which has played a prominent role in RFF's research and analytical contributions during the past quarter-century.

Third, the continent's energy and minerals endowment—if exploitable at all (and in the author's judgment the prospects for this are dim)—constitute at best "resources for the *future*." RFF's interest and participation, therefore, should be self-evident. Indeed, RFF's concern with this subject dates back to its earlier support for Neal Potter's *Natural Resource Potential of the Antarctic*, published in 1969 by the American Geographical Society.

We wish to thank the Carnegie Endowment for International Peace for launching this project and sustaining it in its earlier stages and the Ford Foundation for its support in helping to bring it to a successful conclusion.

Joel Darmstadter, Director
Energy and Materials Division
Resources for the Future

Preface

The purpose of this book is to present a comprehensive picture of the role of Antarctica in world affairs today. Antarctica is the earth's seventh continent and the only one without native human inhabitants. It is the only one whose mineral resources are essentially unexamined and untapped, and the only one whose living resources—found in great abundance in the continent's offshore waters—are barely so. Today, more than three-quarters of the way through the twentieth century, as man moves manufacturing facilities into space and as his probes reach the outer planets, Antarctica remains a great untouched wilderness used almost exclusively for scientific research.

But though Antarctica is the frozen continent, human activities regarding it are hardly standing still. The last decade has seen major changes in the way Antarctica is perceived, used, and governed; the next decade will see more changes. First, there is the growing international interest in Antarctica's resource potential. Also, many more nations and organizations—developing nations, public interest groups, the environmental movement—are demanding a say in the region's administration. A debate on Antarctica is under way at the United Nations. Finally, as this is written, only six years remain until 1991, the earliest date on which a review may be called of the 1961 Antarctic Treaty, which dedicates the region to peace and science and sidesteps the more difficult issues of territorial ownership, resource jurisdiction, and equity for all nations in Antarctica. Time may be running out for the traditional way Antarctica has been governed; thus a stock-taking such as this appears timely.

Of course, there is already an Antarctic literature, most of it highly specialized. There are tales of adventure, of explorers such as Robert F. Scott, Ernest Shackleton, and Richard E. Byrd; there is an extensive scientific literature and a number of articles on international law aspects. And there are picture books. This study differs from most Antarctic writing in that it focuses on the *political,* rather than the legal, technical, or aesthetic dimensions of the

Antarctic question. It tries to state explicitly what have been in some cases unwritten political and diplomatic understandings. A major subtheme is the role of the United States in Antarctica, past, present, and future—terrain not well covered in the literature of American politics, diplomacy, and international security. So this book aims to fill gaps in the literature and serve as an authoritative guide to the problems we face today with respect to Antarctica's future.

The text is organized for both the general reader and the specialist. Chapter 1 is a brief description of the region's geography and history and outlines the issues. The chapters which follow, 2 through 8, each take up a specific aspect of the Antarctic question. Chapter 2 describes the history of U.S. ties to the region, which are little known but vital to understanding the powerful voice the United States has had in Antarctic diplomacy. It relies heavily on original documents in the Polar Section of the National Archives. Chapters 3 and 4 describe the political evolution of the region and the web of political understandings that underpin the treaty. Some of this discussion is based on hitherto unpublished notes of the meetings at which the treaty was negotiated, found in the papers of George Dufek at Syracuse University's George Arents Research Library. Chapters 5 and 6 summarize the living and minerals resource issues, respectively, and attempts by the treaty powers to address them in recent years. Chapter 7 takes up the story of the United States in Antarctica, relating our domestic science program to this evolving international political scene. It also describes policy options for the future. Chapter 8—the last—discusses the interests of Malaysia, India, and other developing countries now asking for a voice in Antarctic governance, the role of the United Nations and the choices the international community faces.

A book such as this has a history of its own. My first acquaintance with the subject came when I was a writer for *Science* magazine and wrote a story about how the Nixon administration, in the emotional aftermath of the 1973 oil embargo, was considering changing U.S. Antarctic policy—and possibly the treaty—to guarantee U.S. companies access to Antarctica's offshore hydrocarbons, then undiscovered and unlooked for. I wrote other articles: the subject was as intriguing as the tip of the proverbial iceberg. Why had Antarctica remained conflict-free all these years? Are scientists really better "citizens of the world" than are members of other professions? Or were there underlying political arrangements which received less publicity, but really made the scientists' role possible? Finally, how likely was it that national governments, in their preoccupation with control of long-term resources, would set the treaty aside to have access to Antarctica's presumed minerals wealth? How stable is the region's peace after all?

I therefore welcomed the opportunity to take a sabbatical from the perennial subject-hopping of journalism to investigate these questions further when the chance was offered by the Carnegie Endowment for International Peace. The Endowment wished to sponsor an in-depth study of the Antarctic question because of its long-standing interest in international peace and how to maintain it. I was also delighted, when the study phase was finished, to accept the sponsorship of Resources for the Future and the Ford Foundation for the manuscript's preparation and publication. RFF's expertise in strategic resource questions and in resource-versus-environment tradeoffs of the sort that may have to be made in Antarctica was particularly helpful.

My personal thanks go to Thomas L. Hughes and Larry L. Fabian, the Endowment's president and secretary, respectively, for their warm and interested support of the project's first phase. I am also deeply grateful to the president of RFF, Emery Castle, and to Milton Russell and Joel Darmstadter, who supervised the project as successive directors of RFF's Energy and Materials Division. Particular thanks go to the RFF staff, to Helen-Marie Streich of the Energy and Materials Division and to Jo Hinkel and Elsa Williams of RFF's editorial and production staff for the many things we all know they did to make the book possible. Rhonda Brown and Michelle Gottleib should also be thanked for their careful research assistance. Finally, I wish to thank my family for their devotion and patience.

Washington, D.C.
September 1985

Deborah Shapley

THE SEVENTH CONTINENT

one

The Setting

Antarctica almost was discovered by the great British navigator and explorer Capt. James Cook, whom the Admiralty had sent out twice in search of the legendary Terra Australis Incognita, and who instead found the east coast of Australia, the New Hebrides, New Caledonia, and the Cook Islands, as well as surveying New Zealand and much of the South Atlantic. Cook came very close to the Antarctic continent on his epic second voyage of 1772–75; he circumnavigated it without sighting land. He was sure it was there, but as a practical man, he dared not "risque...all that had been done during the voyage" by pressing his small ships farther south, because of "[t]hick fogs, snow storms, intense cold, and every other thing that can render navigation dangerous...these difficulties are greatly heightened, by the inexpressibly horrid aspect of the country."[1]

Cook was not the last to remark on the inhospitality of the region. More than one hundred years later, the British explorer Robert Falcon Scott, after finally attaining the South Pole, exclaimed in his diary, "Great God! this is an awful place."[2] And in 1960, when a U.S. Senate committee asked Laurence M. Gould, the dean of American Antarctic scientists, what he liked about Antarctica, Gould said: "For sheer, utter magnificent desolation, it is the most magnificent thing in the world."[3]

Cook had deduced that the southern continent of legend existed because the ice appeared at varying southern latitudes, suggesting it originated with a landmass lying asymmetrically across the pole. Had he the benefit of today's knowledge, he would have known that Antarctica—the earth's seventh continent—is the fifth largest, covering 14 million square kilometers, approximately the area of the United States and Europe combined. The land, especially on the western side, lies partly below sea level, being depressed some 600 meters by the weight of a huge ice cap, estimated to be 24 million cubic kilometers in volume and containing nearly 75 percent of the world's fresh water and 90 percent of its ice.[4]

Cook and later explorers sailed south and, as the Southern Cross rose in the night sky, they encountered many obstacles. In the latitudes of the "roaring forties" strong westerly winds propelled them swiftly across the vast southern seas. Farther south lay the "furious fifties," a belt of heavy seas and storms at 50° to 55° south latitude. Scott, steering south in 1910 in the heavily laden *Terra Nova*, crossed this belt with the ship's boiler room flooded. Some of the dogs and ponies tethered to the decks were throttled by the ship's rolling, and one pony was washed overboard by a big wave, only to be hurtled back again by the next.[5]

Scott's ponies were the victims of one of the most important features of the earth's oceans, the Antarctic Convergence (or Polar Front). The Convergence marks the northern limit of the Southern Ocean (sometimes called the Antarctic Ocean) surrounding the continent. Here the great water masses of the southern Pacific, Atlantic, and Indian oceans collide with the cold Antarctic surface water that constantly is being refreshed by ice melt. The resulting turbulence creates an oceanic barrier that, except for whales, most marine organisms never cross; it contains Antarctica's rich ecosystem in a single region having one-fifth of the world's ocean water.

South of the Convergence lay still more storms, but the pack ice became the explorers' chief hurdle. Pack ice, formed from freezing sea water, is constantly moving. Driven about by the winds, it is capable of slashing the hull of an unlucky or unwary ship. Thus, Antarctic explorers, even in the twentieth century, have preferred wooden ships to steel ones: metal buckles in the ice's terrific squeeze, whereas wood merely bends, allowing the hull to ride up on the ice and be spared—an uneasy truce between Antarctica and its would-be conquerors. The pack ice trapped the little *Belgica* in 1898, forcing the party to winter over, a stay that brought scurvy, depression, and, in some cases, madness to the crew. The Weddell Sea pack trapped Ernest Shackleton's ship *Endurance* in 1915, and finally crushed it, forcing his party ashore. Shackleton himself set off on a journey of more than 2,080 kilometers to get help.[6]

James Clark Ross shared Cook's dream of sailing to the South Pole. In 1840, south of the belt of storms and pack ice, he found another phenomenon: an open, smooth sea bounded on the south by a sheer, flat-topped wall of ice, 60 meters high, and stretching in either direction as far as the eye could see. Ross called it the Great Ice Barrier; and for two successive summers he sailed along it, searching for an opening to the south or a suitable landing place but finding neither. What lay behind? The temperate green land of legendary gold and diamonds? More icy whiteness? Another sea? The secret seemed guarded by a silent volcano at the barrier's western side, eerily smoking although blanketed in snow, which

Ross named Mount Erebus for one of his ships.[7] Another volcanic mountain lay nearby, which he named Terror for the other.

The great white wall was not Antarctica's only "barrier." As large as France, the Ross Ice Shelf, as it is now called, blocked Ross's progress southward. Another huge floating ice shelf, covering the southern part of the sea, was penetrated by James Weddell in 1823. Subsequently, the sea and the 450-kilogram seals that lived on the Antarctic ice pack were named after Weddell, while the ice shelf was named after a later explorer, Finn Ronne. A second major ice shelf in the Weddell Sea was named for the German explorer Wilhelm Filchner, who discovered its hazards in 1911–12, when he tried to put his camp on it, only to have it shear off and fall into the sea, causing the loss of precious gear and supplies.[8]

Antarctica's great ice shelves are but the seaborne part of the huge ice cap overlaying the continent. As the seawater warms its edges, the shelves become unstable and crash into the sea, forming the giant tabular icebergs that ride prevailing currents northward until they melt. As they journey, they become rounded, usually melting by the time they reach the Convergence. But some escape; in 1854 twenty-one merchant ships reported a vast iceberg between 44° and 40° south latitude in the shape of a hook almost 60-kilometers wide and 100-kilometers long. Most likely it was several icebergs locked together, probably the largest mass of floating ice ever reported.[9]

When the explorers finally moved inland from these formidable coasts, they found a land made entirely of ice, except for the nunataks (mountain peaks) protruding through its whiteness. Near the coastline the ice cap rises sharply to thicknesses of 2,000 meters; it is more than 4,770 meters at its thickest point.[10]

It is useful to think of Antarctica as a great ice dome capping the bottom of the world and having an average height three times that of other continents. Geographers, following the suggestion of E. S. Balch, have divided Antarctica in two. East Antarctica is an ancient continental shield that lies mostly in the eastern longitudes. On its western edge it is upfolded, forming the Transantarctic Mountains that bisect the continent. On the western slopes of the mountains, West Antarctica begins; it lies mostly in the western longitudes and includes the milder region of the Antarctic Peninsula. West Antarctica is an archipelago of younger rock and older continental fragments, punctured by the Weddell Sea on the Atlantic coast and by the Ross Sea on the Pacific coast. Glaciologists also consider the continental ice sheet as two pieces; the East Antarctic ice sheet is a single mass, gripping the continent below and probably highly stable. But, slung between the land masses, the West Antarctic ice sheet rests partly on land and partly on water, and

Capt. James Cook's ships harvest fresh water from icebergs for their journey of
far southern exploration.

Weddell Sea pack ice traps Ernest Shackleton's ship *Endurance* in October 1915.

Antarctic spectacle with cliff, man, and
sledge, taken by Robert F. Scott's photographer, Herbert Ponting.

may be unstable enough to melt if the earth's atmosphere warms significantly.[11]

The ice cap is continually forming from snowfall on the dome's high interior, and creeping out and downward to the sea. As the brittle ice bends over the contours of the land hidden below, it breaks into long deep cracks at the surface, perpendicular to its line of flow. These crevasses are sometimes hundreds of meters deep. Their yawning chasms are often masked by snowfall, and explorers, dogs, sleds, and vehicles must thread their way carefully to avoid being swallowed.

The expeditions of Carsten Borchgrevink (1898–1900) and Otto von Nordenskjold (1901–03) were the first to deliberately winter over on the continent. What little exposed rock they found revealed that the country, "doomed by Nature," in Cook's phrase, had once been alive! Capt. C. A. Larsen's expedition found a fossil pine log on Seymour Island in 1892. Edward A. Wilson, returning with Scott from the pole, found leaf tracings in sandstone near the Beardmore Glacier. Coal was found. But how could coal, which comes from forests, and ice coexist? The riddle remained until the 1960s, when the theory of continental drift gained acceptance. Scientists now believe that for millions of years, Antarctica lay over the pole, but was a province of the temperate supercontinent, Gondwana. Not until 115 million years ago did Africa, India, and South America begin to pull away. The present ice cap did not begin to form until after 39 million years ago when the Drake Passage opened, allowing a strong eastward current to beat continually around the continent. This new oceanic wall would seal Antarctica in its prison of cold.[12]

Not only is a journey to Antarctica a trip back in time, into an ice age that early visitors thought once covered the earth; often it seems a journey to another world. Below the Antarctic Circle even the Southern Cross fades from the sky; the sun stops rising and setting and instead crazily circles the horizon. In the austral (southern) spring or fall it sits low in the sky, casting long, dark shadows around the sculpted forms of the ice. At midsummer, in December, it stares relentlessly down, mocking human biological rhythms of day and night. Even the name Antarctica—derived from the Greek *añti,* for "opposed," and *arktos,* the Greek name for the constellation of the bear in the northern sky—implies the region is peculiar.

Its beauty is unearthly. In the thin, clear air of the interior plateau, invisible ice crystals give the sun a beautiful halo of dark red or orange shot through with other colors. When clouds mask the sun, aureoles of orange and red or red and green appear. The cold surface air plays tricks with the light, piling horizons on top of each other; it can make distant, blue-gray mountains seem nearer

and one-dimensional, like huge stage props left by some departed
giant. Fine moisture breaks white light into the colors of the spec-
trum as it floods the ice.[13]

Scott's companion, Wilson, a doctor, zoologist, and water-
colorist, wrote in 1910:

> The sunlight at midnight in the pack is perfectly wonderful. One looks
> out upon endless fields of broken ice, all violet and purple in the low
> shadows, and all gold and orange and rose-red on the broken edges
> which catch the light, while the sky is emerald green and salmon pink,
> and these two beautiful tints are reflected in the pools of absolutely still
> water which here and there lie between the ice floes. Now and again
> one hears a penguin cry out in the stillness....[14]

Equally awesome are the spectacles of the long polar winter,
when the tilt of the earth's axis puts the region south of the Antarctic Circle in darkness. The days grow shorter and shorter;
finally the sun sinks, and the southern stars whirl overhead. On
many nights the auroras wave greenish and yellow, and sometimes
build to frenzied excitement in reds and blues. In the early days it
was suspected that the aurora borealis in the north and the aurora
australis in the south were connected to the earth's magnetism and
to sunspots—but no one knew quite how.

Today we know that the source of the energy in auroras is the
sun. When flares occur on the sun, the great solar wind of protons
and electrons streaming outward becomes charged with extra
electrons. The magnetic field of the earth intercepts the wind and
traps the electrons, which begin bouncing back and forth along
the lines of force of the earth's magnetic field. The energized
electrons are moving vertically with respect to the ground at the
north and south magnetic poles, where they excite particles in the
atmosphere. Producing the characteristic greenish-yellow light of
aurora, excited oxygen atoms then emit crimson light before returning to their original unexcited state. (Man even can create
artificial auroras by injecting energized particles into the ionosphere; nuclear weapons' explosions, for example, can produce
great, reddish auroras.[15]) Study of the aurora borealis in the north
is difficult because of the Arctic's preponderance of floating sea
ice and open sea: Antarctica's continental location makes it a good
laboratory for scientists to observe the upper atmosphere, the
ionosphere, the magnetosphere, and outer space.

But this knowledge came later; the early explorers found aurora australis a source of inspiration as they weathered the rigors
of the polar night. In 1911 Scott thought it had "fluttering ethe-
real life" and wondered why history did not tell of "'aurora wor-
shippers'," so easily could it be considered "the manifestation of
'god' or 'demon'."[16]

Apsley Cherry-Garrard, who with Edward Wilson and Henry R. Bowers, made the first winter overland journey to Cape Crozier to gather Emperor penguin eggs, suffered incredibly. Temperatures reached −70°F and their breath froze over the fronts of their hoods, cutting off air. Perspiration froze on their skin inside their clothes; and an unwitting turn of the head would freeze one's hood in that position, consuming a precious hour while one's companions halted march to hack it loose. Yet one night the waving streams of aurora were so beautiful that Cherry-Garrard was inspired by the lines:

> Here at the roaring loom of Time I ply
> And weave for God the garment thou seest him by.[17]

Perhaps the suspension of ordinary human senses, as well as these eerie and beautiful sights, has fostered the notion that Antarctica has a special connection to other worlds, physical and spiritual.[18] The idea, born in this early literature, has persisted to the present day—from the fact that Antarctic law has been considered a model of law for the moon and outer space, to the fact that astronomers now regard the high Antarctic interior as a unique earthly observation post for studying the interior of the sun.

Antarctic Legacies

Cook had predicted that "the world will derive no benefit" from Antarctica,[19] and his gloomy assessment of 1777 discouraged exploration there for another generation. The explorers of the eighteenth and early nineteenth centuries were concerned with geographic and territorial finds that could add new lands to their empires, bringing home seeds of new crops, or finding gold or lucre, or minerals for the industrial revolution. So they stuck, by and large, to the temperate latitudes. But, periodically, the Antarctic acquired importance and so bequeathed a variety of legacies to the modern world.

Cook had described abundant seals on island beaches in the subantarctic (the area north of 60° south latitude). By 1820, British, Norwegian, and American sealers had established a thriving trade in the South Atlantic based on the little, 160-kilogram fur seals that crowded the beaches of South Georgia and the other islands, barking innocently at the sealers who sidled through the herds with their clubs, leaving a bloodied mass of bodies in their wake. One indication of the butchery is the fact that by 1830 the southern fur seal was nearly extinct.[20]

The discovery of Antarctica was an offshoot of the sealing in dustry. In 1820 Capt. Nathaniel Brown Palmer, then age twenty-one, was in charge of the 14-meter sloop *Hero* on his second annual voyage to the South Atlantic islands with the Stonington, Connecticut, sealing fleet. Sailing southward to scout for beaches for the fleet, Palmer apparently reached Deception Island south of the South Shetlands, which was the best sheltered harbor in the region. On November 17, sailing south from Deception Island, Palmer sighted the ice-clogged coast and cliffs of what he later claimed was the mysterious southern land.[21]

Palmer's motives were economic, but geopolitics was not far behind. The following spring, in the mists off the South Shetland Islands, the *Hero* found itself next to two men-of-war, the *Vostok* and the *Mirnyy,* flying the flag of imperial Russia. They were commanded by Capt. Thaddeus (Fabian) von Bellingshausen, whom the czar had sent south to establish Russia as a global maritime power. Palmer in later life—by then a famous clipper ship captain—would recall that when he boarded von Bellingshausen's ship and told him of his finding, von Bellingshausen called the southern land "Palmer land" in acknowledgment of Palmer's discovery. In modern times, however, the Soviet government would maintain that von Bellingshausen, not Palmer, saw the Antarctic mainland first.[22]

Antarctic sealing died out after 1830, although sealers continued to take the 3,600-kilogram elephant seal—most of which was blubber for oil—from Kerguelen, Heard, and other subantarctic islands. The next wave of interest was commercial and scientific. As imperial trade expanded, more and more steamships plied great circle routes between Europe, Australia, India, and China, around the Cape of Good Hope and Cape Horn. But ordinary sailing compasses were treacherous in the high southern latitudes; the variations in the earth's magnetic field—barely charted for the north—were almost unknown for the south. Friedrich Heinrich Alexander Freiherr von Humboldt—said to be the last man to have all of scientific knowledge in his head—began a systematic study of magnetism using magnetic observatories across Europe, Russia, and East Asia. Johann Karl Friedrich Gauss, likewise interested in studying the north magnetic pole (which lies on an axis inclined at about eleven degrees from that of the geographic poles) predicted the existence of a south geomagnetic pole opposite the northern one.[23]

Gauss inspired three national expeditions to chart Southern Hemisphere magnetism in the 1840s, and to seek the south magnetic pole. None reached it, but each of them contributed significantly to Antarctic exploration. The British expedition, under Ross, found and charted the Ross Sea and the Great Ice Barrier. The American one, under Charles Wilkes, circumnavigated Antarctica and charted the eastern Antarctic coast; in Wilkes's view they proved that Antarctica was a continent. The French explorers, under command of Dumont d'Urville, found a sheltered coast inhabited by hundreds of lively little penguins—which he named the Adélie in honor of his wife.[24] (The south magnetic pole lay inland, and was reached on January 16, 1909, by a group led by Sir Douglas Mawson and Professor T. W. E. David, as part of the British Antarctic Expedition under Ernest Shackleton.[25])

Antarctica regained economic importance in the early twentieth century, when it came to sustain the world's whaling industry. In the austral summer, the Southern Ocean abounded with humpback, blue, fin, and other whales who migrate there to feed. The stocks had been known to whalers for a long time, but they were not of interest until the whalers—following a universal fisherman's rule—had hunted out more accessible herds elsewhere.

In 1904 Capt. C. A. Larsen established the region's first whaling factory at Grytviken on South Georgia. Soon after, other Swedish, Norwegian, American, and British whalers followed, hauling their catch to islands and sheltered coasts where it was flensed or stripped of blubber for transport home. The scale of Antarctic whaling widened when the explosive harpoon cannon and the motorized whale catcher were introduced after 1904. After 1925, when the *Lancing* was fitted with a slipway so the whales could be flensed in the open ocean, the industry became still more efficient and no longer depended on berths in area ports. The Southern Ocean sustained the world's industry for a half-century, sometimes producing ten times as much whale oil as the rest of the world's oceans, until the sudden collapse of the Southern Ocean stocks in the 1960s. Today, there is still some Antarctic whaling, both legal and illegal.[26]

The legacy of the Antarctic has been scientific, geographic, and commercial, but best known is the literature of the generation of explorers who ventured there in the so-called heroic age before World War I. Probably the most articulate group of explorers ever, they have bequeathed us, among other memoirs, Scott's last diary, written as he marched with four others to the South Pole. Arriving on January 17, 1912, he found that Norway's Roald Amundsen had beaten him, getting there on December 14, 1911. Scott kept writing as the entire party trudged back: One member died from a concussion, and another sacrificed himself so the others could survive. Eventually, Scott, Wilson, and Bowers died in their tent only a short distance from a supply depot that could have saved them.

Scott's diary is a paradigm of heroism. Amundsen's account, *The South Pole*—apparently written before he learned of the death of his rival—is laced with his characteristic egotism but contains

Accidental meeting of
Amundsen's ship *Fram* (Nor-
wegian flag astern) with the
Scott expedition's *Terra Nova*.
Through this encounter,
Scott's party learned for the
first time that Amundsen was
trying for the Pole as well.

Roald Amundsen, whose
brother in Norway kept his
accounts while he explored
the Arctic, succeeded in being the first to the South Pole
in 1911–12 with a technically
expert, but ethically controversial expedition.

Disappointment greets
Scott's party at the South
Pole, where they pose for the
camera with Amundsen's
flag signaling his prior
arrival in the background.

many insights about human nature against the trials of polar exploration. Cherry-Garrard's *The Worst Journey in the World* is a bitter, ironic account of the march to Cape Crozier the previous winter. Wilson's writings are especially attuned to Antarctica's biological life and beauty. Shackleton produced *South;* and Mawson wrote *The Home of the Blizzard.*[27]

But they were overshadowed by the towering figure of Fridtjof Nansen, the Norwegian arctic explorer who was a patron of Amundsen and an exemplar to the others venturing to Antarctica. Nansen epitomized the various ideals of polar exploration, embodying the qualities of leadership, bravery, integrity, and spirituality the others all sought—yet he was a practical statesman, having invented a passport for refugees after World War I and won the Nobel Peace Prize. He crossed Greenland and Siberia and navigated the Arctic Ocean. He showed that the complete polar explorer was not only brave and hardy but literate as well. In a sense, he was a von Humboldt of the poles.[28]

The motive for the rash of expeditions to Antarctica from 1900 to 1914 remains unclear despite an extensive literature. Suffice it to say that much of the rest of the globe had been conquered, and the 1890s had seen a burst of exploration of the Arctic. In 1895 the Sixth International Geographical Congress declared Antarctic exploration "the greatest . . . still to be undertaken," whereupon to the Antarctic went Adrien de Gerlache of the *Belgica,* Borchgrevink, and Nordenskjold, as well as the Scot William Bruce and others.[29]

Many of the original territorial and commercial motives for expanding the British Empire had subsided by the first decade of the twentieth century; nonetheless, intrepid Britons were marching into Tibet, scaling Mount Everest, and venturing to Antarctica. In a few years many would enlist in Kitchener's armies to fight in the trenches in World War I, and at least one historian of the empire has drawn analogies between Scott's last expedition and Gallipoli—that is, the British came to lionize heroic defeat as they became more muddled about what victory meant.[30]

Nansen linked polar exploration to the "slackness of varying ages." When nations are rich and comfortable, a few individuals will want to get away to test human endurance against the limits of the known world. The history of polar exploration, he wrote, is

A single mighty epic of the human mind's power of devotion to an idea, right or wrong—a procession of struggling, frost-covered figures in heavy clothes, some erect and powerful, others weak and bent so that they can scarcely drag themselves along before the sledges, many of them emaciated and dying of hunger, cold and scurvy; but all looking out before them towards the unknown, beyond the sunset, where the goal of their struggle is to be found. . . .

But from first to last the history of polar exploration is a single mighty manifestation of the power of the unknown over the mind of man, perhaps greater and more evident here than in any other phase of human life.

[Polar exploration, he continued,] tempered the human will for the conquest of difficulties...furnished a school of manliness and self-conquest in the midst of the slackness of varying ages.[31]

A recent book on Scott's and Amundsen's race to the South Pole indicates that Scott's overriding goal was his promotion in the Royal Navy and that Amundsen's motives were equally practical.[32] It is certainly true that none of these people was shy, and that the stakes were high—if successful, one would achieve world fame and a permanent place in history. But this does not explain all that they did. For example, had Shackleton stayed in England, he could have had a fine career in the merchant marine. Instead, when his plans for the first land crossing of Antarctica were dashed after the *Endurance* sank in the Weddell Sea pack, he got the party over 960 kilometers of ocean to Elephant Island. There, he took his five hardiest men and set out in an open boat to cross the stormiest sea in the world. Nearly drowned and exhausted, they arrived at South Georgia after two weeks at sea; their ship had survived the crossing but nearly smashed on the rocks offshore. Leaving two who were too weak to go on with a man to care for them, Shackleton set out with the remaining two, carrying a primus stove and a cooker, a carpenter's adze, forty-eight matches, 15 meters of alpine rope, and some food. They made a thirty-one-hour march over heretofore untraveled icy mountains until they heard the steam whistle of a whaling factory at the other end of the island. The factory manager, who had heard that the great Shackleton was dead, wept when one of the three terrible, bearded, blackened figures stalking out of the mountains announced that he was Shackleton. After several unsuccessful tries, the entire party was rescued by ship. Not a man had been lost. Shackleton wrote: "We had pierced the veneer of outside things. We had 'suffered, starved, and triumphed, grovelled down yet grasped at glory, grown bigger in the bigness of the whole.' We had seen God in his splendours, heard the text that Nature renders. We had reached the naked soul of man."[33]

Through writings such as these, Antarctica has become linked to the notion of human progress. By now, two or three generations of readers have been uplifted by Scott's inspiring tale. This literature has sparked the belief that if one can reach the ends of the earth and behave well, no matter what adversity looms, humanity will be better off for this example. So one legacy of the heroic age in Antarctica is this idealism, perhaps best expressed in

Departure of Apsley Cherry-Garrard, Edward R. Wilson, and Henry Bowers (*left to right*) on their winter march to Cape Crozier, a trek later called "the worst journey in the world."

Another beautifully composed Ponting photograph gives the flavor of Antarctic travel by dog sledge.

The strain of wintering over in Antarctica shows on the faces of some during Byrd's second expedition, photographed during a midwinter's night celebration.

1909, before the pole had been achieved, by Hugh R. Mill, Britain's outstanding polar geographer:

> The tradition of inaccessibility is a challenge to humanity; and, whether it be the end of the Earth's axis or the summit of a snowy mountain which throws down this challenge, there will never be lacking a few to take it up.... More valuable scientific work and more lucrative commercial returns will accrue... from easier and less fascinating adventures; but the desire to wipe out *terra incognita* appeals more deeply to certain instincts of human nature than either science or trade, and the rough-and-ready test of highest latitude is in one sense a gauge of human progress.[34]

Antarctica Enters a Resource Age

It would be pleasant to dwell further on Antarctica's beauty, early history, and epic human adventures. Even today many opportunities for heroism exist there, although in 1982–83 the continent could be considered almost settled. There are thirty-four year-round stations, with 889 people of thirteen different nationalities (including two Chinese) spending the long, dark austral winter, and this increases to 1,987 in the summer. Virtually all dogsleds have been replaced by helicopters, ski-equipped aircraft, and Sno-Cats. But there is already an extensive literature on Antarctic adventure; it has been described briefly here only to show the richness of the subject beyond the political sphere.

This book focuses on Antarctica's role in international affairs at a critical time when its future is in doubt. Serious crises loom regarding how Antarctica will be governed and what role it will play during the rest of this century and into the next.

Will human beings exploit Antarctica's likely, but undiscovered, offshore hydrocarbons, now that they harvest Antarctica's huge stock of protein-rich krill from the seas each year? How much longer will scientific research be the nations' chief Antarctic activity, now that so many are interested in resources? Who shall decide Antarctica's future? Will it be the twelve nations with historic stakes who negotiated the Antarctic Treaty in 1958–59, and who now administer it, adding only two to their number from 1961 when the treaty went into effect through the fall of 1983? Or will it be decided by nations, such as Malaysia, who have encouraged the United Nations to declare it "the common heritage of mankind," and who, in late 1983, arranged for the UN to consider the matter? What future "swing roles" will be played by India and Brazil, two large, influential developing countries who like to speak for the developing nations outside the treaty, but who nonetheless became the fifteenth and sixteenth full voting treaty

Antarctic Treaty
(summary of basic provisions)

ARTICLE I	Antarctica shall be used for peaceful purposes only. All military measures, including weapons testing, are prohibited. Military personnel and equipment may be used, however, for scientific purposes.
ARTICLE II	Freedom of scientific investigation and cooperation shall continue.
ARTICLE III	Scientific program plans, personnel, observations and results shall be freely exchanged.
ARTICLE IV	The treaty does not recognize, dispute, or establish territorial claims. No new claims shall be asserted while the treaty is in force.
ARTICLE V	Nuclear explosions and disposal of radioactive wastes are prohibited.
ARTICLE VI	All land and ice shelves below 60° South Latitude are included, but high seas are covered under international law.
ARTICLE VII	Treaty-state observers have free access—including aerial observation—to any area and may inspect all stations, installations, and equipment. Advance notice of all activities and of the introduction of military personnel must be given.
ARTICLE VIII	Observers under Article VII and scientific personnel under Article III are under the jurisdiction of their own states.
ARTICLE IX	Treaty states shall meet periodically to exchange information and take measures to further treaty objectives, including the preservation and conservation of living resources. These consultative meetings shall be open to contracting parties that conduct substantial scientific research in the area.
ARTICLE X	Treaty states will discourage activities by any country in Antarctica that are contrary to the treaty.
ARTICLE XI	Disputes are to be settled peacefully by the parties concerned or, ultimately, by the International Court of Justice.
ARTICLE XII	After the expiration of 30 years from the date the treaty enters into force, any member state may request a conference to review the operation of the treaty.
ARTICLE XIII	The treaty is subject to ratification by signatory states and is open for accession by any state that is a member of the UN or is invited by all the member states.
ARTICLE XIV	The United States is the repository of the treaty and is responsible for providing certified copies to signatories and acceding states.

Treaty signatories
Consultative parties

Argentina
Australia
Belgium
Brazil
Chile
Federal Republic of Germany
France
India
Japan
New Zealand
Norway
Poland
Republic of South Africa
Soviet Union
United Kingdom
United States

parties in September 1983? This book intends to explain the background, the underlying stakes, and future possibilities for Antarctica. For Antarctica's future will be determined, not on the violet-
and orange-hued ice pack while a penguin cries in the midnight
stillness—but in Washington, Moscow, Tokyo, London, perhaps
even in New Delhi, Peking, Kuala Lumpur, or the United Nations.

Antarctica is entering a "resource age." It started in the 1970s,
when commercial fishing became significant and resource questions came to dominate the diplomacy of the Antarctic Treaty
consultative parties (in this book called the treaty powers). Critical
decisions will be made in coming years, trading off exploitation
against potential environmental harm well into the twenty-first
century should the minerals of the continent be sought. Such
resources are raising the other nations' stakes in the region—long
thought good only for tests of human character, beautiful scenery,
and scientific research.

The concept of a resource age is artificial but useful. It suggests
that Antarctica passed through an age of exploration, from Cook's
voyage through the last privately financed expeditions after World
War II. The period that began with the 1957–58 International
Geophysical Year (IGY), when Antarctica became an international
scientific laboratory, can be labeled an "age of science." As we shall
see, Antarctic policy in the United States and other countries
changed in the 1950s, veering away from emphasizing exploration
and claims to favor science. The 1950s were a period of ferment in
Antarctic diplomacy, much like today.

The result of that ferment was the Antarctic Treaty of 1961,
which dedicated the region to peace and science and made the
spirit of the IGY permanent. Under the treaty, the twelve nations
having Antarctic programs for the IGY agreed to demilitarization,
to unilateral inspection of all ships, stations, and cargoes, and to a
ban on nuclear wastes and explosions there. Since then, these
provisions have been upheld. Thus the treaty, which entered into
force in June 23, 1961, has allowed Antarctica to live up to H. R.
Mill's vision of it as a gauge of human progress.

The age of science is hardly over. The treaty runs indefinitely,
although the parties may review its operation in 1991. Some of the
science done in the last twenty-five years has been outstanding, but
several new problems need urgent attention, such as whether
global atmospheric warming due to the use of fossil fuels will cause
the West Antarctic ice sheet to melt, raising world sea levels. Antarctica also is useful for studying the interior of the sun. As in the
days of the early explorers, it holds the key to scientific mysteries,
which can be studied if the treaty continues past 1991 in more or
less its present form.

But the treaty has worked as Antarctica's form of government, not because scientific research is inherently more virtuous than other human activity, not because of the preamble's fine words, but because it has held in balance the region's underlying *political* problems—such as the rivalry among seven nations with announced territorial claims, and the balancing interests of the United States and the Soviet Union. International lawyers, who try to understand the Antarctic problem by reading the treaty, often find the document short, ambiguous, and full of holes. They often characterize it as "fragile," but it is in fact "robust" because it well suits the unique political needs of each party. Through it, each territorial claimant, no matter how arcane its case, can hold to its sovereignty position without having to fight about it. It prevents rivals from breaking out and expanding their claims, asserting new ones, or acting contrary to peace, and it holds the web of rival political interests in balance.

Both resources and the prospect of change in 1991 are affecting this time-honored equation for the treaty parties and awakening the interest of outsiders to the club. Therefore, the treaty may have to adjust greatly in order to be as useful in the future as it has been in the past. This book suggests some changes that could be made.

The treaty withstood a severe test in 1982, for example, when Argentina fought a war with Great Britain over its obscure territorial claim to the Falkland Islands on the doorstep of the demilitarized Antarctic Treaty "area" that starts at 60° south latitude. Contrary to some predictions made at the time,[35] the Argentines, after seizing the Falklands and South Georgia, did not cross that invisible international boundary. For Argentina, evidently, war with Great Britain was one thing, but arousing the ire of the twelve other treaty powers, including both superpowers, was another. Indeed, the treaty group proceeded with meetings on sensitive resource issues during those crisis months of 1982. Diplomats later said that, inside the meeting room, one would never guess that two of the nations represented were at war.[36]

The true strength of the treaty may be tested on its thirtieth birthday in 1991, when, under Article XII, the powers may meet and propose changes to it by majority vote, followed by ratification by all the governments. What will 1991 bring? A party, toasts all around, and no substantive discussion? A falling-out among the group because of long-building tensions over resource and other issues? Or, by 1991, will so many new parties—for example, India and China—sit at the meeting table and change the discussion entirely?

The 1991 review may provoke an internal crisis. In addition, the treaty system faces an external challenge from developing nations now debating Antarctica in other forums. The Antarctic Treaty powers, disillusioned with the "common heritage" approach to the unowned deep seabeds, will resist any move by the UN General Assembly to assert jurisdiction over Antarctica. The stage is set for confrontation or conciliation, and what will happen is not yet clear.

The United States in Antarctica

Antarctica's future doubtless will be influenced by the United States in international meetings and "down on the ice" (which is how old Antarctic hands refer to the subfreezing temperatures and hostile conditions of this distant, white land). Both U.S. interests there and the U.S. scientific program face the most serious challenge in their history. For U.S. interests are broad, ranging among access to resources, to protecting the global environment, to checking the Soviet bloc. Yet the content of the U.S. program is confined almost exclusively to university-based small scientific research projects, as though there were no need for the taxpayer to pay more than $100 million per year to go there other than to please a handful of U.S. scientists. The U.S. program is due for a change—perhaps an expansion in its scope and structure—having done little to date to support U.S. interests in resources, environment, and the interests of the developing world.

This book discusses the United States in Antarctica as a secondary theme. For Antarctic history consists not only of oft-told adventure stories, or of the seven nations' quaint territorial claims, or of the advent of science. It includes the ambivalent, deep U.S. involvement in Antarctica for more than 160 years, which led the United States, intentionally or not, to be the chief determiner of events there and chief player in the region's diplomacy. This legacy is particularly relevant now.

The early American sealers were followed by several official attempts to sponsor an expedition, culminating in a serious voyage of exploration led by Wilkes, a character of Ahab-like dimensions. In the twentieth century, expeditions led by private citizens and the U.S. government set out to explore and claim *all* of Antarctica for the United States. It was not clear what the United States would do once it had annexed Antarctica—which is larger than Alaska, indeed, larger than the United States and Mexico combined. Among other things, the explorers believed that the United States in time should benefit from Antarctica's supposed mineral riches.

But the U.S. State Department never promulgated a U.S. claim, even though at the time of the IGY inchoate U.S. "rights" extended over as much as 80 percent of the continent and overlapped the claims of the seven announced claimants. Under the treaty, the United States was careful not to give up the "basis" of these rights, whatever they may be. Ever since, they have been symbolically upheld through continuous U.S. occupation of the geographic South Pole.

The U.S. role is as ambivalent as it is large. Not willing to undergo the painful diplomacy that would lead to a final territorial settlement (the seven claimants are, after all, allies, and the Soviet Union, after 1950, also sought a role), the United States discovered an ingenious solution in the treaty. By removing sovereignty as an issue, the United States was relieved of its own sovereignty dilemma, thus becoming not only the chief architect of the treaty but its chief upholder as well. So the largest potential claimant led the call for peaceful uses, scientific research, and international cooperation.

But this formula does not say how the United States should approach Antarctic resources, nor what it should do about countries, such as Malaysia or India, with different, new demands. The old U.S. formula of peace and science, upholding the basis for U.S. rights, needs redefinition today.

Indeed, U.S. influence in Antarctic affairs may be waning because its program and logistics still rigidly follow the old pattern set during the IGY; ironically, the largest player in Antarctica has been the least willing to change. The United States needs a future vision for Antarctica that is now lacking. What should Antarctica become? A wilderness area? A model of cooperation with developing countries? An internationally managed polar oilfield? These questions must be addressed in open, public debate with Congress, the executive branch, and the scientific community, in order for the United States to play an influential role during the critical time ahead.

Antarctica, apparently so distant from the usual concerns of the policymaker, is connected to many other international issues. It is a geopolitical entity of some importance. It raises arms control issues; NATO allies square off there—albeit politely—against the Soviet bloc; and its fishery—already being exploited—could remedy world hunger. Finally, it offers important opportunities for international cooperation on both science and resource questions. All these concerns must be considered as the nations of the world decide the fate of the earth's seventh continent.

two

The United States in Antarctica: Past

There is an apocryphal story about Richard Evelyn Byrd, the Virginia gentleman turned polar explorer, who more than anyone tied Antarctica's fate to the United States. Standing on the deck of the wooden ship, *Bear of Oakland,* on his third trip to Antarctica, Byrd pointed out to Lt. George Dufek, his navigator, the sheer white barrier of the Ross Ice Shelf. Dufek was making his first trip to Antarctica, although he was destined to return there many times. One can imagine the theatrical Byrd, already world famous for his north and south polar flights, waving a proprietary hand at the great natural spectacle before them. "What do you think of it?" Byrd asked. To which the young naval officer replied, "It's a hell of a lot of ice. But what good is it?"[1]

As told by Paul A. Siple, another famous U.S. Antarctic explorer, the story summarizes the dualism that has characterized U.S. involvement in Antarctica. On one hand, a small group of Americans fell in love with the great white wilderness, and wished to revisit and conquer it and bring it into their country's sphere. On the other hand, many others—including several secretaries of state—wondered what business the United States had in a far-flung place having no direct economic or military value. Thus, the link the explorers forged between Antarctica and the United States persisted, but in a tenuous, shadowy way. For more than a century, U.S. policy on the subject was made *ad hoc* whenever it was made at all.

There was an impulse to explore because new lands promised new resources. Nathaniel Palmer's discovery of the legendary "southern land" was publicized in the 1830s by his sponsor, Edmund Fanning, a Stonington sealing magnate; Fanning and others began lobbying Congress to authorize an expedition south, but they had mixed success. After a series of botched attempts the U.S. Exploring Expedition (1838–42), with a vague but geographically sweeping mandate, was dispatched under Lt. Charles Wilkes. In the twentieth century the lobby was taken up by aviator-explorers—Byrd, Lincoln Ellsworth, Finn Ronne, their relatives, and others—who wanted the government to claim the growing swath of Antarctic territory being discovered.

In the 1930s, the government absorbed the explorers' impulse; Congress was interested, as was President Franklin D. Roosevelt. In 1939 Roosevelt authorized an attempt to occupy and settle Antarctica as a prelude to U.S. possession. After World War II, the U.S. military returned to practice polar logistics and "develop" claims to Antarctica by dropping brass cannisters with signed documents asserting ownership. Thus, during the twentieth century the United States has flown more planes, mapped and photographed more territory, and sent more expeditions than any other nation. In 1960 Sen. Ernest Gruening of Alaska would testify that Americans—and only Americans—had seen approximately 80 percent of the continent.[2]

It all was designed to lead to a high-level declaration of a U.S. claim, although for years discussions inside the State Department centered on which parts of Antarctica should be claimed and under what legal theory such a claim could be upheld. As other countries announced their Antarctic territorial claims, usually on the basis of discovery or the sector theory used in the Arctic, the difficulties of a final territorial disposition grew. Ironically, as the area of a potential U.S. claim grew with successive expeditions, U.S. public silence on the subject deepened.

The reasons for this silence and for official reluctance to annex Antarctica deserve discussion. Since the purchase of Alaska in 1867 the United States has balked from outright annexation of foreign territory as being incompatible with American democratic ideals that reject the appearance of colonialism. In the past, when such opportunities for annexation have presented themselves, as in the case of Cuba or the Philippines, the United States chose not to do so. Instead, it "liberated" them from Spain and Japan, respectively. Another factor is the public's reluctance, despite the backing of specific groups, to support exploits on "foreign" soil. Thus, the ambivalence that runs through U.S. Antarctic history about whether Antarctica is "ours" or not or whether it is domestic or foreign territory stems from a larger dilemma between the obvious opportunities for expansion afforded by U.S. power and its image of itself as an antidote to imperialism.

Finally, Antarctica—like other far-flung places such as Micronesia and the Philippines—has never held the attention of the public, Congress, or the president, for very long. Many serious attempts to establish a coherent Antarctic policy were interrupted and eclipsed by more pressing events, such as wars. This intermittent attention meant that at many junctures the United States had no Antarctic policy at all. Official statements were so few and infrequent that even in the early 1970s, when asked for the U.S.

position on sovereignty in Antarctica, the State Department would quote a brief statement made by Secretary of State Charles Evans Hughes to the Republican Publicity Association in 1924.

This chapter describes the little-known heritage of the United States in Antarctica through the end of December 1956, when Siple's team completed the Amundsen–Scott base at the South Pole. This accomplishment symbolized U.S. interests in the continent and was meant to preserve U.S. dominance there during the eighteen-month International Geophysical Year (IGY) that began on July 1, 1957. Although Byrd and other explorers wrote individual memoirs, a detailed account has not been published elsewhere. A quasi-official history of American involvement was compiled by historian Kenneth J. Bertrand, but his account omits the diplomatic deliberations and internal government discussions. Parts of the account which follows are based on recently de-classified materials, particularly the list of U.S. claims shown on the maps in this chapter.[3] Most of it has been pieced together from established sources, but, even so, the full story cannot be told until more of the official record, particularly that of the 1950s, is released.

This chapter illustrates the nationalistic and expansionist motives that were the basis of U.S. involvement in Antarctica, as well as one pillar of the U.S. role there. The second pillar—the international, peacekeeping, and scientific roles—emerged in the 1950s with President Dwight D. Eisenhower's decision to downplay the "national program" of the military and the explorers and to highlight scientific cooperation—without, however, ceding the basis of U.S. territorial rights. The consequences of this shift and the 1959 Antarctic Treaty, which codified this stance in international law, will be described later.

By the 1960s, the issue of U.S. claims there became moot—except that under the treaty, and even today the United States retains the "basis" for the enormous territorial claim laid down by the U.S. expeditions. Lawyers can argue about whether the huge, unannounced U.S. claim has any standing in law, and can evaluate its merits vis-à-vis the announced claims of the other nations. It may well have little legal weight, but then announced claims of others (see chapter 3) are dubious too.

Today, the validity of the U.S. claim is less important than the political message it conveys. We will see how, by the late 1940s, the United States almost by accident became *the* power in Antarctica and the region's unquestioned leader in international diplomacy. This story sheds light, then, on how the United States came to its position of leadership for the region, which in turn is necessary for understanding the dilemma that it faces there today. And any domestic debate about the U.S. stake in Antarctica—whether it is "ours" or "foreign," about whether we are there for national and economic motives, or for loftier, peacekeeping ones—must take account of this legacy.

The "First" Discovery of Antarctica

There is a three-way tug-of-war among the United States, Great Britain, and the Soviet Union over whose mariners discovered the Antarctic continent. Aside from its symbolic overtones, the debate has some significance in law, since discovery is sometimes considered a basis for territorial possession. Some evidence supports the claim that an American, Nathaniel Brown Palmer of Stonington, Connecticut, made the first sighting of the continent in November 1820. But it is possible that a Briton, Edward Bransfield, who sailed near the mainland the previous January, sighted it first. But Bransfield's logbook is lost, making the British claim hard to substantiate. Likewise it is difficult to document the Soviet government's assertion that Adm. Thaddeus von Bellingshausen, sailing the southern seas for the czar in 1819–21, saw the continent first. Von Bellingshausen's log makes no such claim, and maps drawn after his return to Russia appear to credit Palmer's discovery.

Capt. James Cook, in his earlier voyage, had not seen a southern landmass, but his account of the voyage nonetheless sparked commercial interest in southern seals. By 1820, rival British and American sealers were working the beaches of the southern coast of South America from Buenos Aires to Valparaiso. The vessels traveled in fleets, taking all the seals they could find and constantly moving, scouting for fresh, undiscovered beaches.[4] The fleets passed one another in fair weather and foul, sometimes showing colors and tipping off one another to lucky finds, but sometimes passing silently, incognito.

Edouard A. Stackpole of Nantucket, an authority on the period, believes that most sealers operating around the Falkland Islands and the South Shetland Islands knew that a continent lay to the south. A clear sky would suddenly fog over, tremendous gales would suddenly sweep up from the south; such weather was unlikely if only open water lay beyond the horizon. But while looking for land, they could easily be confused in finding it; the fog, the singularly high, steep peaks of the islands and lifting of over-horizon features by mirage—as well as the secretiveness of the sealing trade—heightened their confusion about what was floating ice, islands, or the legendary "southern land."[5]

In any event, Captain Nat, as Palmer was called, was known to most sealers in the area in 1820 as a clever, sharp-eyed seaman. He had returned with the Stonington fleet that season as commander

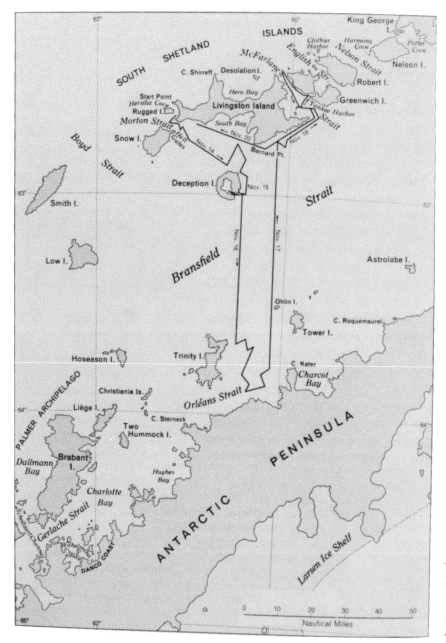

The cruise of Nathaniel Palmer's *Hero* on November 16–17, 1820. Allegedly, Palmer was the first to recognize the Antarctic continental mainland.

of the *Hero,* a 14-meter sloop with a mate and crew of four. Given the harsh weather and seas of the region, it is remarkable that the *Hero* drew only 2 meters and was smaller than any of Columbus's ships. It was also shorter than the *Gypsy Moth,* which Sir Francis Chichester sailed alone around the world in 1966–67.[6]

Palmer's job was to find additional sealing beaches, so he headed south from the South Shetland Islands in November. He reached Deception Island—a partly sunken volcano so named because a ship could sail almost all the way around it before seeing the narrow entrance to its sheltered harbor, a captain's dream in those stormy seas. Leaving Deception Island, Palmer set off to the southeast on November 16. The weather was clear, giving him visibility across the 88 kilometers of water to the cliffs opposite which rise to more than 300 meters near the coast and to 1,800 meters inland. Palmer seems to have crossed from Deception toward the mainland. At midnight on the 17—during the brief "night" of that time of year—he hove next to Trinity Island, just off the mainland. At 4:00 A.M. he "[d]iscovered a Strait," his logbook says, "Tending SSW & NNE—it was Literally filled with Ice and the shore inaccessible." This is widely interpreted to have been the mouth of the Orleans Strait (so named later by Dumont D'Urville) that separates Trinity Island from the mainland. But Palmer's log says he "thought it not Prudent to venture in"; he probably followed the shore, which was "every where Perpendicular," to the north. Bearing northwest, he was back at "Freeseland" (now called Livingston Island), in the South Shetlands by 2:00 A.M. on November 18.[7]

Palmer apparently believed that he had found the southern "land," a continent instead of an island, and publicized his discovery as such. A map describing it as "Palmer Land" was issued five months later; the sponsors of Stonington fleet described it in *Voyages Round the World,* published in 1833. Palmer's niece, Mrs. Richard Fanning Loper, published Palmer's own account based presumably on his log and recollections.[8] American maps for years labeled the peninsula the Palmer Peninsula although by international agreement it is now called the Antarctic Peninsula.

But Palmer's priority has been questioned. Stackpole has written that Palmer saw only the south coast of Livingston Island, in the South Shetlands. The British argue that Bransfield, aboard the *Williams,* saw the snow-capped peaks, black rock faces, and stony beaches of the mainland near the Orleans Strait in January of 1820.[9] Most historians of the period agree that Bransfield sailed the strait dividing the South Shetland Islands from the mainland that now bears his name, and spotted islands later noted by Palmer. But without Bransfield's log it is impossible to document what southern coast he saw and whether he recognized it as the great "southern land." In 1949 the USSR All-Union Geographic Society would claim that von Bellingshausen had discovered the continent during his circumnavigation of Antarctica, although von Bellingshausen's log makes no such claim.[10]

There is little doubt that Palmer saw the continent again in a subsequent cruise in January 1821. His chance encounter with von Bellingshausen in the fog off the South Shetlands provides much grist for folklore and historians. The American version is colorfully told in Mrs. Loper's account. In it, Palmer in his sealskin coat

Nathaniel Palmer's
14-meter sloop *Hero* meets
von Bellingshausen's men-of-
war *Vostok* and *Mirnyy* off the
South Shetland Islands.

boards the *Vostok* and proceeds to astonish the elderly, courteous,
and cautious von Bellingshausen, who is surrounded by officers in
full-dress uniform.[11]

Von Bellingshausen, who had been naming the South Shetland
Islands after such napoleonic defeats as Borodino and Waterloo to
please the czar, was upset to learn that the British had discovered
them long before and had already taken seals there. When Palmer
told him of "what I had discovered"—a reference to Palmer Land,
presumably—

> He rose much agitated, begging I would produce my logbook and
> chart, with which request I complied. . . . He examined them carefully
> without comment, then rose from his seat saying, "What do I see, and
> what do I hear from a boy in his teens—that he is commander of a tiny
> boat of the size of a launch of my frigate, has pushed his way to the
> pole through storm and ice and sought the point I, in command of one
> of the best appointed fleets . . . have for three long, weary, anxious
> years searched day and night for. . . . What shall I say to my master?"[12]

Palmer said that von Bellingshausen acknowledged Palmer's priority, a view bolstered by the fact that Ivan F. Krusenshtern of the
Russian Imperial Navy published a map of the Pacific Ocean after
von Bellingshausen's return showing "Palmer Land" south of the
islands. Nonetheless, von Bellingshausen's log makes no mention
of Palmer's announcement of his discovery of the land.[13]

There were many vessels in the region during those years; the
logbooks of many are still missing; and it is conceivable that some

other vessel's captain sighted Antarctica first. Stackpole was given a book belonging to descendents of Christopher Burdick of the *Huntress* of Nantucket. Under clippings pasted in the family scrapbook was Burdick's log of his journey to the far south and his landing on February 7, 1821, at Hughes Bay south of the Orleans Strait, accompanied by Capt. John Davis in the *Cecilia* out of New Haven. This is the first documented landing on the mainland, although according to second-hand accounts,[14] Captain McFarlane in the British ship *Dragon* may have landed earlier.

The Wilkes Expedition

Palmer's discovery sparked two subsequent expeditions to the south, the second of which was the U.S. Exploring Expedition of 1838–42 led by Wilkes.

The push came from Fanning, who believed that further exploration of the far south was too expensive to be done privately. Another promoter was Jeremiah N. Reynolds of Wilmington, Ohio. Whereas Fanning's motives were commercial, Reynolds shared the beliefs of one John Cleves Symmes, Jr., an Ohio promoter who traveled to small towns lecturing on a variety of "scientific" subjects. A pet theory of Symmes's was that of the "holes in the poles," namely, that at the poles the earth bent inwards. If one dared sail into these caverns, one would find another world. Reynolds interpreted this odd belief to mean that the United States should explore the far south at once.[15]

By 1828, because of the efforts of Fanning and Reynolds, President John Quincy Adams and Secretary of the Navy Samuel L. Southard favored such an expedition. Congress authorized only one ship to go to the Pacific and the "South Seas," however, stipulating that the voyage had to be paid for from existing funds.[16]

Reynolds began compiling data on the southern regions from New England ship captains, and he prepared two vessels, a crew, and a scientific staff (including Wilkes, who was to be the expedition astronomer). But upon taking office, President Andrew Jackson quashed the scheme in order to save government funds. Subsequently, Reynolds transformed the project into a private venture whose members would include Fanning and Palmer, among others.[17]

The expedition is known as the Palmer–Pendleton Expedition of 1829–31. It was important for the work of a young scientist, James Eights. Although Eights's work remained unknown until the twentieth century, it is now accepted that his keen observations of the geology, flora, and fauna of the South Shetland Islands preempted the work of better known naturalists who came later. For example, Eights found the first fossil in the Antarctic region, a discovery formerly attributed to Capt. C. A. Larsen in 1892. Eights also found a unique ten-legged spider (*Decolopoda australis*) whose existence was doubted until it was found again by William Bruce in 1903. Some of Eights's geologic analyses anticipated those of Charles Darwin.[18]

But for all his brilliance, Eights was a shooting star. After he turned over his samples to the Albany Institute upon his return, he did nothing of consequence and died a disappointed man. Nonetheless, the scientific tradition he started in the Antarctic would bear fruit more than 100 years later.[19]

By 1836, Congress had approved an official expedition to the south; evidently Fanning and Reynolds had won their point after all. That year, $300,000 was authorized for an exploring expedition "to the Pacific Ocean and the South Seas" to aid commerce and navigation and "to extend the bounds of science." But the final approval and outfitting of the expedition became highly controversial. As a result, Reynolds was not included, and Wilkes was picked as commander at the last minute by the navy.[20]

Wilkes's official instructions emphasized "the important interests of our commerce embarked in the whale-fisheries." He had approximately two years to circumnavigate the Southern Ocean—though it would take an extra year—and to push farther south than had Cook or the British captain James Weddell. He was "to determine the existence of all doubtful islands and shoals" there. Thereafter he was to explore the Pacific and the west coast of North America. Eights was initially accepted, then dropped from the group; but other scientists went and produced important results.

The expedition was dogged with problems from the start. The ships were in poor condition: The *Peacock* sailed poorly and wrecked; the slower *Relief* had to be left behind; the *Sea Gull* was lost in a gale; and the *Flying Fish* had to be sold before the expedition was over. This left the *Vincennes* and the *Porpoise;* their sides, punctured with gunholes, afforded little protection against the heavy southern sea swells and Antarctic blizzards.[21]

Finally, there was Lieutenant Wilkes himself, who, after leaving port, promoted himself to captain against navy rules; his arbitrary nature and harsh punishments made the voyage a torment for the crew. He accomplished his mission successfully, but on his return he found a court-martial instead of a hero's welcome. Thus, the expedition's achievements were eclipsed by Wilkes's peculiar nature.[22]

(Recently, a scholar of Herman Melville's works, David Jaffe, found the copy of the *Narrative of the U.S. Exploring Expedition,*

Charles Wilkes, leader of the
first dedicated U.S. Antarctic
expedition, as he appears in
his *Narrative*, published on
his return. This picture could
have influenced Herman
Melville's description of
Captain Ahab in *Moby Dick*.

Ko-Towatowa, a Maori on
Wilkes's expedition, as he is
portrayed in the *Narrative*.
He may have been the model
for Queequeg, the Indian
harpoonist in *Moby Dick*.

which Wilkes wrote in his own defense in 1845, which had been
acquired by Melville in 1847. Soon after, Melville began writing
Moby Dick. Jaffe concludes, on the basis of Melville's notes on the
volume, that Wilkes was the model for Captain Ahab, or at least
for the mystical, obsessive Ahab who emerges in the later part of
Moby Dick. They have similar biographies, and many incidents of
Ahab's voyage resemble those that occurred on Wilkes's. Even the
description of Ahab's terrible scar could be based on the frontis-
piece of the *Narrative*, which shows Wilkes with a long, dark
shadow along one side of his face. There were analogies with the
"blackness" of Ahab's and Wilkes's natures, with their accidental
encounters with happier ships. Even Queequeg, the Indian harpoonist in *Moby Dick*, could be modeled on Ko-Towatowa, a Maori
on the Wilkes voyage, whose portrait is one of the *Narrative*'s more
exotic illustrations. Finally, Ahab's consuming wish to find the
great white whale could be drawn from Wilkes's obsessive, dangerous charting of the great, white Antarctic continent.[23])

Wilkes hunted—and found—Antarctica. Whatever his literary
legacy, he succeeded, despite terrible weather and lack of sheltered harbors, in following the coast of East Antarctica for 2,400
kilometers, making sightings and sketching and charting coasts

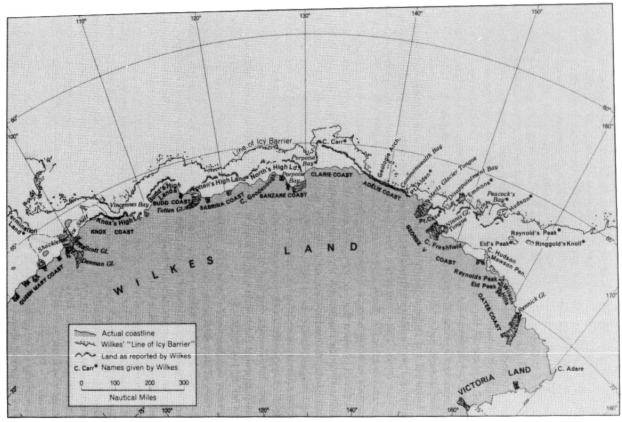

Wilkes's journey along the East Antarctic coastline mapped many features for the first time. Here, his sightings are compared with later, more accurate ones. Contrary to the views of well-known geographers, Wilkes concluded that Antarctica was a continent and not an archipelago.

and islands. The expedition took on political significance as well, for Sir James Clark Ross was shortly to venture south for the British, and Dumont D'Urville was going there for the French. For three successive summers Wilkes plunged his ships southward to the ice-ridden waters around the Antarctic continent. He saw enough of the coast to conclude that Antarctica was indeed a continent and not an archipelago—a proposition he would defend, to his credit, against other better-known authorities such as Ross.[24] Wilkes's survey and its general accuracy, despite many disputed sightings, gave the United States a record of "discovery" in East Antarctica that later explorers and State Department geographers would not forget. A portion of East Antarctica is named Wilkes Land in his honor.

Two other noteworthy U.S. efforts in the Antarctic took place in the nineteenth century. One was the discovery of Heard Island, at 53° 06′ south latitude and 73° 30′ east longitude, on November 25, 1853, by Capt. John J. Heard of Boston in the bark *Oriental*. Heard was testing a great circle route through the southern Indian

Ocean when he saw the island but did not land. Six weeks later, a Briton, Capt. Edwin A. McDonald in the *Samarang,* also spotted Heard Island.

Heard Island's huge colonies of elephant seals soon provided a lucrative stock for American sealers. The efforts of Heard and his widow to secure an official government claim to the island were fruitless. As late as the 1930s the U.S. government considered claiming Heard Island for its strategic location, but a British claim based on McDonald's sighting held, and the island now belongs to Australia.[25]

Matthew Fontaine Maury, an important nineteenth-century American scientist who systematically compiled world weather and oceanic data for the U.S. Navy, was interested in the far south too. In compiling his key works, *Sailing Directions* and the *Physical Geography of the Sea,* which were essential to captains of the period, Maury became concerned about the lack of data for the far southern region. In 1861 he wrote letters to eight European governments proposing an international program of polar exploration. His injunction illustrates the view of nineteenth-century science and the close relationship between science and geographic exploration of the time: "Remember that the earth was made for man; that all knowledge is profitable; that no discoveries have conferred more honour and glory in the age in which they were made, or been more beneficial to the world, than geographical discoveries."

Maury's plan was interrupted by the Civil War. A Virginian raised in Tennessee, Maury resigned his post as chief of the navy's hydrological office to join Confederate ranks in April 1861. But if Wilkes was dogged by his defects, Maury continued to be blessed by his strengths; he had thrown over his career for the Confederate cause but still was an international figure. Later both the czar's brother and Napoleon III would invite Maury to live in their respective countries, and Hugh R. Mill, the British geographer, paid him handsome tribute. Indeed, Maury's plan for an international expedition led in 1882–83 to the first International Polar Year, carried out mostly in the Arctic.[26]

The Turn of the Century

The United States was relatively inactive in the Antarctic at the turn of the century and during the heroic age when several European expeditions visited Antarctica. Nonetheless, there were some U.S. contributions. One was the visit to South Georgia in 1912–13 of Robert Cushman Murphy, a twenty-five-year-old American naturalist, who arrived on the brig *Daisy* of New Bedford. Sketches he made and collections he gathered during his three months ashore helped to establish his international fame as an ornithologist.[27] Another American, Frederick Cook, was the ship's doctor on the *Belgica*'s ill-fated expedition of 1897–99. The ship was trapped accidentally in the ice and had to winter over. Cook helped the crew to survive their unexpected trial by giving them fresh penguin and seal meat to fight off scurvy. Later, he would achieve notoriety by falsely claiming he had reached the North Pole on another expedition. [28]

During World War I, the *Carnegie*, a nonmagnetic ship run by the Carnegie Institution of Washington, undertook a magnetic survey of the Southern Ocean. It took 118 days to circumnavigate Antarctica from the time the ship left Lyttleton, New Zealand, until its return. There was precipitation on one hundred of those days and gales on fifty-two, half of which reached hurricane force. As for the ice, J. P. Ault recalled, "It was like trying to sail down Broadway with all the skyscrapers gone wild and drifting around in our pathway."[29]

American Aviation in Antarctica

World War I marked a turning point in Antarctic history and in America's role there. In 1916 Shackleton returned to a Europe at war, for practical purposes ending the heroic age of Antarctic exploration, although he would make one final expedition there. Thereafter, the European nations were too consumed with the war and its aftermath to be active in the distant, far south. The "slackness of varying ages," which Fridtjof Nansen had said made polar exploration possible, was taken up, and the incredible human energies Antarctica consumed were applied back home.

Yet World War I gave rise to aviation, which would define the next chapter in Antarctica. Aviation made the United States, hitherto an occasional player in the far south, the dominant national presence there. The growth of U.S. Antarctic exploration by air was almost accidental, an offshoot of the nascent U.S. aerospace industry which sought to whet public enthusiasm for and confidence in aviation after the war by sponsoring spectacular flying feats.[30] Thus, the United States happened to have the sponsors, the money, the donated aircraft, and a band of ambitious pilot-explorers eager to use Antarctica as a fitting fairground for their publicity-seeking exploits.

On May 9, 1926, Byrd announced that he, along with pilot Floyd Bennett, had successfully flown over the North Pole.[31] Byrd also had wanted to make the first nonstop flight across the Atlantic and had leased Roosevelt Field for the purpose; but he allowed

THE UNITED STATES IN ANTARCTICA: PAST

Aircraft for the first Byrd expedition were ferried over the ocean on ships. They were a revolutionary technology for Antarctic exploration.

The elderly Roald Amundsen (*left*) congratulates Richard E. Byrd and Floyd Bennett, who have just returned from their North Pole flight.

Charles Lindbergh to use it for the takeoff of what proved to be Lindbergh's successful transatlantic crossing in 1927. A few days later, Byrd and pilot Bernt Balchen flew the Atlantic, landing in Normandy before reaching Paris. And on November 16, 1928, Sir Hubert Wilkins, an Australian backed by William Randolph Hearst among others, made the first airplane flight in the Antarctic, around Deception Island.[32]

Meanwhile, Byrd was approaching the Ross Sea by ship with a

Ford trimotor plane, the *Floyd Bennett*, lashed to the deck; he planned to fly over the South Pole. The well-known journalist, Charles J. V. Murphy, Byrd's friend and sometime ghost writer, once called Byrd a "mystic," implying his troubled side.[33] But Byrd's romance with Antarctica, his love of exploits and kindness to his men, his boosterism, his political clout, all these would lash the fate of Antarctica to the United States—a tether which still binds today.

Byrd's 1929 South Pole flight brought America to the forefront of Antarctic exploration. After his North Pole flight, Byrd had proposed a similar flight over the South Pole in an after-dinner talk with Roald Amundsen. As retold by Byrd in *Little America,* Amundsen's reaction sounds apocryphal and makes Antarctic explorers sound like a medieval guild of masters and apprentices who passed their craft down from generation to generation. Reportedly Amundsen said: "A big job, but it can be done. You have the right idea. The old order is changing. Aircraft is the new vehicle for exploration. It is the only machine that can beat the Antarctic."[34]

Late in 1928 Byrd returned to beat the Antarctic, with airplanes, snowmobiles, radio, the new Fairchild K-3 aerial camera, and plans to explore on a hitherto untried scale. He established his base, as Amundsen had, at the Bay of Whales. On November 29, with Balchen as pilot, Harold June as co-pilot, and Ashley McKinley as photographer, and with Byrd navigating, crammed in the back at the chart-table, he sent messages to Balchen on a trolley, shouting instructions over the noise of the engines. Having dumped 250 pounds of food to lighten the plane so it would lift over the "Hump" in order to reach the polar plateau, they finally flew over the South Pole.[35]

As Byrd recalled:

> For a few seconds we stood over the spot where Amundsen had stood, December 14, 1911; and where Scott had also stood, 34 days later, reading the note which Amundsen had left for him.... There was nothing now to mark that scene; only a white desolation and solitude disturbed by the sound of our engines.... And that, in brief, is all there is to tell about the South Pole. One gets there, and that is about all there is for the telling. It is the effort to get there that counts.[36]

The Byrd expedition of 1928–30 was the first of a series of Antarctic expeditions in which Byrd and others, on behalf of the United States, saw, mapped, and claimed more land than any other nation's expedition. They photographed approximately 150,000 square miles of Antarctica from the air. On the south polar flight they went as far in nine and one-half hours as

Amundsen had gone in fourteen weeks. On one flight, for example, Byrd flew east from the Bay of Whales to the interior and saw a previously unseen mountainous region which he named Marie Byrd Land for his wife, claiming the land west of 150° west longitude for the United States.[37]

Flying in Antarctica was dangerous. The extreme cold often made the airplane's equipment inoperable, and the oil in the engine could freeze upon landing. Landings were difficult: the glaring austral summer sun blanked out shadows on the white ground, so a place that looked smooth from the air could be riddled with sastrugi (snow ridges) and treacherous for landing. Quick-rising Antarctic mists or snowstorms within seconds could suspend a pilot in "a world of milk,"[38] with no horizon and no distinction between safe white cloud and fatal snowy mountainside.

Navigation proved difficult, too. Magnetic compasses were useless, and winds constantly pushed a plane off course so that pilots had to use dead reckoning and sight. Blizzards not only blew one off course, but wrecked planes tethered on the ground. In 1929, when Laurence M. Gould, Byrd's second-in-command, with two companions made a "simple" two-hour flight from Little America to the mountains, a blizzard wrecked their moored plane, dashing with it many of the expeditions's plans.[39]

But the advantages of flight were legion. From the ground Shackleton and the other heroic age explorers could see a swath about 5 kilometers wide. The accuracy of their mapping depended on sketches and the occasional photograph, but a pilot in a plane 600 meters above ground could see 96 kilometers away. With a good aerial camera and good ground control, he could map a strip 192 kilometers wide along his flight path. If the plane climbed to 1,525 meters above ground, he could thus "discover" new lands up to 149 kilometers on either side of the flight path.[40]

Heroic age explorers could march 25 kilometers per day with dog-drawn sledges and 16 kilometers per day if they manhauled them. They depended utterly on caches of food and stove fuel located approximately 112 kilometers apart stored there previously.[41] But a plane could drop food and fuel without landing, thus improving the safety and range of later flights or overland journeys. The first Byrd expedition started the engine of Antarctic aviation, but the enormous U.S. role in Antarctic exploration and discovery—its political legacy—was in part the accidental result of technology, namely, the ever-greater ranges of ski-equipped aircraft used on successive expeditions.

The first Byrd expedition also made overland journeys of importance. Gould led an overland party to the Queen Maud Mountains at the southern edge of the Ross Ice Shelf and found them a continuation of the ranges of Victoria Land. Thus it was estab-

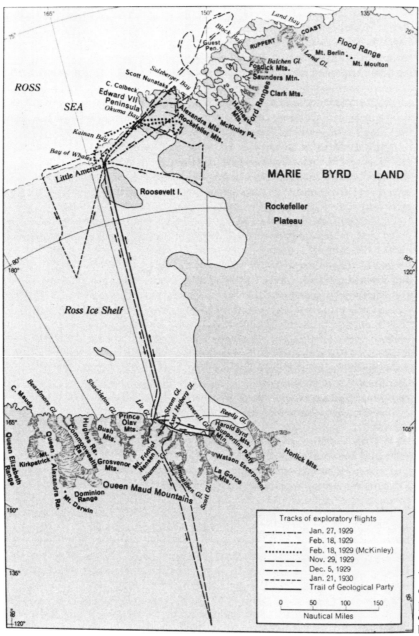

The first Byrd Antarctic expedition, showing flight paths and the trail of the overland geologic party led by Gould.

lished that Antarctica is bisected by a single great range, now called the Transantarctic Mountains.[42]

Byrd saw his Antarctic expeditions as America itself in microcosm; he named his base Little America. His financial backers were a *Who's Who* of national life at the time: the Rockefellers, Edsel Ford, Tidewater Oil, the *New York Times,* and myriad other influential individuals and organizations. The *Times* started a tradition of sending a correspondent along on Antarctic adventures. A national contest was held to select an outstanding Boy

The Ross Ice Shelf, a great natural spectacle, also proves to be a hazard for the first Byrd expedition when a large chunk of ice falls into the Bay of Whales near Little America.

Scout to accompany the expedition. The winner, Paul A. Siple, went to Antarctica at age nineteen in 1928 and became, with Gould, a leader of the next generation of American "Antarcticans."[43] Peculiarly American were Byrd's instructions to his party when, on a later expedition, he left them to winter over alone:

> Every man in this camp has a right to be treated fairly and squarely, and the officers are requested to hold this fact in mind. In a sense our status is primitive.... We have no class distinctions as in civilization. What a man is back home does not count at Little America. He who may have failed back there has his chance to make good here; and he will not be judged by the position he holds so much as by the way he plays the game and does his job, however humble it may be....[44]

The first Byrd expedition was popular back in the Depression-struck United States. Byrd's powerful brother, Harry Flood Byrd, saw that Byrd received a naval promotion, though he had retired from active duty in 1916 after an injury. By a special act of Congress, Byrd was promoted to rear admiral (retired) in 1930. The positive public response to his expedition enabled Byrd to raise funds for a second expedition from, among others, Edsel Ford, chocolate drop manufacturer William Horlick, brewer Jacob Ruppert, and the National Geographic Society. Newspaper and broadcast rights helped fill the till.[45] As part of the preparation, a Byrd family associate, Sen. Millard E. Tydings of Maryland, introduced Senate Resolution 310 to "authorize and direct the President to lay claim to all areas in the Antarctic which have been discovered or explored by American citizens."[46]

Byrd's second expedition aimed at extending the explorations of the first by using the nucleus of polar explorers he had already trained. But in the tradition of many polar explorers, Byrd wrote that his return was "not so much a spontaneous thought as a maturing compulsion bred by the work of my first expedition."[47]

The second Byrd Antarctic expedition of 1933–35 mapped 1.16 million square kilometers. Byrd returned with the *Bear of Oakland* and *Jacob Ruppert* to the Bay of Whales and rebuilt Little America. From there, exploratory flights to the east and west mapped and extended the area discovered earlier. The expedition found that the Ross Ice Shelf, where it was camped, conceals an ice rise, which it named Roosevelt Island.[48]

One of their aims was to make continuous meteorological observations from the Antarctic interior through the austral winter, since every previous expedition had nestled along the coastline, buffeted by weather originating in the interior. With overland tractor hauls, Byrd's men built "Advance Base," but weather conditions and other problems forced them to construct their camp only 160 kilometers from Little America. Because the impending winter cut construction time short, the base was smaller than originally planned. Byrd therefore decided to winter alone in the underground 2.7- by 3.9-meter hut built into the ice shelf, noting, "I could not bring myself to ask a subordinate to take the job." Two men could not winter together without getting dangerously on one another's nerves. Moreover, he added, it "was an experience I hungered for, as soon as I grasped the possibilities."[49]

During his solo attempt at wintering over, Byrd injured himself and was poisoned by gas from a faulty stove, but concealed his condition in radio reports to Little America. He finally was rescued by three men who stayed with him in the hut for two months before they could leave. This tremendous feat of human endurance compares with Shackleton's attempt in 1909 to *run* toward the South Pole during the last day on which his party could keep pushing south and still have enough supplies for the return journey.[50] It compares with Scott's decision, when his motorized sledges and ponies had failed to haul the 270-pound sledges up the steep, deeply crevassed Beardmore Glacier to reach the pole in 1911–12. Along with Scott's diary, Byrd's book, *Alone,* is a classic story of heroism.

Americans struggling with the Depression could hear Byrd's voice patched in by ham operators on their radios and Columbia Broadcasting System broadcasts from Little America. Journalist Murphy was along, producing news copy and helping in other ways. When, for example, the Virginia gentleman alone under the ice in Antarctica found he did not know how to cook flapjacks, Murphy arranged contact with the chef at the Waldorf Hotel in

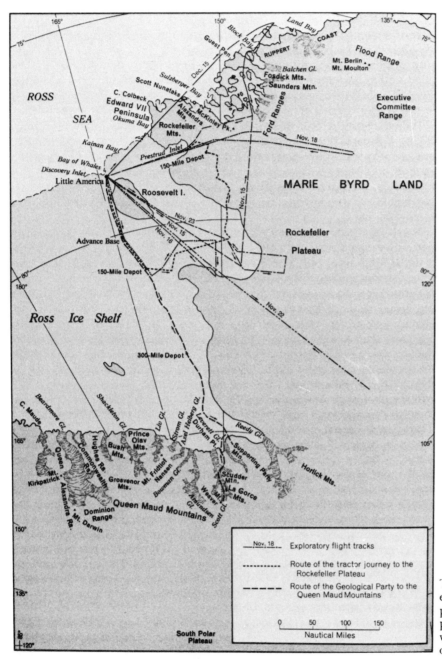

The second Byrd Antarctic expedition, showing flight paths, the trails of overland parties, and "Advance Base," where Byrd alone wintered over.

New York—an event which must have produced great excitement in the kitchen as well as amusing news stories for an otherwise grim world.[51]

Byrd would return to the Antarctic often, as head of the next expedition and as a titular figure on others. Toward the end he was resented by others who wanted to be "Mr. Antarctica" themselves. His stay alone under the ice shelf reinforced his mysticism, and some of his later pronouncements would prove embarrassing to the government, which by then was taking Antarctic policy more seriously. He continued to be the freewheeling "Admiral of the Antarctic," a role incompatible with the bureaucratic formalities of later government expeditions.[52]

The whole Byrd story is far from told; and, some of his achievements have been questioned. After one navy expedition, journalist Drew Pearson reported that Byrd had fought with the official photographer who argued Byrd had not seen some Antarctic territory he claimed to have seen. After Byrd's death in 1957, Balchen, his pilot on the transatlantic and South Pole flights, prepared to publish calculations showing that Byrd and Floyd Bennett could not have made it to the North Pole in 1926. Historian Richard Montague has written that the Byrd family, angered by the impeding accusation, pressured Balchen's publisher to delete the offending passages. Montague includes these passages and the story of the censorship in his own book, *Oceans, Poles and Airmen.*[53] But neither Balchen nor anyone else has questioned that Byrd got as close to the South Pole as was possible, given the crude state of navigation at the time.

But Byrd's political legacy was enormous in the swath of territory that his Antarctic expeditions had found and mapped. He trained a generation of polar experts who carried on even as his star faded: Finn Ronne, whose father, Martin Ronne, had been a sailmaker for Amundsen; Siple, whose book, *A Boy Scout with Byrd*, and poise on a subsequent international tour won him lasting fame; Richard Black, a navy man who held several important Antarctic commands; Thomas Poulter, a physicist; Amory H. Waite, an electronics specialist; Peter Demas; Henry Dater; Carl Eklund; and geologist Gould, the only one whose scientific credentials were extensive enough (he later became president of Carleton College) to continue as a leader in Antarctica's age of science and the IGY.

But this group, once it returned from Antarctica, was unable to turn its vague vision of an "American" Antarctica into official policy. Polar explorers are not team players, except sometimes "down on the ice." The ego required to lead a group through the austral winter does not sit well on government committees or other official settings. Moreover, the group was riddled with rivalries and jealousy, and once back home, its members dispersed. Siple worked for the army on polar clothing and equipment and later headed army research; Black returned to the navy; Ronne appears to have had loose ties to the Central Intelligence Agency; Waite was part of the army signal corps; and Dater became historian of the U.S. Antarctic Projects Office. While unable to mount a coherent Antarctic policy back home, they nonetheless convinced their separate employers of Antarctica's importance. Indirectly, they influenced U.S. Antarctic policy for decades.

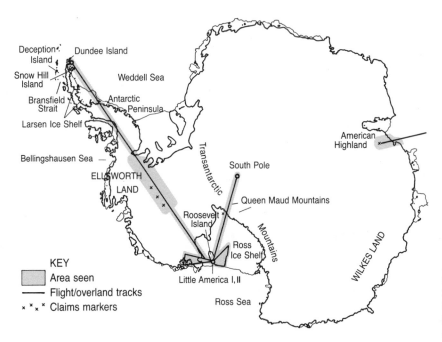

Area seen and claims
markers put down by the
first and second Byrd
expeditions and the first and
second Ellsworth
expeditions.

Ellsworth's Expeditions

Lincoln Ellsworth was born rich enough to inherit a coal fortune
and a Swiss castle. He dabbled in civil engineering and biology,
and flew over the North Pole with Amundsen in a dirigible in
1926. He saw Antarctica as the American West of a century before,
envisioning himself as a "pathfinder" of aviation. He idolized
Wyatt Earp and kept Earp's cartridge belt at his bedside and with
him in his plane. Earp, he wrote, "entered the Western scene at the
beginning of its most vivid era, and when he put away his gun that
era was ended."[54]

In 1935 Ellsworth went to the Antarctic Peninsula. With pilot
Herbert Hollick-Kenyon, he flew down the length of the peninsula and across West Antarctica to the Ross Ice Shelf, crossing
much territory not previously seen. On the flight he claimed some
906,500 square kilometers between 80° and 120° west longitude
for the United States. One indication of how the airplane
revolutionized Antarctic exploration was the fact that the two traveled by air for twelve days over 3,520 kilometers of previously
unseen territory, yet still knew where they were. But when they
were forced to land finally, they spent eleven days wandering
around, manhauling their supplies, until they found the abandoned Little America base. It was only 26 kilometers from where
they had landed.[55]

Subsequently, Ellsworth reinforced U.S. claims in East Antarctica. In 1938 he returned in the ship *Wyatt Earp,* planning to fly

inland to the pole from Enderby Land, in the sector claimed by Australia in 1933. Ellsworth's ship hove as close to shore as possible, enabling him to examine the geology of some of the offshore islands. But the pack ice, the absence of a good coastal base site, and an emergency—involving a crewman who fell overboard—dictated a quick return to civilization. Before turning northward, however, Ellsworth made one flight inland on January 11, 1939, flying south to the 72° longitude meridian, where he dropped a claims marker and flag, claiming an area of approximately 207,200 square kilometers for the United States and naming it American Highland.[56]

The State Department and the Explorers

Ironically, as aviators were discovering more and more Antarctic territory, spokesmen for the U.S. Department of State, in their rare comments on the subject, downplayed discovery as a sufficient basis for territorial claims in Antarctica. During the 1920s and 1930s, official U.S. policy moved in the direction of a new doctrine of "constructive occupation," that is, the revisiting of places previously seen and mapped, as the proper basis for sovereignty. This legal evolution was secret, however; for the most part it took place in intragovernmental memos written in response to pressure from the explorers. But however chancy, it resulted in President Roosevelt's 1939 plan to permanently occupy Antarctica as the basis for an outright U.S. claim to as much of the continent as practicable.

The first policy statement had been made by Secretary of State Charles Evans Hughes in 1924, who was reacting to an erroneous story in the *Rochester Herald* concerning Amundsen's forthcoming flight over the North Pole. The story discussed whether an American along on the flight might make claims for the United States to previously unseen territory there. Apparently, Hughes had been advised that the discovery doctrine, under which European nations had taken possession of the New World and Africa, was no longer sufficient. He said:

> ... Rights similar to those which in earlier centuries were based upon the acts of a discoverer, followed by occupation or settlement consummated at long and uncertain periods thereafter, are not capable of being acquired at the present time. Today, if an explorer is able to ascertain the existence of lands still unknown to civilization, his act of so-called discovery, coupled with a formal taking of possession, would have no significance, save as he might herald the advent of the settler; and where for climatic or other reasons actual settlement would be an impossibility, as in the case of the Polar regions, such conduct on his part would afford frail support for a reasonable claim of sovereignty.[57]

Hughes reiterated this view in May of the same year in reply to one Anson W. Prescott, secretary of the Republican Publicity Association, who, for reasons that are not clear from the surviving record, wanted to know why the United States had not claimed that part of East Antarctica discovered by Wilkes. Hughes replied:

> It is the opinion of the Department that the discovery of lands unknown to civilization, even when coupled with a formal taking of possession, does not support a valid claim of sovereignty unless the discovery is followed by an actual settlement of the discovered country. In the absence of an act of Congress assertative in a domestic sense of dominion over Wilkes Land this Department would be reluctant to declare that the United States possessed a right of sovereignty over that territory.[58]

While this formula seemed to detract from a U.S. claim based on the discoveries of Palmer and Wilkes, it had the advantage of undercutting other nations' claims, particularly those based on the heroic age discoveries, without further State Department action. It also passed to Congress the burden of deciding whether the United States should assert formal ownership in Antarctica, a gauntlet which, despite the Tydings resolution, Congress never picked up.

But Byrd believed the government should assert ownership there. "I am the mayor of this place," he once wrote,[59] and he seemed to think of himself as the custodian of Antarctica until his government took over.

Byrd and Ellsworth were both world-famous figures with powerful family connections; they did not hesitate to pressure the State Department to arrange Antarctic matters to suit their convenience. Byrd's brother Harry came to the Senate in 1933; perhaps Roosevelt's deputies thought that some of Harry Byrd's opposition to the New Deal could be softened if they went along with his brother's wishes regarding Antarctica.

By the 1930s, the State Department allowed acts of administration to take place on Antarctic expeditions; these would later be invoked to support the doctrine of "constructive occupation." When the New Zealand government—which played host to Byrd's expeditions before they plunged southward—presented orders to one of Byrd's lieutenants not to slaughter seals in the Ross Dependency, the order simply was not transmitted to Byrd. Thus, the expedition proceeded to take 200 seals as needed for dog food and oil, politely ignoring New Zealand–British jurisdiction over the area. Postmaster General James Farley sent a postmaster along on the second Byrd expedition to cancel mail at Little America. When the British government protested this challenge to its sovereignty over the area, the British ambassador added that he did not expect and did not want a formal reply.[60] As Europe moved toward war in the late 1930s, the British and U.S. governments were not about to quarrel over stamps in Antarctica.

Harry Byrd wanted the secretary of state to arrange for the government of Norway to promote to captain a Norwegian commander on his brother's Antarctic expedition. He wanted a French ship to pick up Byrd upon his return to Tahiti. Ellsworth's family likewise bullied the State Department to back up his wish to make claims in Antarctica. Bernon S. Prentice, Ellsworth's brother-in-law, asked how the department would react to Ellsworth's plan to claim land for the United States between 80° to 140° west longitude on his transantarctic flight. Undersecretary of State William Phillips telephoned Prentice, saying the proposed claim "would not embarrass the Department."[61]

State Department records reveal an amusing scene between Ellsworth's other brother-in-law Jacob Ulmer and Samuel W. Boggs, the geographer of the department. For decades Boggs was the department's expert on Antarctic geography and claims. The record shows, through his 1933 study of the polar regions and its revisions, Boggs's evident belief that Antarctica was rich in resources and a fair prize. But when Ulmer visited Boggs in 1938 to ask what lands Ellsworth should claim, Boggs repeated the official formula that claims would not "embarrass" the department. Ulmer apparently took Boggs to task for this, saying the British Foreign Office would never be so lukewarm toward its explorers trying to expand British-owned territory.[62]

Ulmer's lobbying worked nonetheless. En route to Antarctica, Ellsworth stopped at Capetown, South Africa, where the U.S. consul gave him secret official instructions on August 30, 1938. Cordell Hull sent these instructions to the consul regarding his meeting with Ellsworth:

> ... You are requested to inform him, in strict confidence, that it seems appropriate for him to assert claims in the name of the United States as an American citizen...regardless of whether or not it lies within a sector or sphere of influence already claimed by any other country. It is, of course, preferable, that such claims shall relate to territories not already claimed by another country. Reassertion of American claims to territory visited by American explorers several decades ago would seem to be appropriate if he should desire to explore such areas. You may suggest the possibility of dropping notes or personal proclamations....[63]

But, just as Phillips had telephoned rather than written to Prentice, Hull was not about to leave a record of official endorsement. Hull's instructions added:

> It should be made clear to Ellsworth that he should not indicate or imply advance knowledge or approval of the Government of the

Finn Ronne, son of
Amundsen's sailmaker
Martin Ronne, led
several U.S. expeditions
to Antarctica.

Polar exploits drew U.S. public attention in the 1930s.
The enthusiasm shown dur-
ing this tickertape parade for
Byrd after his North and
South Pole flights fore-
shadowed that directed toward space exploration during the 1960s.

United States but that he should leave it for this Government to adopt its own course of action.

You may inform Mr. Ellsworth that if he should care to communicate to the Secretary of State a report of his expedition, particularly in relation to areas visited and claims asserted, the Department would be pleased to receive such a report.[64]

Ellsworth obviously followed these instructions on his expedition. When he returned, he met with Hull, on April 19, 1939, to present the claims he had made. Hull received them without comment. Ellsworth also showed Boggs some fossil rocks he had taken on his first expedition from the Snow Hill Islands off the peninsula, and told Boggs that they indicated the likelihood of finding petroleum in Antarctica.[65]

Other nations had been asserting claims in the 1920s and 1930s—without, however, sending expeditions there: Great Britain in 1908, New Zealand in 1923, France in 1924, and Australia in 1933. In 1939 Norway asserted a claim just as the Third Reich was dispatching an expedition to an area of the Antarctic coast that Norwegian whalers had used. How much of this activity was a response to the U.S. expeditions, and how much the U.S. expeditions were a response to the claims of the others, is hard to determine. Nonetheless, the pace of diplomatic activity was quickening. In 1938 President Roosevelt asked for a study of the issue. Hugh S. Cumming, Jr., of the Division of European Affairs at the Department of State, drafted a reply. Responsibility for Antarctica had been given to this division, apparently because of the number of European nations who were asserting rights there.

Cumming's memo of July 1938 expressed concern that the government was too "passive" about its rights in Antarctica. It read: "A naked reservation of American rights now would probably have little weight in an ultimate settlement when balanced against the concrete acts of other nations."[66] Instead, Cumming noted, the U.S. government should be "active" in disputing the claims of other nations, and try to achieve sovereignty in Antarctica through "constructive occupation." The memo defined this as discovery, subsequent exploration, formal claim of possession, and carrying out administrative acts. It is one of the few documents available that spells out the legal basis of subsequent U.S. activity. Because these deliberations were classified, however (Ellsworth's instructions were not declassified until 1954), there is little discussion in the available legal literature about the nature of "constructive occupation" or its validity in law against other theories of sovereignty in Antarctica.

Cumming's memo was the basis for Roosevelt's decision of January 7, 1939, to mount an official government expedition whose aim would be establishment of U.S. rights in Antarctica. Under-secretary of State Sumner Welles forwarded the memo to the president over his own signature on January 6, 1939.[67] Earlier, in May of 1938, Black, who was working for the Division of Territories and Island Possessions of the Interior Department, wrote to Ernest Gruening, its director, outlining U.S. interests in Antarctica and proposing an expedition led by himself and Ronne. Gruening, who apparently liked the idea, told the president about it.[68]

Approached from two fronts, Roosevelt approved the Cumming-Welles memo and sent Cumming and Boggs on a confidential visit to Byrd to ask his help in carrying out the plan. Byrd had been planning a third private expedition, but instead offered his services gratis to the president. Soon after he was invited to visit Roosevelt at the White House and their friendship was launched.[69] Byrd quickly marshaled his polar colleagues to plan what would be the 1939–41 expedition of the U.S. Antarctic Service.

The U.S. Antarctic expeditions were to be run by the U.S. Antarctic Service, located in the Department of the Interior, which oversees other U.S.-claimed territories. The aim of the expedition would be to permanently occupy previously claimed regions and to revisit them frequently in order to maintain sovereignty against other nations' claims. Siple even mentioned "colonizing" Antarctica.[70]

But the goal of sovereignty was to remain secret so as not to alarm rival claimants and spur an outright territorial race there. Roosevelt's instructions to the U.S. Antarctic Service expedition said: "Members of the Service may take any appropriate steps such as dropping written claims from airplanes, depositing such writing in cairns, et cetera, which might assist in supporting a sovereignty claim by the United States Government. Careful record shall be kept of the circumstances surrounding each such act. No public announcement of such act shall, however, be made without specific authority in each case from the Secretary of State."[71]

Apparently, Roosevelt shared the concern of the Norwegian government which feared the possibility of the Third Reich using Antarctica as a wartime base. Thus, the new, "active" U.S. policy also had a security goal—to preserve the southern flank. (Indeed, German raiders based on subantarctic islands would do considerable damage during World War II.) Moreover, he seemed to be thinking of the Monroe Doctrine and at one point suggested that the U.S. claim be asserted on behalf of all American republics. The notice of the expedition that was sent to Latin governments was a masterpiece of ambiguity about what the United States would claim in Antarctica, and for whom.[72]

The U.S. Antarctic Service Expedition of 1939–41

The expedition of 1939–41 established two "permanent" bases: East Base at Marguerite Bay on the peninsula, which was led by Richard Black; and West Base, also called Little America III, on the Bay of Whales.[73]

Contributions came from private donors such as Charles R. Walgreen, the drug store magnate, William Horlick of the malted drops, Walter S. Kohler who made bathroom equipment, Justin W. Dart, and others. The most curious donation was the Snow Cruiser, a giant, 17-meter-long vehicle—longer than Captain Nat's *Hero*. It could sleep four people, had a darkroom and a laboratory, and carried an airplane on its roof. Conceived by Thomas Poulter and designed by the Research Foundation of the Armour Institute, of which Poulter was scientific director, the Snow Cruiser was to be a mobile base. Its front wheels (3 meters in diameter) were designed to retract as the vehicle started across a crevasse and to lower again as the back wheels retracted and followed over. But like Scott's motorized sledges of 1911–12, and numerous other machines designed to "beat" the Antarctic, the Snow Cruiser failed. Not only was it too heavy for the ice and snow, it broke the ramp as it came off the ship. The party spent a week moving this behemoth from the ship to West Base, but following a secret attempt to use it the next spring, it was abandoned.[74]

The expedition carried out wider-ranging explorations than previous explorers had and was also more systematic at leaving and recording claims. The longitude, latitude, time, and date, along with signatures, were recorded on claims sheets that were enclosed in brass cylinders and dropped from the air or buried in cairns. Copies were retained to put on file, presumably, with the secretary of state.

The expedition's discoveries were marked in a new raft of features on the map: the Walgreen Coast, Mount Siple, Mount Tricorn, and, deep in Marie Byrd Land, the Executive Committee Range, named for the interagency group which formed the U.S. Antarctic Survey's board of directors. Ronne made a 1,920-kilometer sledge journey down the west side of the peninsula from East Base, where Ronne was commander, and found that Alexander I Land—discovered and named by Bellingshausen—was only an island, thus disproving any Russian claim that he had discovered the continent based on that sighting.[75]

But like so many previous efforts to establish a coherent Antarctic policy and presence, the U.S. Antarctic Survey and the policy course it represented were cut short by the outbreak of World War II. A more immediate factor was the hostility of Congress which, in its rush to appropriate funds for the expedition in 1939, had not realized that the government sought an ongoing, permanent presence requiring more money every year. Congressmen began asking whether private companies had enjoyed favoritism donating dog food, cigarettes, toilet paper, and even mittens to the expedition. Some asked whether Byrd was profiting from his appointment as commander of the expedition. Byrd rushed back from Antarctica to fight for his expedition's budget, but to no avail. Although the charges were never proved, Congress cut funds drastically, and the men in Antarctica had to evacuate and come home. The expedition's reports were never compiled and published, although some individuals managed to file reports before going on to their wartime jobs.[76]

Postwar Expeditions: Bigger Than Ever

World War II transformed the United States' role in the world. It emerged from the war as the preeminent military and industrial power, and as the sole possessor of the atomic bomb, and so faced new, uncomfortable responsibilities toward the devastated European nations and the growing threat of the Soviet Union. This change transformed all aspects of U.S. foreign policy, including that for Antarctica.

From 1946 on, U.S. activity in Antarctica could no longer be a one-sided national adventure. Instead, Antarctica became a single element in a complex of relations with Europe, New Zealand, and Australia, and the Soviet Union. Government expeditions continued to go there, and helped to build the case for U.S sovereignty. But in government councils back in Washington, outright U.S. ownership became less of a final and more of a first step toward reaching a settlement with other interested nations. The search for an agreement would lead to Eisenhower's 1956 decision to downplay U.S. mapping and claims in favor of supporting science "down on the ice" and, ultimately, would lead to the 1959 Antarctic Treaty.

This new, tempered official attitude also stemmed from the eerie postwar atmosphere. When the president of Chile visited Antarctica and returned with ore possibly containing uranium, he sparked rumors of vast Antarctic uranium deposits and speculation about who would control them. It was also rumored that the United States would test atomic weapons in Antarctica or use nuclear explosions to break up the ice cap so as to reach the presumed mineral wealth below.[77] Such speculation put pressure on U.S. officials to proceed cautiously to avoid inflaming matters, and to seek a peaceful settlement.

So the United States moved from being a quixotic claimant to a regional power broker by virtue of its superior logistics, its knowledge of the peculiarities of the Antarctic region, and the enthu-

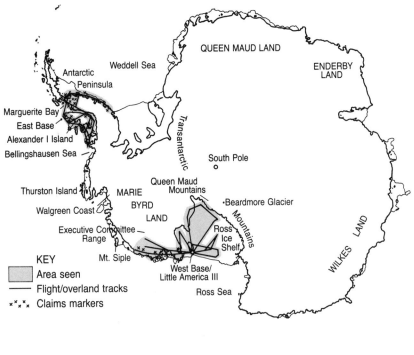

Area seen and claims
markers put down by the
U.S. Antarctic Service
Expedition.

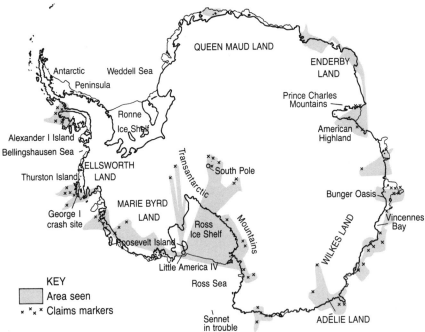

Area seen and claims
markers put down by U.S.
Operation Highjump.

siasms of its native "Antarcticans." This evolution took place in
phases. First, from 1946 to 1949, the State Department tried to
quiet tensions caused by the dispute between the British, Argentines, and Chileans over the Antarctic Peninsula. The department
wanted to announce a U.S. claim as the opening move in negotiations, but this approach failed. A draft condominium proposal

circulated by the United States was rejected quickly by the other claimants, hoping for something better. Then in 1950, the Soviet Union put the others on notice that it wished to participate in any Antarctic settlement, thus changing the entire diplomatic equation. Both events sent U.S. policy back to the drawing board. A ten-year mapping and claims effort, called the National Program, was launched. However, in 1956 during preparations for the IGY, President Eisenhower subordinated this effort to the needs of the scientists for less politically sensitive, more internationally cooperative activities in Antarctica. The emergence of scientists as a new constituency for U.S. Antarctic policy ended the nationalist, expansionist phase of Antarctic activities.

But first came Operation Highjump, the code name for the U.S. Naval Antarctic Developments Project of 1946–47, the single most massive assault on Antarctica undertaken by any nation before or since. In all, 4,700 servicemen participated, along with 51 scientists and observers. One motive was to give the men something to do—not to mention the 13 ships and the tons of equipment—including "weasels," caterpillar tractors, Jeeps, tracked landing vehicles, planes, and helicopters. Thus, it helped the navy solve the problem of postwar demobilization. The military rationale behind the operation was to acquire polar logistical experience, since U.S. forces might have to fight a war against the Soviet Union in the far north. Indeed, two 1946 exercises, Operations Nanook and Frostbite, were carried out in the Arctic, but they had to be kept to a small scale for political reasons. Antarctica offered the military—itself overblown because of the war—a fitting, large-scale exercise ground. Finally, as Lisle Rose, a historian of Highjump I, has written by the winter of 1946–47 the navy needed a showy project to demonstrate its unique importance back in Washington, where the battle over service unification was raging. Byrd also played a role: He had spent the war as a special assistant to Adm. Ernest J. King, the chief of naval operations; afterward he lobbied King's replacement, Adm. Chester Nimitz, and Navy Secretary Forrestal for another Antarctic expedition.[78]

Byrd had the title of officer-in-charge, but the expedition's real commander was Rear Adm. R. H. Cruzen. Cruzen had commanded the *Bear* during the 1939–41 U.S. Antarctic Service expedition and Operation Nanook. Highjump I experienced numerous problems "down on the ice" partly due to the use of so many people unfamiliar with Antarctic conditions. Three men died when the patrol plane George-1 crashed, the first deaths to occur on any Byrd expedition. The submarine *Sennet* was sent with the fleet, but it proved to be a disaster in the pack ice; because it dared not submerge, the ice piled up on the sloping bow. When an icebreaker moved alongside to cut the stricken sub free from the ice, it nearly crushed the delicate, low hull. The *Sennet* was towed into open water, and that was the last time anyone seriously considered sending a submarine into the Antarctic pack. One of Highjump's triumphs, however, was that the icebreakers and helicopters performed admirably. Almost as much as the airplane, they would give the United States an extraordinary advantage in future Antarctic operations.

The expeditions sighted and photographed an estimated 60 percent of the Antarctic coastline, of which about 25 percent previously had been unseen by man. Not only did the expedition revisit West Antarctica, the area most often explored by previous U.S. expeditions, but it also revisited East Antarctica explored by Wilkes and Ellsworth.[79]

Highjump I's survey program seemed designed to help U.S. claims meet Cumming's test of constructive occupation. Acting Secretary of State Dean Acheson notified the secretary of the navy on December 14, 1946, as the expedition was sailing southward, that it would be acceptable for members of the expedition to claim not only previously unseen land but land claimed by other nations. Acheson urged the expedition to "take appropriate steps such as depositing written claims in cairns, dropping from airplanes containers enclosing such written claims, etc., which might assist in supporting a claim of sovereignty by the United States Government and . . . to keep a careful record of the circumstances surrounding each act. . . . No public announcement with respect to these activities shall be made. . . ."[80] As the Central Intelligence Agency would write in a 1948 report, "No suggestion was made for any form of semipermanent occupation. If 'constructive occupation' had still been thought necessary, this principle was defined as including the policy of basing claims on repeated revisiting of places previously discovered."[81] In all, sixty-eight claims markers were dropped during the expedition.

The Windmill and Ronne Expeditions

Highjump I failed in some ways. It photographed 777,000 square kilometers, but this was merely one-fourth of the expedition's goal. Much of the aerial photography lacked adequate ground control.[82] Consequently, during the following season, the navy sent a follow-up expedition, code-named Operation Windmill, to obtain ground control points and continue mapping. Acting Secretary of State Robert Lovett reaffirmed Acheson's policy on claims; twelve claims markers were dropped in the course of the expedition. Among the high points was an attempt to reach a rumored warm spot—an understandable daydream of many a shivering Antarctic explorer—on the East Antarctic coast. Seen by Lt. Cmdr. David E. Bunger from a plane the season before, it had been nicknamed Bunger's Oasis. The expedition's icebreakers, plowing in toward the so-called oasis, indeed found a warm stretch of coast, but it was surrounded by an apron of pack ice and inaccessible from the sea. However, the expedition discovered nearby the Windmill Islands, an excellent landing place in Vincennes Bay, probably seen by Wilkes a century before.[83]

The 1947–48 season saw the last private American expedition to Antarctica, led by Ronne. He reopened the U.S. Antarctic Service's East Base in Marguerite Bay on the peninsula. In a diplomatic minuet that was typical of other occasions when expeditions of rival claimants met "down on the ice," one Major Pierce-Butler, leading a neighboring British expedition, objected when Ronne hoisted the Stars and Stripes over his camp (Ronne was canceling mail there as well). Ronne sent this note of reply: "As an American expedition reoccupying this base on Stonington Island, we have reflown the American flag on the American-built flagpole at the American camp."[84] Indeed, U.S. expeditions had orders not to discuss the U.S. sovereignty position with foreigners in Antarctica, and also were warned not to needlessly irritate rival expeditions. Ronne's reply, and his later good relations with the British group, indicated how Americans in Antarctica would deflect discussion of the issue in public.

Ronne's expedition filled in the presumed American claim in Antarctica from the peninsula—seen and claimed on several prior U.S. expeditions—eastward across the Weddell Sea toward Coats Land on the other side. His flight on December 12, 1947, traced the previously unseen front of the Weddell Sea ice shelf, well south of where Shackleton's ship *Endurance* had sunk years before. Ronne claimed the region for the United States. In all, he said the expedition mapped 1,165,500 square kilometers, half of it not seen previously.[85]

An International Settlement Rebuffed

Postwar U.S. Antarctic diplomacy was the result of the interest of other nations who were more vocal publicly about their claims. After the European powers asserted claims, Argentina jumped in during the 1940s, actively disputing British sovereignty over the peninsula region. Chile, with similar anticolonial emotion, followed suit. Both nations announced their apparently eternal and mutually contradictory claims to the peninsula; and taking advantage of Great Britain's preoccupations elsewhere, they sent expeditions to harass the British outposts in the region.

State Department records indicate that these three rivals preferred to argue with one another than with the United States which, they knew full well, had at least as strong a claim to the same territory and would be a formidable foe if it chose to defend its Antarctic claim. So the three challenged each other, rather than challenging the fourth, publicly silent claimant. This state of affairs was fine with the U.S. State Department, which declined to take sides with the Latin nations against Great Britain or vice versa, let alone challenge all three at once.

The international dispute over ownership and the big U.S. expeditions made other nations nervous. In the late 1940s, Belgium, Norway, Australia, and New Zealand all took steps to remind the others of their interests in the region. Even South Africa took possession of the subantarctic Prince Edward and Marion islands, another indication of the growing international concern.[86]

A 1946 State Department study had concluded that if the U.S. Antarctic claim were put forward and the issue submitted to arbitration, the outcome would be uncertain. In 1947, when Secretary of State James Byrnes was asked at a press conference if the United States planned an international conference on the Antarctic question, he remarked that with so many more important problems in the world, Antarctica hardly warranted a conference of its own.[87]

Ironically, it was Ronne, from lonely East Base on Marguerite Bay in the dead of the austral winter, who sparked a serious government quest for a solution. From there, on June 2, 1947, Ronne sent a telegram to George C. Marshall, the new secretary of state, suggesting a ten-year program of U.S. mapping and claims, and outlining a possible international settlement.[88] The telegram was circulated to the U.S. Army, the Air Force, the Navy, and the Departments of Commerce and the Interior.

The State Department, meanwhile, was talking with the British, and mentioned the possibility of making Antarctica an international trust territory. The British objected, arguing that the Soviet Union would have to be involved, thus the security of the southern portal of the Drake Passage around Cape Horn would be jeopardized. It also warned against encouraging the kind of "neutrality" Chile and Argentina had shown in World Wars I and II. The State Department reassured the British government that no arrangement would be made including the Soviet Union. In fact, the U.S. military was adamantly opposed to any solution involving the Soviets, but did not object to a trusteeship *per se*.[89]

In June 1948, armed with a Central Intelligence Agency report that updated and expanded Boggs's 1933 study, the State Department prepared to go ahead with a trusteeship for Antarctica. The president approved a plan by which the United States would announce its claim first and then move toward a trusteeship negotiation. But the British objected, saying that they preferred an eight-way condominium, one of the earlier options. The department revised its plan and proposed a condominium instead, but the proposal flopped. Five governments rejected it outright;

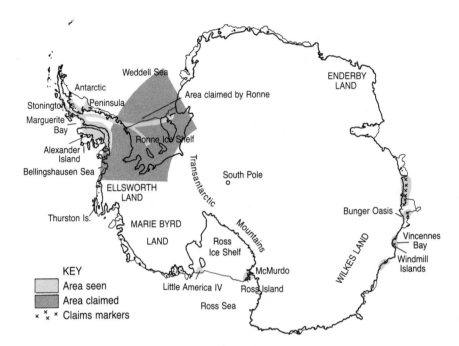

KEY
Area seen
Area claimed
× × × Claims markers

Area seen and claims markers put down by Operation Windmill and the Ronne Expedition.

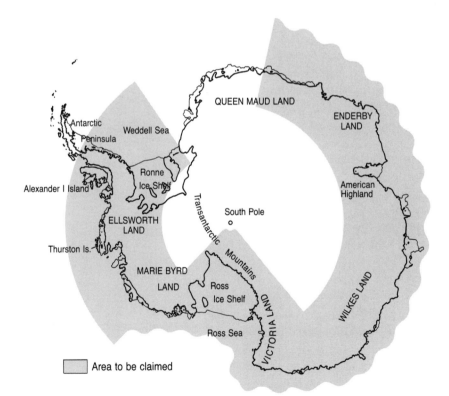

Area to be claimed

Draft U.S. claims to Antarctica drawn up by the State Department in 1948.

As the "Admiral of the Antarctic," Byrd and his circle of fellow explorers would influence U.S. Antarctic policy through the 1957–58 International Geophysical Year.

the other two allowed only that it could be a basis for discussion. In fact, national feeling about Antarctica had just been aroused; it could hardly be eliminated by a stroke of the U.S. pen. Finally, adverse public reaction to the U.S. plan in Latin America caused the department to postpone plans to announce its claim later that year.[90]

This more "active" U.S. policy included plans for Highjump II, which would "assault" Antarctica once again. As approved in April 1949, Byrd would be in charge and Dufek and the others would have key roles; 158 officers, 1,093 men, scientists, and news correspondents would go to the Antarctic for the 1949–50 season.[91] But this time the "Antarcticans" were unlucky. Harry Flood Byrd had opposed President Truman's bid for reelection in 1948, something which Truman would not forget. Truman, who could not retaliate directly against the senator, reportedly quipped that there were "too many birds" in Washington. In August 1949, the president got back at Harry by canceling Richard's pet project, Highjump II. The secretary of defense, looking for defense budget cuts, concurred. Thus, so long as Truman was president any chance for further U.S. Antarctic expeditions was doomed.[92]

Soon after, time ran out for the United States to obtain an all-Western settlement for Antarctica. The stepped-up national interest in the region was noted in the Soviet Union, and in June

1950 it sent diplomatic notes citing von Bellingshausen's voyage, whaling, and other Soviet interests, and saying: "The Government of the USSR cannot agree that such a question as that of the regime of the Antarctic be decided without its participation.... [It] cannot recognize as legal any decision regarding the regime of the Antarctic taken without its participation."[93]

The National Program, 1950–56

These setbacks put U.S. Antarctic policy back to square one. Acheson, rebuffed on his condominium proposal, decided to continue diplomatic discussions on the basis of a 1948 proposal by Chilean law professor, Julio Escudero Guzman, for a moratorium on the Antarctic claims dispute. But now, the U.S. negotiating position would have to be secure against a challenge from the Soviet Union, since Soviet propaganda was asserting von Bellingshausen's prior "discovery" of the Antarctic continent. A July 1951 policy statement listed four U.S. goals for the region: (1) to eliminate international friction on claims; (2) to undertake systematic scientific investigation of the area; (3) to ensure that U.S. enemies would not control any part of Antarctica; and (4) to uphold U.S. "Arctic" [presumably meaning Antarctic] interests.[94]

As to the relative merits of U.S. and Soviet claims, one study quoted the State Department's view: "The 'eminent, very extensive, and long-continued activities of official and private American parties in Antarctica' placed the United States in a 'wholly different status' from any which could successfully be claimed by the Soviet Union on the basis of Russian activities of 127 years ago."[95]

The National Program was the government's plan to document this assertion through a ten-year effort. Siple's book, *90° South,* is the main source of information at present. In 1950 the State Department "suddenly" decided to reconsider making a U.S. claim, but Siple told the officials:

> "... You haven't got a single document in your files to show what the United States has done...."All that existed was a mass of personal literature in the form of a number of books and scientific articles, diaries, letters, specimens and photographs, none of which had been examined and authenticated officially. Even the vast amount of data gathered by the U.S. Antarctic Service had not been collected and printed....
>
> "All right," said State Department officials, "let's put things in order now." A note of urgency hung in the atmosphere.
>
> "Everything is going forward at an ever-accelerating rate," Admiral Byrd said and smiled.[96]

Committees were set up to consolidate U.S. rights and carry out the mapping planned for Highjump II. There was a mapping committee, of which Siple was a member, with two subcommittees called "Where" and "How." Byrd was "only adviser to the Chief of Naval Operations but he had no intention of taking a back seat to others."

Another committee gathered the historical record. This was the Technical Advisory Committee on Antarctica (TANT). Chaired by the perennial Boggs, it included Siple and McKinley (who had done aerial mapping on Byrd's South Pole flight and subsequent expeditions), archivist Herman Friis, and representatives of the State Department, CIA, the Weather Bureau, the Interior Department, and of the Army, Navy, and Air Force. In 1953 the State Department set up an *ad hoc* interdepartmental committee to coordinate Antarctic policy and claims. One deadline set for the claim announcement was April 1954.[97]

A major effort was made to collect all records relating to a possible American claim in Antarctica. Paul Schratz, then a commander in the navy who was Byrd's representative to TANT, recalls that as part of the effort, a young naval officer was dispatched to New England libraries and museums to find out about the activities of nineteenth-century sealing captains in Antarctica. Much of the historical work discussed earlier dates from this period. Schratz recalls that the officer found old logs of American captains, and that the government promptly classified them. There is no documentation of such an incident, but others involved with TANT say it may have happened.[98] In any event, it is possible that TANT uncovered new historical material.

A draft U.S. claim drawn up at this time included the periphery of the continent from 35° west longitude around to 13° east longitude, excluding the very tip of the peninsula north of 68° south latitude. No rationale for this is included in the document. The claim was not put forward because Byrd advised that the "blank" areas on the map could be filled in and claimed by the next expedition.[99] Indeed, the flight paths of aircraft on the subsequent Deepfreeze expeditions and locations of markers dropped suggest that this is just what they were doing.

The "Marriage of Convenience" for the IGY

But as TANT and the other committees got under way, a rival group was making known its wish to use Antarctica for a different purpose, that of scientific research. At a dinner party held by James A. Van Allen in 1950 in Silver Spring, Maryland, some prominent scientists conceived of the idea of using the upcoming 1957–58 Third Polar Year (following on those of 1882–83 and

1932–33) as a chance to corral not only other nations, but the new postwar technology (such as sounding rockets, improved balloons, and the like) to investigate earth geophysics, a science just blossoming. The "year" would coincide with a solar "maximum," a period when solar activity would be impacting on the earth's ionosphere and could be observed from both poles. Because the Arctic was mostly an ice-clogged ocean, observations at that pole would be difficult, but the relatively accessible continent of Antarctica offered an excellent potential "laboratory" for the international effort.[100]

The plan, which soon broadened to include observations from all over the earth's surface and encompassed several other disciplines, was rechristened the International Geophysical Year (IGY). It also became something of a *cause* to the organizers, who sought to include as many different countries as possible, regardless of their political stripe, as a way of showing how science could advance international understanding (the Soviet Union and both Chinas were among the sixty-seven participating countries). But the key U.S. organizers included some of those with strong loyalties to Antarctica, such as Lloyd V. Berkner, a radio engineer on one of Byrd's first expeditions, who had written a 1949 National Academy of Sciences report on the need for scientific research in Antarctica and who had been at Van Allen's dinner party. There was also Gould, whose scientific credentials were excellent. President of Carleton College since 1945, Gould headed the USNC's Antarctic Committee.[101] In addition, the group had a powerful institutional base in the National Academy of Sciences' U.S. National Committee for the International Geophysical Year.

At first, the scientists hardly believed they would obtain government funding for this highly ambitious project, especially the Antarctic effort, so they considered renting Danish icebreakers to transport their expeditions. According to several accounts, initially the government was reluctant to underwrite the project. It was unused to funding huge scientific undertakings and, moreover, the thought of cooperating with the Soviet Union in the early 1950s seemed peculiar.

But gradually, within the National Security Council and other forums, the scientists won their case. Eisenhower became persuaded of the importance of a major U.S. commitment to the IGY generally, including the Antarctic phase. Of course, because the National Program was classified, as were the existence of TANT, the "Why" and "How" committees, and the plans for "constructive occupation" of the Antarctic continent, so many scientists involved with the IGY were unaware of the government's plans. Siple somehow bridged both sides of what he called this "Antarctic ice fence": he worked on the National Program and also served on Gould's IGY Antarctic committee. He wrote: "It occurred to me early that

the two separate ventures of science and national interest could be
loosely linked by a marriage of convenience. Other countries, such
as Belgium and Argentina, planned to do just this, while the Norwegians, the French, and the New Zealanders made a point of
planning their IGY bases in territories they claimed."[102]

But Siple, who felt he had been brushed aside by the scientists,
clearly had difficulty visualizing a nonpolitical Antarctica. Strad-
dling the "ice fence" was not easy: "Some of the scientific group
would have been aghast at any suggestion that the dedicated international program join hands with the nationalistic interests. On
the other hand, the national program mapmakers were reluctant
to consider IGY as of any possible value in their own programs
and goals."[103]

The "marriage of convenience" is reflected in the scant records
of President Eisenhower's Antarctic policy decisions during this
period. In July 1954, Eisenhower approved NSC 5424/1, a policy
based on a National Security Council staff study that combined
both elements. It judged Antarctica's strategic and resource value
to be slight. Nevertheless, it recommended that the United States
should assure its rights, deny the region to probable enemies, and
assure access to any natural resources found there. U.S. policy
should reassert U.S. territorial rights at the appropriate time,
while showing readiness to negotiate a settlement with the other
claimants. Thus mapping, exploration, and "permanent stations"
were needed to support this policy.[104]

At the same time, the policy gave the scientists what they
needed. Antarctica's present value was judged to be scientific, and
the policy urged support for scientific programs and other international efforts that would lower tensions there. When he approved the policy, Eisenhower also agreed to request funds for
U.S. Antarctic activities for the IGY. The National Science Foundation, founded in 1950 to support U.S. science, would serve as
fiscal agent, while the quasi-governmental National Academy of
Sciences (that is, its IGY committees) would decide how to spend
it. Almost magically, after this, the scientists found the military—
whose "Antarcticans" sniffed a return to Antarctica in the air—
very willing to consider new bases, ship routes, and flight programs. Many were unaware of the ulterior motives involved, so it
was, for most participants, a secret wedding.

The Mappers Overstep

But the National Program came to an abrupt end, even though its
essential duties of mapping and establishing claims would continue to be advocated by the military. In 1954 Boggs died, and

with him went much of the State Department's institutional memory concerning U.S. national interests in Antarctica. The TANT committee reorganized, turning its work over to Bertrand, the professional historian and geographer who finally published a history of the expeditions—in sanitized form—in 1971.[105]

Siple's book relates that the National Program "fell to pieces" in early 1956, when the overly enthusiastic mappers in the "How" group asked for $56 million to make a map of Antarctica on a finer scale than some maps of the United States. When the president saw this request, he was both surprised and irritated. "What are we actually committed for down there?" he asked. Alan Waterman, director of the National Science Foundation, replied that the government was committed only to the IGY scientific program. "[W]hen no one contradicted Waterman, the President said we would carry out the IGY program and nothing else," Siple wrote. The budget sent to Congress had no provision for an Antarctic mapping program. Nonetheless, Siple notes that the reaction of Byrd, by then in ill health, was to support the IGY as the "opening wedge for a renewal of the National Program."[106]

This precise story cannot be confirmed, but something like it must have happened, for NSC 5424/1 was revised by Eisenhower personally in January 1956, when the references to "permanent stations" and an ongoing mapping program were deleted. Instead, the stations and the ongoing program were to be "for scientific purposes only." Instead of emphasizing eventual assertion of a U.S. territorial claim, the revised policy emphasized that the United States would seek agreement with "free world claimants to Antarctic territory" under which U.S. rights would be reserved, and freedom of exploration and scientific investigation carried out.[107]

A similar policy shift occurred in the orders given to the three official expeditions sent to Antarctica for the IGY: the reconnoitering trip of the icebreaker *Atka,* sent in 1954–55 to look for suitable base sites; Deepfreeze I, which left the following year in order to build the bases; and Deepfreeze II, which was to provide logistical support for the IGY itself. The *Atka* expedition's guidance, issued November 2, 1954, said that its purpose was not only for reconnoitering for the upcoming IGY, but for the "preservation and enhancement" of U.S. rights. In the past, members of U.S. expeditions had made claims, and members of the *Atka* expedition should "feel free to do the same according to the attached form, retaining appropriate record of the same." Members need not keep their claims secret but should emphasize that "they are not claims formally made by the U.S. government" although the government might announce a formal claim in the future. As to the legal basis of U.S. sovereignty in Antarctica: "[A]ll kinds of

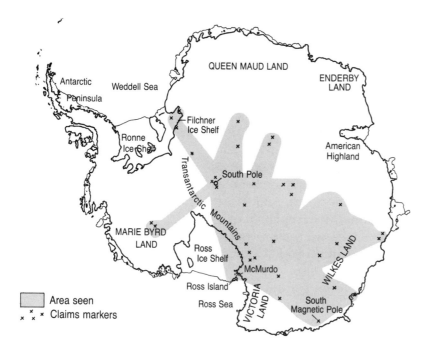

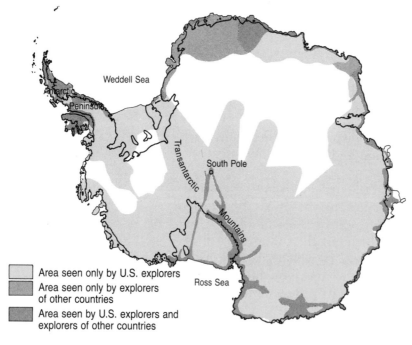

Area seen and claims markers put down by Operation Deepfreeze I and II.

U.S. government map released in 1960 at the ratification of the Antarctic Treaty, showing that 80 percent of the Antarctic continental area had been seen only by American explorers, while a smaller part had been seen by American and other explorers and a still smaller part seen only by explorers of other nations.

national activities may have some significance as grounds for potential claims to the areas where they occur. Examples include discoveries, maintenance of stations, detailed exploration and scientific observations, all of which are of extreme value if they result in physical evidence of a public nature...." As for encountering foreigners, the guidance recommended minuets like the one Ronne had danced with the British earlier: "In the past such meetings have generally been friendly, although some have led to unpleasantness. No firm rules can be prescribed, and it may be necessary to use ingenuity to avoid incidents, while at the same time not relinquishing the right to freedom of movement."[108]

The orders for Deepfreeze I, issued the following year, included instructions to conduct mapping operations as recommended by TANT and to establish permanent stations in support of U.S. rights. But the orders for Deepfreeze II, issued a year later, after Eisenhower's decision, required that operations support the U.S. IGY program. No mention was made of U.S. "rights."

But the idea of an "American" Antarctica died hard, presidential policy or no. Henry Dater's list of the location of claims markers shows that Deepfreeze II, as well as Deepfreeze I, dropped claims. Ronne even put one down after the IGY had begun in 1957.[109] Possibly similar reinforcement of the "prior" claim—which is allowed under the Antarctic Treaty—has continued since then, although there are no public references to such activity.

The United States at 90° South

By the time the IGY began on July 1, 1957, U.S. Antarctic policy had become an amalgam of nationalistic elements subordinated to international cooperation in the name of science. The age of science—and public belief that Antarctica was "international"—had begun. Even so, Siple and Navy Seabees had just finished building a station at the geographic South Pole, the heart of the continent and the point at which the sector claims met. This station symbolized the previous American interest in acquiring territorial rights to the entire continent, standing there in silent rebuke to the others, who had neither the money nor logistical expertise to build it. So if an image of the nationalist, expansionist legacy of the United States in Antarctica exists, it is this: the Navy's best Seabees, who had trained for every possible contingency; the big silver and red Air Force Globemaster airplanes that left McMurdo and made the five-hour flight to the Pole; the pallets of building equipment, scientific instruments, rations, humans and dogs falling into loose Antarctic snows while men in parkas ran around waving and shouting.

No men had stood at the South Pole since Scott in January 1912. But from the grave, Scott made his contribution. When Admiral Dufek decided that the snow around the Pole would be too soft to permit landing and takeoff of heavy, ski-equipped aircraft, one ingenious subordinate dug out the famous picture of Scott's party at the Pole with Amundsen's flag in the background. By examining how far their boots had sunk in, it was determined that the snow was hard enough to proceed. There were other reminders of Scott's awful journey. The first team was accidentally dropped 13 kilometers from the true Pole. Immediately short of breath and weakened because of the high altitude, they nonetheless dog-sledged and navigated with sunsights in the traditional manner until they found the exact spot. Once there, they kept moving the "Pole" around as the sightings were refined; meanwhile the prepared symbol for the Pole, a gay orange and black barber pole, sported from the roof of a new-built garage.[110]

What was the United States doing there, sending a team to build real buildings at the heart of the earth's most hostile continent? As Dufek had said of Antarctica years before, what good was it? The nationalist expansionist effort had failed, basically, since the imperfect U.S. claim was never promulgated by the government back home. The group of "Antarcticans" who had worked so hard to annex Antarctica to the United States—though for what purpose still was not clear—was being pushed aside by the new group of scientists with very different concerns. One thing had been accomplished: by virtue of its extensive visits, use of technology, map-making, and expertise about the region, by having the huge potential claim in its diplomatic back pocket, and by being interested in keeping the peace there, the United States had become key to Antarctica's disposition, during the IGY and for years after.

three

Other National Interests

A widespread misunderstanding about Antarctica is that it is a political blank, a *tabula rasa* with no memory of human passage, for which political institutions can be built at will, just as a well-prepared team can pop up at a station on the white expanse of the South Pole.

But as U.S. diplomats found in the late 1940s when they tried to arrange an international settlement for the seventh continent, Antarctica is hardly a glistening white gift waiting to be presented to the world community by some benevolent power. Besides the United States, many other nations have sent explorers trekking across the ice and snow, threading their way up glaciers, and chipping away at rocky outcrops. Many nations' airplanes have cast shadows on the whiteness, dropping claims markers and flags. No less important is the diplomatic brouhaha over territorial rights that has been heard in capitals and embassies around the world ever since Great Britain, in 1908, announced the first formal claim to land there.

The web of disputes and alliances between the nations involved in Antarctica has shifted often. It changed as other European powers followed Britain's suit, announcing and recognizing one another's formal claims in the 1920s and 1930s. It shifted again in the 1940s, when Chile and Argentina became vocal about their sovereignty and asserted their positions with shows of force: some Argentines even fired over the heads of a British landing party in 1952. It shifted again in 1949–50 when the Soviet government, citing von Bellingshausen's voyage of 1819–21, said it wished to be part of any disposition of Antarctica.

This chapter traces the interests of other nations in Antarctica, including the eleven with enough historic interest to mount scientific programs in the region for the International Geophysical Year (IGY) which later negotiated the Antarctic Treaty. It also discusses the historic interests of West Germany and Poland, the two nations to have attained voting, or consultative, status under the treaty from 1961, when it entered into force, until developing world interest prompted India and Brazil to join in 1983. An understanding of this web of competing political interests of the historically interested countries is essential to an understanding of the Antarctic Treaty, which works by holding them in balance. The chapter also describes the IGY and how as a result of it, the original twelve nations came to negotiate the treaty.

Territorial Claims and International Law

Antarctic claimant nations, in general, vigorously uphold their "rights" at home and in international meetings. But in Antarctica itself, while their national expeditions try to do nothing to undermine their sponsors' political positions, the explorers' rule of self-help also obtains, enabling supposedly rival groups to cooperate and even to enjoy one another's company. Disputes over territorial claims in Antarctica have a unique quality that can be illustrated by two anecdotes.

One concerns the British ship, *John Biscoe,* which brought men to rebuild the British base at Hope Bay on the peninsula in 1952. An Argentine party was already there, however, and fired over the heads of the British landing party. Soon after, the Argentine government, while not renouncing its claim to Hope Bay or other territory, said the incident had been an error. Subsequently, the British party landed and built its base.[1]

The second involves E. W. Kevin Walton, another Briton, who encountered a Chilean party at the base on Neny Island in Marguerite Bay. The Chilean shipmaster disputed British authority there, Walton wrote, but followed up his protests with invitations to "dinner and pictures."[2]

No shots have been fired lately among groups in Antarctica, but rival claims continue. An indication of the national emotions that can underlie such claims was shown in 1982 when the ancient dispute between Great Britain and Argentina over sovereignty to the Falkland Islands erupted into war. The rest of the world had ignored the dispute and the intermittent negotiations it engendered; it seemed an anachronism in the late twentieth century.[3] Nonetheless, the two countries fought a war over the islands, showing what deep emotions can underlie quaint legalisms. Indeed, in the United States, public awareness of its historic tie to Antarctica has declined in recent decades; however, in some other nations the public is more aware of the Antarctic connection. Australian and Argentine politicians seriously view "their" Antarctic territory as an extension of the territory of the nation. And in Great Britain and France, both historic and current ties to Antarctica are well known.[4]

One doctrine used by competing claimants is *discovery.* The British invoke it concerning Edward Bransfield's supposed first sighting of the Antarctic mainland in January or February of 1820, and the first sightings of James Clark Ross, James Weddell, and the heroic age expeditions. But even discovery in the nineteenth century did not give title. The U.S. Secretary of State Charles Evans Hughes said as much in 1924, when he noted that discovery gave title only if it heralded the "advent of the settler."[5]

Hughes was invoking *effective occupation,* a doctrine used often in the eighteenth and nineteenth centuries. If a nation showed it could control a region administratively and defend it from rival claimants by force if need be, it was considered to be the owner. But following Hughes's statement, several international arbitrations moderated this doctrine for remote areas.

In 1928 the Netherlands was awarded the tiny island of Palmas (or Miangas) in the Pacific because, as Judge Max Huber has written, "Manifestations of territorial sovereignty assume, it is true, different forms, according to conditions of time and place. Although continuous in principle, sovereignty cannot be exercised in fact at every moment on every point of a territory."[6]

And in the 1932 arbitration over Clipperton Island, a desolate guano-covered rock located north of the Galapagos Islands, the arbitrator—King Victor Emmanuel III of Italy—allowed that occasional administrative acts were a basis of sovereignty. According to a respected account by Philip Jessup and Howard Taubenfeld, the ruling allowed sovereignty if it was "reasonable . . . in view of the extent of the territory claimed, its nature, and the uses to which it is adapted and is put. . . ."[7] In 1933 the Permanent Court of International Justice disallowed Norway's claim to part of Eastern Greenland, and ruled for Denmark, because it had shown a continuous display of authority with the intent of acting as sovereign.[8]

Besides effective occupation, there is *constructive occupation,* as outlined by Hugh S. Cumming, Jr., of the Department of State in 1938, and proposed to President Franklin D. Roosevelt by Under Secretary of State Sumner Welles, as the basis for a U.S. sovereignty claim in January 1939. Constructive occupation was defined first as discovery, followed by "subsequent exploration by air or land, coupled with a formal claim to possession," as well as "other acts short of actual and permanent settlement."[9] Roosevelt tried to implement this policy with the additional aim of permanent settlement through the U.S. Antarctic Service expedition of 1939–41; later U.S. expeditions carried it out through revisiting and making claims, which were, however, not publicly announced. So although the United States proposed the doctrine and seemed to adopt it as official policy, it is probable that its own claim may

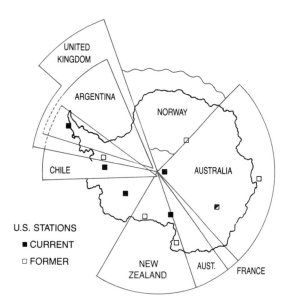

The seven announced claims to Antarctica. Six have sector claims converging at the Pole.

not meet the standard set by the doctrine, since "formal claim to possession" was made by setting down claims markers in Antarctica but never promulgated publicly from Washington.

Six Antarctic claimants (all but Norway) invoke the sector principle. It was probably initiated by a Canadian, Senator Poirier, as the basis of Canadian sovereignty over Arctic islands and waters. Reasoning that these are a "region of attraction" to the Canadian landmass, Poirier proposed that two lines, extended from either end of the main landmass, should run due north until they converge at the geographic North Pole. Everything inside this wedge, he said, should belong to Canada. Canada's minister of the interior detailed this position in 1925.[10]

The first official use of the sector principle was made by Great Britain when it announced its Antarctic claim in 1908; in 1926 a Soviet statement invoked the principle to assert Soviet sovereignty over a wedge of the Arctic extending to the North Pole.[11]

Supposedly, the sector principle is based on concepts of *continuity* and *contiguity* under which some coastal nations claim islands offshore. Yet Antarctic claimants invoke the sector principle to parts of the Antarctic continent that are across the water—700 miles away in the case of Chile and Argentina and 2,200 miles away in the case of New Zealand. Moreover, Antarctica is a geologically distinct continent from these landmasses, whereas the Arctic islands are geologically continuous with the Asian and North American landmasses lying south of them.[12]

Great Britain

The British interest in the polar regions, north and south, was an outgrowth of empire and of the advance of British science and geography in the eighteenth and nineteenth centuries. The Royal Geographic Society urged scientists to push back as far as possible the limits of the known world, while the empire—ruled and served by sea—had a proprietary reflex concerning any islands and harbors, no matter how remote, that guarded sea lanes and straits.

The British see their sovereignty in the South Atlantic–Antarctic region as falling into three categories. The first is the northern part, and it comprises the Falkland Islands well to the north of the Antarctic area which starts at 60° south latitude. Then there are the South Sandwich Islands and South Georgia, just north of 60° south latitude, and the South Shetland and South Orkney Islands just below 60° south latitude. All these islands were British-discovered and used by British sealers and whaling-station masters. The third element is the peninsula and coasts, and some inland areas of the Antarctic continent itself, that have been explored by British subjects.

The British claim to the Falkland Islands is legally distinct from its claims to the other two elements. Capt. John Byron (grandfather of the poet) hoisted the British flag there in 1765 because the islands controlled the ship route around Cape Horn. Earlier, Louis Antoine de Bougainville had established an outpost for France on another part of the islands, and in 1766 he transferred the French title to Spain. In 1770 Spain challenged England's possession of the Falkland Islands, nearly causing a war, but the British continued to occupy them from 1833 onward. Thereafter, in the frequent disputes with Argentina, which inherited the title from Spain, the British colonists of these windswept, sheep-dotted hills repeatedly stated their preference for Great Britain as their ruler. Their wishes were a key factor in the British government's decision to oust the Argentines by force after they invaded in 1982.[13]

As whaling in the region increased at the beginning of the twentieth century, the governor of the Falkland Islands was empowered to issue licenses to whaling-station operators and to collect revenues. Indeed, the reason the government issued Letters Patent in 1908, extending the Falkland Islands Dependencies southward over Antarctica, was to reinforce the system of licenses.[14]

In 1908 the claim was drawn to include not only all the islands, but all land south of 50° south latitude in a sector extending down to the Pole. However, the claim thereby included part of the Chilean and Argentine mainland. This error was corrected in 1917; the area was redefined as two adjacent wedges, one starting north of the Falkland Islands at 50° south latitude and one beginning south of the South American mainland.[15] The claim soon would become unimportant economically, for after 1925 the whalers were able to flense their catches in the open ocean and no

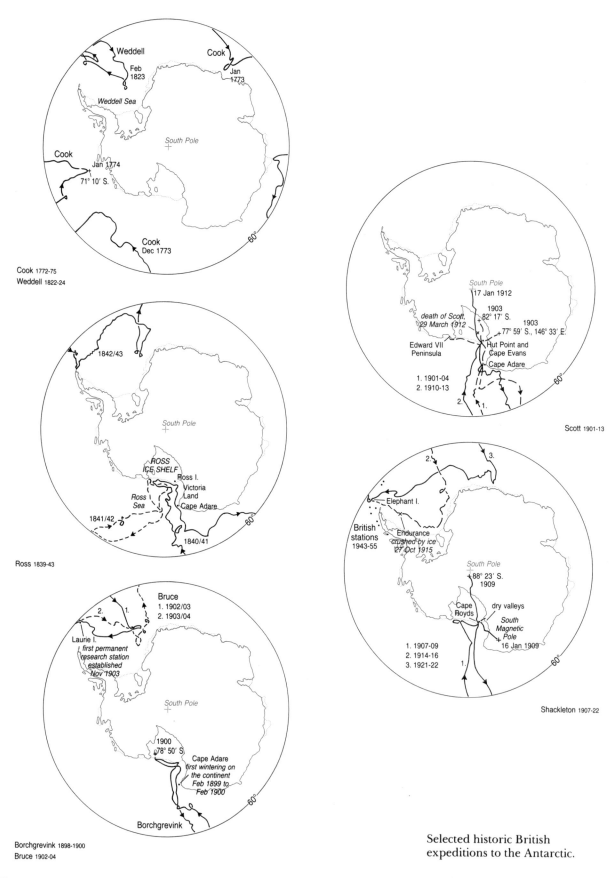

Cook 1772-75
Weddell 1822-24

Ross 1839-43

Borchgrevink 1898-1900
Bruce 1902-04

Scott 1901-13

Shackleton 1907-22

Selected historic British
expeditions to the Antarctic.

Within the first map:

Weddell
Cook

Feb
1823

Weddell Sea

Cook

Cook

Jan
1773

South Pole

Jan 1774
71° 10' S.

Cook
Dec 1773

60°

Within the second map:

1842/43

South Pole

ROSS
ICE SHELF

Ross I.

Victoria
Land

Ross
Sea

Cape Adare

1841/42

1840/41

60°

Within the third map:

Bruce
1. 1902/03
2. 1903/04

2.   1.

Laurie I.
first permanent
research station
established
Nov 1903

South Pole

1900
78° 50' S

Cape Adare
first wintering on
the continent
Feb 1899 to
Feb 1900

Borchgrevink

60°

Within the fourth map:

South Pole
17 Jan 1912

1903
82° 17' S.

death of Scott,
29 March 1912

1903
77° 59' S., 146° 33' E.

Edward VII
Peninsula

Hut Point and
Cape Evans

Cape Adare

1. 1901-04
2. 1910-13

2.   1.

Within the fifth map:

2.   3.

Elephant I.

British
stations
1943-55

Endurance
crushed by ice
27 Oct 1915

South Pole
88° 23' S.
1909

Cape
Royds

dry valleys

South
Magnetic
Pole
16 Jan 1909

1. 1907-09
2. 1914-16
3. 1921-22

1.

60°

longer depended on shore stations. The first sign of Argentine objections to British sovereignty came in 1925, when Argentina filed for an internationally issued radio license for one of the island stations.[16]

Tacit recognition of British sovereignty by other nations eroded in the 1940s with the Chilean and Argentine challenges. The emerging dispute and the German naval threat prompted Operation Tabarin to assure British occupation of key points in the region. Named for a Paris nightclub, it was run by J. W. S. Marr, who would make a major contribution to the study of Antarctic krill. In 1944 the operation was renamed the Falkland Islands Dependencies Survey (FIDS); its explorer-scientists stayed on in the region to conduct research and uphold British rights in the increasingly acrimonious sovereignty dispute.[17]

In 1947, 1951, 1953, and 1954, Great Britain suggested that the three parties take the issue to the World Court, but Chile and Argentina declined. Then, in 1955, Great Britain applied unilaterally to the court for arbitration, but the case was dropped after Chile and Argentina claimed that the court did not have jurisdiction.[18] The dispute continues today, a paler form of that which arose over the Falkland Islands.

In 1953 management of the FIDS was transferred from the governor of the Falkland Islands to a separate scientific bureau headed by explorer Vivian Fuchs. Fuchs led the Commonwealth Trans-Antarctic Expedition during the IGY, during which time the survey's acting director was Raymond Priestley, who had been on the Scott expedition of 1910–13. Priestley extended the survey's ties to British university science, and in 1962 it was renamed the British Antarctic Survey (BAS). It was at first part of the Foreign Office—a remnant of the colonial past—but in 1967 the survey was incorporated into the Natural Environment Research Council, one of the five councils that sponsor British scientific research.[19]

Although threadbare, the British effort in Antarctica has been characterized by a high degree of continuity. Those, such as Frank Debenham, Sir Charles Wright, Priestley, and Marr, whose first polar experience was in the heroic age, continued with posts in the British Antarctic Survey and the Scott Polar Research Institute at Cambridge University. The next generation includes Fuchs, N. A. McIntosh, and Brian Roberts, who went on the British Graham Land Expedition of 1934–37. Roberts was both director of the Scott Polar Research Institute and chief Foreign Office diplomat for Antarctic matters. After his retirement from the Foreign Office, he was succeeded by his long-term second, John Heap. Richard M. Laws, a distinguished Antarctic scientist, became director of the British Antarctic Survey in 1973. In addition to running the two ships, six bases, and the scientific staff of the British Antarctic Survey program, Laws is also an influential representative at international meetings, particularly those dealing with Antarctic marine living resources. And, when the House of Lords discusses Antarctica today, Lord Shackleton, Baron of Burley and son of the explorer, makes his views known.[20]

New Zealand

British subjects discovered much of the coast of Antarctica south of New Zealand, so in 1923, as part of its assertion of sovereignty, the British government by Order-in-Council claimed a wedge stretching across the Ross Sea and running inland to the South Pole. It called this area the Ross Dependency and assigned it to New Zealand; thereafter Norwegian whalers working in those waters were to apply to New Zealand for licenses.[21]

Hilly, green, temperate, and above all friendly, New Zealand has assisted many Antarctic expeditions. Adm. Richard E. Byrd once remarked that Nature must have had in mind compensating Antarctic explorers for their difficulties when she endowed New Zealand with such generous people. Lyttleton, Wellington, Christchurch, and Dunedin have seen polar ships coming and going, often stopping to refuel and enjoy the benefits of civilization. When someone has been found unsuitable at the last minute, often a New Zealander has been chosen as a substitute. A famous statue of Robert Falcon Scott adorns one of Christchurch's principal public parks. Christchurch's Canterbury Museum has a large collection to which many Antarctic explorers have contributed. This helps to explain New Zealand's awareness of "its" Antarctic territory, as well as its traditionally cooperative stance in Antarctic diplomacy. In 1956 Prime Minister Walter Nash proposed making Antarctica a "world territory" under UN control. In 1975 the government suggested making Antarctica a world park, although subsequent governments have tended to favor orderly minerals exploration instead.[22]

New Zealand's own contributions have been slim, however, down on the ice. It did not mount its own expedition there until joining a commonwealth expedition in 1929–31. Called BANZARE for British, Australian, and New Zealand Antarctic Research Expedition, it rediscovered parts of the East Antarctic coast, surveyed by air, investigated whaling grounds, and collected oceanographic data.[23]

France

Fearing still more British claims, in 1924 the government of France reacted by asserting one of its own to the coast of Terre

Adélie, which had been seen by Dumont d'Urville in 1840 but would not be revisited by the French until 1949. The claim was defined as the coast lying between 136° and 142° east longitude and 66° and 67° south latitude. After Australia, in 1933, had claimed two huge sectors on either side, France, in 1938, redefined its claim to include a sector extending from 60° south latitude inward to the South Pole.[24]

Although the basis of the French claim is discovery, France has never claimed other coasts found by Frenchmen, such as the Orléans Strait sighted by d'Urville. In 1903–05 Jean B. Charcot in the *Français* and, again, in 1908–10, in the *Pourquoi Pas?* braved the pack ice west of the peninsula, charted the Biscoe Islands, sighted Peter I and Alexander I islands, and also discovered Marguerite Bay and other areas. Probably a wish to cooperate with Great Britain on the matter of claims (France recognizes British sovereignty in Antarctica) dissuaded the government from asserting claims based on these voyages.[25]

France's attitude toward its Antarctic claim has been bound up with its interest in its subantarctic islands of Kerguelen, Crozet, Amsterdam, and Saint Paul, where French sovereignty is undisputed. At first the islands and Terre Adélie were administered by the governor general of Madagascar, but in 1954 their administration was moved to the Ministry of Overseas France in Paris.[26] Despite their relative inactivity in Antarctica, the French have stuck avidly by their Antarctic rights, presumably to protect the benefits they receive from owning Kerguelen and Crozet, which have proved to be rich fishing grounds, and any resources found in or offshore Terre Adélie.

Australia

Australia bases its rights to two enormous sectors of Antarctica not only on the discoveries of various explorers from England, but also on those of an Australian—Sir Douglas Mawson—who ranks with the heroic age explorers in stature. But Mawson was younger, and his expeditions extended well into the modern age; in addition, he wrote important books on Antarctic research and believed in its minerals potential.

Spurred by the previous British claims elsewhere in Antarctica—and by Byrd's first expedition and the American plans for others—the Australian government asked Great Britain to assert an Antarctic claim on its behalf. This was done by Order-in-Council in 1933. The western limit of the Australian claim at 45° east longitude was based, so to speak, on an explorer's handshake. Mawson, who was instrumental in the BANZARE expedition of 1929–31, had worked his way along the coast toward the region where a Norwegian expedition led by Hjalmar Riser-Larsen was operating. When the two met up they agreed that Mawson's territory for further exploration would be east of 45° east longitude and Riser-Larsen's would be west of it. When the Order-in-Council fixed this boundary as the limit of the Australian claim, that country won the better deal, so to speak, as some of the coast in Australia's claim had been discovered by Norwegians.[27]

The Australian parliament quickly passed the Australian Antarctic Territory Acceptance Act in 1933. The debate in parliament stressed Antarctica's minerals potential and income from whaling grounds; analogies were made with the U.S. annexation of Alaska.[28] Henceforth, this would be symptomatic of Australia's fierce nationalism about "its" Antarctic territory.

Norway

In 1938 Great Britain, Australia, New Zealand, and France mutually recognized one another's territorial claims in Antarctica. Norway was the last of the European nations to throw its hat into the ring. Although a Frenchman had discovered Bouvet Island in the South Atlantic back in 1739, Norwegian whalers used it extensively in the early twentieth century, and in 1928 Norway asserted ownership to this strategically located island and called it Bouvetøya. In 1931 Norway made claim to Peter I Island, off West Antarctica, which was used by its whaling captains. In 1939—alarmed by the imminent expedition of the German Reich to the coast of Queen Maud Land—Norway asserted sovereignty over the land between the limits of the British claim at 20° west longitude in West Antarctica around to 45° east longitude in East Antarctica, the Riser-Larsen–Mawson boundary. Norwegian claims to this coast and to the two islands were based less on discovery than on their occupation and use by its captains. Elsewhere, Norway preferred to have its captains pay the license fees sought by other claimants rather than to quarrel over sovereignty. Therefore, Norway recognized the claims of the four other European powers.[29]

Indeed, Norway had—and has today—far more vital interests in the Arctic. It could not afford to prejudice its position there by whatever position it took in Antarctica. So the most famous discovery of all—Roald Amundsen's first at the South Pole in December 1911, when he claimed the south polar plateau for King Haakon VII—was not followed up by his government. If Norway were to claim a sector in Antarctica extending inland to King Haakon VII's Plateau (as Amundsen had named it) at the Pole, it would validate the sector principle, which the Soviets assert to Norway's disadvantage in the Arctic. Thus, Norway worded its Antarctic claim to include only the land and hinterland, with no fixed southern boundary.[30]

Germany

Germany has been active in Antarctica historically although it has never asserted a claim. Its history there, as well as its current interest in Antarctic resources, caused the Federal Republic of Germany (West Germany) to become a full consultative member of the Antarctic Treaty in 1981.

A series of German expeditions to Antarctica culminated in the Third Reich's near-claim to part of Queen Maud Land, which its 1939 expedition named New Schwabenland. In the late nineteenth century, Georg von Neumayer had been a key figure; he sparked British scientific interest in the Antarctic, and influenced the decision to send a German expedition, led by Eduard Dallman, to the peninsular region in 1873–74. Geography professor Erich von Drygalski, in 1901–03, sailed south in the *Gauss*. After the ship became imprisoned in the ice, von Drygalski bravely trekked to shore and surveyed the area from a balloon. He saw as far as some mountains located on an island now named for him. Wilhelm Filchner, on his 1911–13 expedition, attempted to build his camp on the Filchner Ice Shelf in the eastern Weddell Sea. Two German expeditions visited Antarctica in the 1920s, and German scientists frequently participated in the expeditions of other nations as well.[31]

The Third Reich's expedition was led by Capt. Alfred Ritscher. As part of Hermann Goering's plan to organize a whaling effort to relieve a severe fat shortage in Germany at that time, the party sailed from Hamburg in December 1938, in the *Schwabenland*, a catapult-equipped ship, with two "flying boats" aboard. It stayed off the coast of Queen Maud Land for three weeks, launching exploratory flights that dropped spearlike rods with swastikas, flags, and claims sheets. There is no evidence the party landed on the continent itself, however. The would-be German claim extended from 4.2° west longitude to 14° east longitude.[32]

But the Reich never formally took up its claim, and in the meantime Norway had claimed the same territory for itself. Any rights Germany might have based on its earlier expeditions were renounced in the Treaty of Versailles after World War I. Following its defeat in World War II, Germany was not required to make a specific renunciation of Antarctic claims. What the postwar German government thought is not known; officially, it has not said anything on the subject.[33]

The Federal Republic of Germany had no Antarctic program during the IGY. In 1978, however, spurred by an interest in Antarctic resources and science, the government agreed it was in the West German interest to achieve consultative status. A budget of DM330 million was allotted for 1980–85, for an Antarctic station, an ice-strengthened ship, and a polar institute. Besides the polar institute, other West German institutes dealing with geology and fisheries are conducting Antarctic research, giving that nation a prominent role in Antarctic diplomacy today.[34] The German Democratic Republic (East Germany) is an acceding power to the treaty and, for some of the same reasons, may well try to become a full voting member soon.

Argentina

The Argentine claim, asserted vocally by the Peron government after it came to power in 1947, is a wide sector claim whose eastern limit is beyond the eastern limit of the Argentine landmass. The basis in law is the Papal Bull, *Inter Caetera*, of 1493, under which Pope Alexander VI gave Spain the entire world west of the 46th meridian, and the Treaty of Tordesillas in 1494, under which Spain and Portugal divided up the Western Hemisphere. Under the legal doctrine of *uti possedis juris,* Argentina alleges it succeeded to the title when it achieved independence from Spain in 1816—even though no one had sighted Antarctica at that time.

Argentina also invokes continuity and contiguity, since the mountains of the peninsula are thought to be a continuation of the Andean chain. Continuous occupation is also invoked, because in 1904 the Scotsman Dr. William Bruce, having built a weather station on Laurie Island, turned it over to Argentina to operate, which it has done ever since.[35] The Peron government built and occupied bases—which provided postal services and appointed magistrates as acts of administration—close to the British ones in the South Shetland Islands and on the peninsula, but the heightened tensions caused by the dispute led to a 1948 agreement between the two nations not to send warships south of 60° south latitude. Military vessels were allowed to resupply stations however. In the 1947 meeting that led to the Inter-American Treaty of Reciprocal Assistance, known as the Rio Treaty, Chile and Argentina included provisions extending the area to be protected southward to the South Pole.[36]

Argentina is the most active of all the claimants in carrying out symbolic acts to assert sovereignty. No maps of Argentina are printed that do not also show Argentina's Antarctic claim. The Argentine cabinet has met in Antarctica and transacted business there, and in 1978 at Esperanza Base on Hope Bay a baby was born, named Emilio Marcos Palma. He was promptly declared an Argentine citizen.[37]

Chile

Chile proclaimed its sovereignty over part of Antarctica in 1940. By presidential decree "all lands, islands, islets, reefs, pack-ice, etc., already known and to be discovered" from 53° to 90° west longitude were designated as Chilean Antarctic Territory. Chile's claim overlaps those of Argentina and Great Britain and extends into Ellsworth Land, discovered and mapped by U.S. expeditions. The overlap between the Chilean and Argentine claims, however, seems to bother neither. In 1947 the two nations agreed to cooperate in the Antarctic. Eschewing logic in favor of friendship (or, more likely, common anti-British feeling), the agreement declared that both of their titles to their Antarctic claims are "indisputable."[38]

Chile traces its "indisputable" title back to *Inter Caetera,* the Treaty of Tordesillas, and to Chile's inheritance of the Spanish title at independence. Geologic continuity is also invoked. One Chilean author has proposed that Chile and Argentina divide their territory in Antarctica along the watershed line of the mountainous spine stretching across the length of the peninsula—a principle that has been used to resolve boundary disputes in the Andes. If implemented, this would give Chile the western half of all the islands of the Scotia Arc and move Chile's eastern limits farther to the east than they are now. The dispute between Chile and Argentina over ownership of the Beagle Channel running through Tierra del Fuego (which has producing oil and gas wells) has not interfered with cooperation farther south in Antarctica.[39]

Chile, like Argentina, carries out administrative acts in the Antarctic to preserve its claim: in 1948 President Gonzalez Videla visited Antarctica in a highly publized trip. Today Chile cancels stamps and carries out other official acts. It has stated that its 200-mile territorial sea off its own coasts also applies to "its" Antarctic area, an issue which so far has not enmeshed other nations' fishermen in the Antarctic. But Chile has a tradition of statesmanship in the Antarctic, too: a Chilean law professor, Julio Escudero Guzman, proposed the *modus vivendi* on claims that became the basis for the Antarctic Treaty.[40]

The Republic of South Africa

Although it "looks" across the ocean toward Antarctica, South Africa never has sought a territorial claim there. Its principal concern has been with the security of its southern flank, which, in the late 1940s, led Prime Minister Smuts's government to take possession of the Prince Edward Islands, which actually had been discovered by Marion du Fresne of France in 1772 and rediscovered later by Captain Cook. Cook named the largest of them Marion in recognition of du Fresne's prior discovery.[41]

South Africa has a long-standing practical interest in Antarctic meteorology; it runs weather stations on Marion Island and, for the IGY, operated them on the subantarctic islands of Tristan de

Cunha and Gough. It took over Norway's IGY base in 1959–60, renaming it the South African National Research Expedition (SANAE). Its research there is also mostly meteorologic.[42]

Belgium

Belgium has always wanted a say in Antarctic affairs because of the role of the *Belgica* expedition led by Lt. Adrien de Gerlache in 1897–99. Some Belgians have favored a territorial claim based on the discoveries of de Gerlache, but Belgium has neither the resources nor the interest, it appears, in adding its iron to the already smoldering fire of competing peninsular claims. Nonetheless, many features on the map credit the expedition—such as Cape Gerlache and the Belgium Mountains. In 1948, after hearing about the U.S. proposal to the seven claimants to form an international regime, Belgium sent an offical note to the U.S. State Department saying it should be consulted in any future arrangements for Antarctica. In response, the U.S. government sent the Belgian government a copy of the ill-fated plan for an Antarctic condominium. Later, Belgium played a major role in organizing and carrying out the IGY, and operated a station during that time. Thus it got its wish of being included in the treaty negotiations that followed.[43]

Japan

Given Japan's long reach as a seafaring nation, Antarctica is relatively close to home, lying 13,600 kilometers from Yokohama. Japan's history in Antarctica is remarkably similar to that of the European powers; it sent an expedition during the heroic age and, in the 1930s, seems to have had imperial designs on the frozen continent. In 1911–12 Lt. Choku Shirase led an expedition to Antarctica to try for the South Pole. Unable to penetrate the pack ice in time to establish a base for the winter, he camped on the ice shelf east of Amundsen's base, Framheim, and encountered the Norwegians while sledging. With little to do and not enough time to try for the Pole before winter, Shirase returned to Japan. Later, when Byrd visited that part of the ice shelf and gave it American names, he renamed one feature Okuma Bay upon learning of the Japanese priority. The map also shows Shirase Glacier and the Shirase Coast.[44]

One sign of the Japanese government's interest in a claim in the late 1930s was the fact that when Chile asserted a claim in 1940, the Japanese government protested. But if the empire had planned to assert its own claim, the war intervened. At the end of World War II, upon Australia's urging, Japan was required to renounce all claim or right or title "in connection with any part of the Antarctic area" under the Treaty of Peace signed in San Francisco in 1951.[45]

The principal Japanese activity in the Antarctic has been whaling and, in the last decade, krill fishing. For the IGY, Japanese scientists mounted a model effort at their station, Showa. Since then, they have retained a strong scientific program, especially in upper atmosphere physics and meteorite research. Most recently, Japan has made broad surveys of parts of the Antarctic continental shelf.

Poland

Poland has a long Antarctic tradition but never hinted at making a claim there. Henryk Arctowski, a Pole, sailed on the *Belgica* expedition. In the 1920s Antoni Branislaw Dobrowolski, a Polish glaciologist, made important scientific contributions. Poland sent scientists to the Soviet station, Oazis, during 1959–60, and in 1959 the Polish government asked to be included in the Washington conference that concluded the Antarctic Treaty. But the request was denied because Poland had not had its own program in Antarctica for the IGY, which was the basis for inclusion in the treaty negotiations.

In the 1970s Polish interest in Antarctica revived, and krill-fishing fleets followed those of the Soviet Union and Japan into the Antarctic. Meanwhile, the government mounted a scientific program, establishing two scientific stations named Arctowski and Dobrowolski. In 1977 Poland became the first nation to achieve consultative status under the treaty since its adoption.[46]

The Soviet Union

When, in 1950, the Soviet government sent a diplomatic note announcing it wished to be part of any Antarctic settlement (see chapter 2), the Western powers already interested in the region began a debate over Soviet motives that continues today. Then, as now, there were two schools of thought: first, that Soviet Antarctic policy is driven by Soviet expansionism and, second, that the Soviet regime wished to be taken seriously by the other powers, to be seen as equal to the United States in diplomatic clout, technology, and in science. Soviet spokesmen have done little to shed light on their motives: they have made only a few gruff official statements. But their presence in Antarctica from the time of the IGY on— backed by the enormous resources and expertise of their Arctic program—is a key factor in Antarctic diplomacy.

The Soviet government has not announced a specific claim to

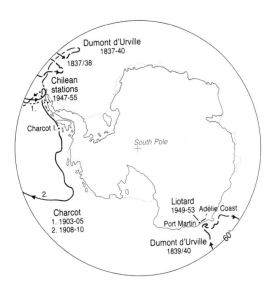

Historic French Antarctic
expeditions and Chilean
station.

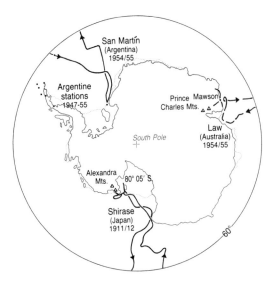

Historic Australian
expeditions, including joint
efforts. Expeditions by
Argentina and Japan.

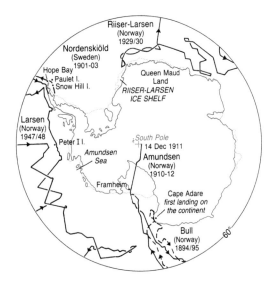

Historic Norwegian
expeditions, and that led by
Sweden's von Nordenskjöld
(Nordenskiöld).

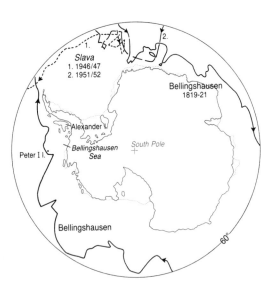

Route taken by the
Russian-sponsored von
Bellingshausen expedition of
1819–21.

Antarctic territory, although beginning in 1949 it made references to Soviet "rights" based on von Bellingshausen's discovery of the Antarctic continent.[47] Other nations acknowledge von Bellingshausen's discovery of Peter I Island and Alexander I "Land"—the latter shown by the U.S. Antarctic Service expedition to be an island—but his log made no claim to discovery of the continent. As we saw in chapter 2, a Russian atlas published on his return seemed to hand that laurel to the American Nathaniel Palmer.

Moreover, in the Arctic the Soviet Union does not recognize rights based on discovery, but instead advocates a sector principle. In 1926 a Soviet sector in the Arctic was announced covering Arctic islands, many of which were discovered, in fact, by Norwegians. Some Soviet commentators have explained this discrepancy by arguing that the sector principle makes more sense in the Arctic, where the region of attraction is close to the main landmass, but should not apply in the Antarctic, which is a continent and distant from the other landmasses. But other Soviet commentators have justified the Soviet Arctic sector principle on that asserted by Great Britain in the Antarctic, thus giving implicit recognition to the principle's validity in the Antarctic, after all.[48]

This contradictory legal discussion is less useful than a look at Soviet base patterns in Antarctica. For the IGY, it installed bases on the East Antarctic coast and a series of bases going inland to the deep interior, forming a wedge in the heart of the Australian claim. When the Soviet Antarctic program revived in the late 1960s and 1970s, bases were installed in a ringlike pattern around the continental periphery; several are located as near as possible to the boundaries between other nations' claims. Thus, the government originally may have considered a sector-type claim based on effective occupation and later launched on a more ambitious plan for a large coastal "presence," particularly near sensitive areas and areas of resource interest. The implications of the Soviet presence today are discussed in chapter 7.

Then, as now, bureaucracy is the most obvious explanation for the large resources, in terms of numbers of ships and men, the Soviet Union sends to Antarctica. Much of the Soviet landmass and the mouths of its major rivers lie north of 60° north latitude. There are few overland transportation routes between the raw materials of the east and the industrialized, European west of the country. Historically, the job of the Arctic research institutes—the Chief Administration of the Northern Sea Route and the Chief Administration of the Hydrometeorological Service—was to keep the northern sea routes open for ship traffic for vital economic functions. The 1956 study of national interests in Antarctica by the CIA, mentioned in chapter 2, notes that these units were contributing personnel to the IGY. Since then, the importance of these sea routes has grown. In 1979 the institute—renamed the Arctic and Antarctic Research Institute—had an estimated 2,000 employees and many ice-strengthened ships. So the steady growth of Soviet personnel in the austral summer and winter may result from a need to have something for all those glaciologists, engineers, and Arctic geologists to do, especially during the Arctic winter, which occurs when there is continual daylight in Antarctica.[49]

Regardless of Soviet motives and Western fears, most of the evidence shows that the Soviet individuals who participated in the IGY were mainly interested in doing science on a cooperative basis with Western scientists and anxious that their contributions be respected. Among the first to travel abroad after the death of Stalin, they seemed to consider themselves fortunate to attend the international meetings and to visit other nations' academies of science. So their presence alone was a significant event. Small wonder then, that when the IGY took place in Antarctica without political incident, western scientists boasted that international, cooperative science could thaw the Cold War.[50]

The Idea of the IGY

By the early 1950s, the web of interlocking national rivalries in Antarctica had come to resemble a diplomatic perpetual motion machine. Great Britain, France, Australia, New Zealand, and Norway recognized one another's claims, but their mutual agreement could not be extended because Chile and Argentina disputed Great Britain's claim and had competing claims of their own. None of these took account of the huge potential U.S. claim, which the government had been on the verge of announcing, on and off, for decades, but which remained inchoate because the United States had declined to be drawn in to the disputes of the other seven.

One obvious solution was for the United States to limit itself to the unclaimed sector between 90° and 150° west longitude, which included Marie Byrd Land and the part of West Antarctica discovered and mapped mostly by Americans. Apparently, the U.S. government did not want to limit itself to that forbidding region, with few harbors or coastal base sites. After 1950, the Soviet Union added another challenge; an eight-way settlement among the Western powers could have provoked a Soviet challenge. And Norway and New Zealand, which had no bases of their own in Antarctica, were nervous that a Soviet expedition might camp in their claimed territory.[51]

The diplomatic cable traffic could have clattered on forever.

Instead, a new element was introduced that showed the rivals the way to a solution. This was the International Geophysical Year—the eighteen-month "year" that ran from July 1, 1957, to December 31, 1958—conceived and implemented through the cooperative efforts of scientists of many nations. Among its other programs, it used Antarctica as an international scientific laboratory. The idea that many nations could do scientific research in Antarctica without raising territorial problems was new; a precondition was that the scientists be allowed to manage the effort themselves with a minimal amount of political interference, and that they be permitted to settle any problems that arose among themselves. It was a powerful, new concept in Antarctic diplomacy, and one the IGY group later would eulogize.

But at the time, preparations for the IGY were frantic, politically difficult, and exciting. British, American, French, and Belgian scientists were the nexus of the organizing effort. They were well-organized, arrogant, and often acted spontaneously to hold at arm's length any political factors which would create problems "down on the ice" in Antarctica. During the IGY, when twelve nations operated fifty-five stations and several thousand personnel in the Antarctic, they began to think of ways to institutionalize this innovation as working friendships began to form and all went well.[52]

A few incidents from the IGY Antarctic program illustrate why this idea was so successful and why scientific research, then and now, is a useful way to organize human activities in Antarctica. The "year" was really the Third Polar Year, the First International Polar Year having been 1882–83, when twelve countries installed approximately forty stations in the Arctic. The Second International Polar Year in 1932–33 had been a modest effort because of the worldwide depression and political tensions in Europe at the time. Both polar years made significant contributions (the second identified separate reflecting layers in the ionosphere, which made possible the growth of the radio industry, for example), but neither used the Antarctic in a serious way.[53]

The dinner party that gave birth to the plans for the third "year" was given in 1950 by James A. Van Allen (who would become famous for the IGY's best-known discovery, the Van Allen belts of radiation encircling the earth), in honor of Sydney Chapman, then Sedleian Professor of Natural Philosophy at Queen's College, Oxford. Chapman had built a model of the physical relationships between the earth and the sun.[54] At the party, Lloyd V. Berkner proposed a Third International Polar Year to be held twenty-five years after the second, in 1957–58. This would also coincide with the next predicted "maximum" of solar activity. The sun's activity—electrical, physical, magnetic—could be observed, along with its effects on the earth's ionosphere, which turns inward to the ground at the magnetic North and South Poles, and thus would serve to amplify and verify Chapman's model. The scientists' idea was adopted quickly by the various scientific organizations that could contribute to or sponsor it. The Mixed Commission on the Ionosphere approved it that same year; and the International Council of Scientific Unions—the central organ of all interested scientific societies—became the "year's" official sponsor.

The World Meteorological Organization thought the observations should be global and not limited to the polar regions, so the "year" was renamed the International Geophysical Year. At a 1954 planning meeting, it was decided to emphasize two areas: outer space, because new rocket and balloon technology, and possibly manmade earth satellites, could reach this frontier; and Antarctica, because it was almost unknown to science, yet within technological reach.[55]

The Political Innovations of the IGY

The *de facto* result of having a clique of scientists from many nations interested in Antarctic scientific cooperation was a new pattern for Antarctic diplomacy.

The first contribution of the IGY was a "gentleman's agreement" that scientific stations and expeditions there would not have the political significance they had had in the past. They were to be there only for the agreed scientific program and would have no legal significance. Thus a country whose sector was the site of other nations' scientific bases should not feel its sovereignty threatened, and the scientists could move freely around the continent and offshore waters. The idea originated with Georges Laclavère, the strong-willed chairman of the Antarctic Committee of the Comité Spécial de l'Année Géophysique Internationale (CSAGI), the central organizing committee of the IGY internationally. The effect of the "gentleman's agreement" was to give the scientists the upper hand over their political counterparts in determining where bases should be, although, not surprisingly, the seven announced claimant countries put scientific stations in their claimed sectors.[56]

The geographic South Pole remained the prestige location, the symbolic heart of the continent, and a place not visited by man since Scott's party reached it in January of 1912. All the sector claims converged there, so it had legal significance as well. IGY planners were sensitive to the implications—whoever occupied the South Pole would dominate the world's perceptions of who controlled Antarctica. So a key decision was made in July 1955, at a planning meeting held in the ornate old Observatoire in Paris, when, on short notice, Vladimir Beloussov and a delegation of Soviets arrived.

At the time, eleven nations had announced plans for Antarctic stations—all of them located on the coast except for those planned by the United States for the polar plateau and in Marie Byrd Land, where the weather of the region was believed to originate. Laclavère asked the Soviets where they planned to go. Beloussov replied that there would be three Soviet Antarctic stations: one at the geographic South Pole, one on the Princess Astrid Coast (or the Knox Coast) in East Antarctica, and one in between.

With perhaps a wink at his American colleagues, Laclavère replied that the Americans already planned to have a station at the geographic South Pole and it would be scientifically repetitious to have two. On the other hand, he added, there was a more difficult location at the Pole of Inaccessibility, so named because it is equally difficult to reach from all parts of the coast. It was near the geomagnetic South Pole, the southern end of the axis of the earth's magnetic field. Man had never seen this region, and so far, no station was planned for it. Fine, said Beloussov, according to the account published by Walter Sullivan, who was covering the meeting for *The New York Times*, "We do not insist on the Geographic Pole." Later, the Soviets agreed to locate their inland station at this "other" pole.[57]

Apparently, only after this meeting did the U.S. delegation, led by Laurence M. Gould, receive final approval for the U.S. South Pole base. Both superpowers took seriously, both then and now, their occupation of these respective "hearts" of Antarctica. The Americans stay on at considerable human and financial cost. The Soviets, who raised their flag over Sovietskaya, at the Pole of Inaccessibility in February 1958, abandoned it and their other inland base, Vostok, after the IGY. But later they reopened Vostok and use it to this day.

Having deftly avoided one confrontation, Laclavère faced another. Belgium and Japan wanted to put scientific bases at Bunger's Oasis, the supposed "warm spot" on the East Antarctic coast found by the Highjump I expedition. But the Soviets would need a station nearby to stage their overland treks to "their" pole. And just along the coast at Mawson would be the Australians, representing a government that was not happy at the prospect of even neutral Soviet bases on "its" Antarctic territory. A recurring nightmare for the IGY planners was what would happen if the gentlemen's agreement broke down, especially during the long austral winter when tempers are shortest and base crews under the most strain. The possibility of incidents had to be minimized.[58]

Belgium and Japan agreed to go elsewhere—to other gaps in

the coast of Queen Maud Land and Enderby Land, respectively—helping to complete the ring of bases the scientists desired. A second Australian base (Davis) would be located down the coast from the Soviet coastal one (Mirny). The United States wanted to be there too, for scientific reasons and as a political buffer.

As it happened, Gould had dined one evening with Sullivan, who had been to Antarctica before for the *Times,* once on Highjump I and later on the exploratory cruise of the *Atka.* He had been reading the records of Operation Windmill, which had followed Highjump I, and told Gould that the Windmill expedition had found a sheltered spot nearby, in Vincennes Bay (see chapter 2). Gould decided the United States would go there and Laclavère agreed. It was later named Wilkes base.[59]

A similar problem developed on the other side of Antarctica, along the Weddell Sea, where the British planned two coastal bases and the Argentines had one. As recently as 1952, the Argentines had fired over the heads of the British landing party in Antarctica. Would the bases remain "neutral?" The United States felt it also should be there and planned to build a station as near to the others as could be justified in terms of science. Fortunately, there were good scientific reasons for another base along the edge of the auroral zone there.[60]

As its second political contribution, the IGY provided some international administration for the region: vital services such as communications, weather forecasting, and rescue were undertaken. These beginnings are important, for subsequently they have served to reduce the risks to others—tourists, mountain climbers, expeditions of other nations—in Antarctica. In fact, no expedition to Antarctica would dare send a ship into the ice pack, or build bases, or move inland (let alone, in future, exploit mineral resources) without this assistance, for to do so would invite failure and possible loss of life and national prestige. Chapter 8 will take up the legal meaning of this so-called system of administration for the region.

The best known feature of the system is the Special Committee on Antarctic Research (SCAR), organized by the scientists when it was clear that Antarctic research would continue after the IGY. SCAR first met in February 1958 at The Hague in order to coordinate the various national research programs. One reason for its success then and in the years since is that its members largely are veterans of the skirmishes between national interests and apolitical science of the IGY. As such, they are used to finding ingenious solutions to the many unexpected, awkward situations that can arise in Antarctica. Chapters 5 and 6 describe SCAR in the 1970s: it has helped to ease the transition of some Antarctic research programs from being purely scientific, as they were during and

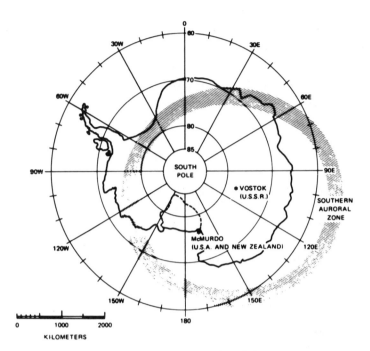

Surface area intersected by the auroral oval that was a principal subject of IGY research.

Stations established in Antarctica for the International Geophysical Year (IGY).

after the IGY, to being concerned with living and mineral resources and environmental impacts. Just as it has helped to set international priorities in basic Antarctic science, SCAR now helps set research priorities on resource and environmental questions.

In 1961 the committee was renamed the Scientific Committee on Antarctic Research; its offices were located at the Scott Polar Research Institute in Cambridge, England. For years SCAR ran on a meager budget of $15,000 annually in the form of contributions from the participating nations, yet its impact was enormous. Indeed, as early as 1959 when negotiations for the Antarctic Treaty were being finished, the framers felt no need of a separate administrative body to administer Antarctica or Antarctic science; they already had a *de facto* one in SCAR.[61]

A less well-known innovation was Antarctic Weather Central and its communications net. The fifty-five stations and many field parties in Antarctica for the IGY offered a unique opportunity to gather simultaneous meteorological data from many points in the region. In addition, systematic weather data would be needed for the IGY on a large scale, which justified the meteorological service. Moreover, since many visitors to the region for the IGY were not polar veterans, everyone would be safer with a good communications net. However, a completely centralized network was impossible; the auroral zone in the middle of the continent wreaks havoc with direct cross-continental communications. So a decentralized system would have to do.

But a political nightmare loomed, and it seemed best that bases and trail parties not depend for such vital assistance on potential political rivals. So the system was designed accordingly and nicknamed the "mother-daughter" system. All the stations fed weather information at regular intervals to Weather Central at Little America. "Daughter" stations communicated with their own field parties and listened for emergency broadcasts from them, while "mother" stations, with more powerful equipment, watched over their "daughters." The network was constructed so that no "daughter" was dependent on a politically unsympathetic "mother."[62]

A final innovation of the IGY was the practice of having personnel from other stations serve tours at Antarctic Weather Central during the "year," including the austral winter. This led to an arrangement whereby someone from the U.S. Weather Bureau wintered over at the Soviet base Mirny while a Soviet meteorologist, along with meteorologists from other nations, wintered over at Little America. This practice has continued, so that today there is a Soviet guest at a U.S. station each austral winter, while an American winters over each year with the Soviets. Sullivan wrote that "nothing could have done more to allay fears and suspicions [during the IGY] than the series of personnel exchanges."[63]

Although marred by Sputnik, by political wrangles over the roles of Taiwan and the People's Republic of China, and by an attempt at censorship by the Soviet authorities during one key meeting in Moscow, the IGY was an enormous success internationally. It became the model for a number of other cooperative, international "Big Science" programs of the 1960s, and established the influence of the informal network of senior scientists worldwide. And, by all accounts, the success of the IGY Antarctic program became the model for a new political institution erected for Antarctic governance, the Antarctic Treaty.

four

The Antarctic Treaty

Where the scientists trod, the diplomats followed. The immediate impetus for the Antarctic Treaty was Australia's concern that the Soviet Union, which had stations in its sector, would stay on and announce a sector claim after the International Geophysical Year (IGY) had ended. At a meeting of Australia, New Zealand, and the United States (ANZUS) in September 1955, Australian Minister R. G. Casey said that Soviet plans for regular flights to Antarctica were "most menacing." When U.S. Secretary of State John Foster Dulles visited Australia in March 1957, Casey asked him to do something about Antarctica's political situation for the period following the IGY.[1]

Returning to Washington, Dulles looked for someone in the State Department to deal with the Antarctic situation. Samuel W. Boggs, the department's long-time Antarctic expert, had died in 1954, and since then the TANT committee's work and the Antarctic studies sparked by Boggs had lost momentum. Dulles chose Paul C. Daniels, a retired diplomat with ambassadorial rank who was familiar with Latin America. Daniels began by consulting with other government agencies and the British government.

But the true ancestor of the treaty was the Escudero Declaration, proposed by Chile in 1948. It advocated at least a five-year moratorium on the Antarctic sovereignty dispute while the parties concentrated on scientific research. It also endorsed free access to the continent and political neutrality for expeditions. The declaration was wordy and abstract, the product of a Chilean law professor, Julio Escudero Guzman, but it suddenly became relevant when the IGY demonstrated what such an arrangement would be like in practice. Indeed, after his condominium plan had been rebuffed in 1948, Acheson decided to make Escudero's text the basis for further talks.[2]

Two paths were open to the United States after the IGY. One was to announce a U.S. claim as the opening move in a final territorial settlement. (A variant was proposed by Adm. George Dufek, in June 1957, to Under Secretary of State Christian Herter. A favorite of the explorers, Dufek's plan mandated that the United States, Australia, and New Zealand form a condominium encompassing their sectors and Marie Byrd Land. Later, France and Norway would be allowed to join, followed by Great Britain, Chile, and Argentina if they settled their differences.[3])

The other way was Escudero's. When President Eisenhower revised U.S. policy in January 1956, he abandoned the goal of a final territorial claim to Antarctica and decided to emphasize international cooperation in seeking a settlement. Word of Daniels's talks with the British leaked in the press, and consequently the talks were broadened to include all nations having IGY Antarctic programs. On May 2, 1958, President Eisenhower sent a formal letter of invitation to the eleven other countries, outlining the principles of an agreement, including Escudero's *modus vivendi* on claims.[4]

Representatives of the twelve met biweekly, from June of 1958 until well into 1959, in the board room of the National Academy of Sciences in Washington.[5] They failed to meet their self-imposed deadline of December 31, 1958, when the IGY officially ended, but did not miss it by much. The main elements of the treaty were in place by early 1959, and it was signed at the Washington Conference on December 1, 1959. It was immediately hailed as an international breakthrough advancing the cause of world peace.

Why had the informal sessions at the Academy and the six-week Washington Conference produced the agreement that had eluded diplomats and politicians since the late 1940s? Certainly the IGY contributed, by breathing life into Escudero's airy phrases, but another reason was *Sputnik*. The first orbiting artificial satellite was launched successfully by the Soviet Union on October 4, 1957 (ironically, while leading Soviet space scientists were in Washington on IGY business). *Sputnik* demonstrated to the astonished Allies that the Soviets had the capability to send rockets over intercontinental distances. Besides its impact on Western defense strategy and foreign policy, let alone its repercussions for the IGY satellite program, *Sputnik* made Southern Hemisphere allies nervous that the Soviets might install missiles in their Antarctic bases, putting their countries within range. Suddenly, their internecine disputes in Antarctica were dwarfed by the common fear of a Soviet military presence there. A formal accord among nations, including the Soviet Union, renouncing military activities there seemed the only way to forestall a costly arms race.

A third factor was India. In 1956, and again in 1958, India had suggested that the United Nations take up Antarctica. The claimant states successfully dissuaded India from pursuing the move further, but during the months the group met at the Academy, a UN debate seemed imminent.[6]

The Text of the Treaty

The Antarctic Treaty was signed on December 1, 1959, and took effect on June 23, 1961. Later chapters will describe how the treaty powers are using it as the keystone for agreements on minerals and living resources and how governments that are nonparties want to sweep the treaty away on the grounds that the original signers had no right to negotiate it in the first place. A critical test of the Antarctic Treaty will come in 1991, the date when a review of its operation can be called and the world community—nations involved in the treaty and probably those outside it—will decide how it should continue. The treaty, then, is the key to Antarctica's future.

A discussion of the treaty and how it evolved in the negotiations follows (see Appendix A for the treaty text). The discussion is based on previously unpublished notes from each of the shirt-sleeve meetings held in the Academy board room that were taken by a member of the U.S. delegation and sent to Rear Adm. George Dufek, the U.S. Antarctic projects officer.[7] Although the set of memorandums is incomplete, the documents nonetheless shed light on how the parts of the treaty were drafted and evolved. They reveal the thinking of the group, especially concerning its right to negotiate such a treaty and its relationship to the world community. They include a draft article the group considered, promising states which were nonparties to the treaty equal treatment. But the group did not adopt the proposal—leaving unsolved today the relationship of the treaty to the world community.

On balance, the treaty represents a massive compromise among historic Antarctic rivals—and an unfinished political deal. Its Preamble and arms control, inspection, and scientific provisions are wonderfully clear and explicit. Other articles, such as the one on territorial claims, are ambiguous and designed to accommodate opposing legal and political positions. Still others—such as the provisions for future administration and the scope of the treaty—are hasty and almost unfinished. Other key issues—the rights of nonsignatories, determination of legal jurisdiction, and the disposition of Antarctica's resources—are handled only by implication, if at all.

The Preamble of the treaty shows that a new era in Antarctica has been reached. In it the parties recognize that the peace of the region could be assured by continuing the innovations of the IGY: "Recognizing that it is in the interest of all mankind that Antarctica shall continue forever to be used exclusively for peaceful purposes and shall not become the scene or object of international discord."

The parties acknowledge the "substantial contributions to scientific knowledge" of the IGY and say they are "convinced that the establishment of a firm foundation for the continuation and development of such cooperation on the basis of freedom of scientific investigation in Antarctica as applied during the International Geophysical Year accords with the interests of science and the progress of all mankind."

Finally, the signatories acknowledge the United Nations, concluding that "a treaty ensuring the use of Antarctica for peaceful purposes only and the continuance of international harmony in Antarctica will further the purposes and principles embodied in the Charter of the United Nations."

Article I(1) demilitarizes the region, stating that "Antarctica shall be used for peaceful purposes only. There shall be prohibited, *inter alia,* any measures of a military nature, such as the establishment of military bases and fortifications, the carrying out of military maneuvers, as well as the testing of any type of weapons." However, military personnel or equipment may be used for scientific research "or for any other peaceful purpose" [Article I(2)]. According to the memoranda, at their second meeting Article I was quickly adopted by all—including the Soviet Union.[8]

Article II provides that scientific research shall continue throughout the region, as it did during the IGY. Article III explains how the parties should cooperate in this undertaking through advance notification of expeditions and by exchange of personnel and data among expeditions. New Zealand was chosen to draft the language of this article, perhaps because of its geographic and historic position as a staging area for Antarctic expeditions.

To further acknowledge the United Nations, the statement was added: "Every encouragement shall be given [to] cooperative working relations" with the United Nations [Article III(2)]. This gave a nod in the direction of that body without suggesting that the group was subordinate to it—a suggestion that ardent claimants such as Argentina and Chile opposed.[9]

Article IV is the treaty's linchpin. In it, the parties agree to disagree on territorial sovereignty. In effect, it was an admission that the problem preoccupying them for three decades was insoluble. Because this *modus vivendi* had been a precondition to the negotiation, it was accepted promptly. The language is a terse rephrasing of the parallel provision in the 1948 Escudero declaration. It embodies the "gentleman's agreement" of the IGY to immunize expeditions and stations from the sovereignty dispute. At the same time it allows countries to maintain their traditional sovereignty positions, so long as they do not cause the "international discord" prohibited in the Preamble. The article's full text reads as follows.

Nothing contained in the present Treaty shall be interpreted as:

(a) a renunciation by any Contracting Party of previously asserted rights of or claims to territorial sovereignty in Antarctica;

(b) a renunciation or diminution by any Contracting Party of any basis of claim to territorial sovereignty in Antarctica which it may have whether as a result of its activities or those of its nationals in Antarctica, or otherwise;

(c) prejudicing the position of any Contracting Party as regards its recognition or nonrecognition of any other State's right of claim or basis of claim to territorial sovereignty in Antarctica.

No acts or activities taking place while the present Treaty is in force shall constitute a basis for asserting, supporting or denying a claim to territorial sovereignty in Antarctica or create any rights of sovereignty in Antarctica. No new claim, or enlargement of an existing claim, to territorial sovereignty in Antarctica shall be asserted while the present Treaty is in force.

Article IV is so ambiguous that it has produced an outpouring of legal literature too tortured to repeat here. Generally, there have been two lines of interpretation. One is that Article IV is the first step toward *terra communis,* that is, co-ownership of Antarctica by the parties in a some kind of condominium. The second interpretation is that Article IV leads toward *terra nullius,* that is, renunciation of ownership by the parties. One conclusion drawn from this second interpretation is that Antarctica either has, or is aquiring through time, the same legal status as the unowned floors of the world's oceans or the surface of the moon.[10]

Several papers speculate on legal status of the ice cap, some saying it is part of the high seas, because it originates with un-owned snow falling over the high Antarctic interior. Others argue that the ice cap is *terra firma,* since for practical purposes it is immobile (a snowflake falling near the South Pole takes 100,000 years to creep down to the sea). The lawyers discuss the status of the big tabular icebergs which, even if they were once *terra firma,* may be international by the time they "calve" and drift northward on the high seas. Another legal mindbender is the status of the giant ice shelves, which—because they are part of the icecap—still could be *terra firma,* and perhaps belong to states claiming the adjoining land. But, asks one paper, what then is the status of the salt on which the ice shelves float? Taken solely as legal proposi-tions, the arguments seem endless: Fast ice, which attaches to the shore, may be part of the continent, whereas annual sea ice, which refreezes each year, may be part of the high seas.[11]

Unfortunately, most of the international law papers primarily refer to one another. They ignore obvious realities such as the competing web of national political interests that Article IV—however odd—holds in balance. Discussions of an Antarctic condominium usually overlook the fact that the powers have rejected any suggestion that they are merging ownership—a condominium would contradict *all* of their respective sovereignty positions and would force each to sacrifice the advantage it gains from its particular "rights." Likewise, papers arguing that Antarctica is the "common heritage of mankind" overlook the fact that the powers will never renounce the *possibility* of ownership (except Japan, which had to do so after World War II).[12] To some of the powers, it may seem absurd to read, for example, that the ice cap over which Robert Scott and Roald Amundsen trudged and planted national flags is, after all, an ocean.

Article IV proves unsatisfactory to readers unfamiliar with Antarctic politics; indeed, its obvious textual problems have led many legal writers to conclude that it is meaningless. But Article IV becomes understandable, even clear, in its political context: it stabilizes the web of competing national interests among claimants, between claimants and nonclaimants, between the superpowers, and between rich nations able to maintain Antarctic presences and smaller ones, like Belgium and South Africa, which have barely done so. The treaty's preservation of the political balance, far more than its inspiring language, is the true reason it governs Antarctica so successfully.

Article IV does, however, raise two important issues for future activities and minerals development there. One is the lawyers' question of whether a document such as the Antarctic Treaty can stop "acts or activities taking place while the present treaty is in force" from affecting sovereignty. If the treaty were to break down, what would prevent the United States from asserting that its continuous presence at the South Pole since 1956 is evidence of "rights" to the central continental area—as the explorers envisioned? Would not the Soviet Union make the same claim for its ring of stations around the Antarctic coast, even though all but two of them were built after the treaty went into effect? The language of Article IV and other parts of the treaty tries to remove the legal implications of stations and expeditions; from a practical point of view, it is most useful. Nonetheless, if political circumstances were to test it severely, it might not hold up, and a turf fight, with its accompanying legal ballyhoo, might erupt once again.[13]

Another issue unresolved by Article IV is who holds title to mineral resources. It offers no solution if, for example, New Zealand, (a claimant), the United States (with the basis of claims), and Japan (which does not recognize the claims of others and makes none of its own) all want title to petroleum pulled from the Ross Sea floor. Even if an administrative regime could solve this problem by vesting title to the petroleum to a developer who meets certain requirements, a court might find it hard to agree that title never existed until the regime came into being.[14] Currently, the treaty powers are trying to solve this problem (see chapter 6), but the memorandums of the 1958 meetings in the National Academy board room do not indicate that this issue was discussed.

The treaty's arms control features are found in Article I on demilitarization, and in Article V, which prohibits any nuclear explosions in the area or the storage of nuclear wastes in Antarctica [Article V(1)]. For the latter case, an exception is allowed if a subsequent "international agreement" on nuclear explosions or storage is agreed to by all the consultative parties [Article V(2)]. The ban on nuclear explosions was suggested at the Washington Conference by Argentina, which was concerned, apparently, about fallout from Antarctic nuclear tests blowing north to South America. Since Argentina was also a claimant that might have trouble getting the treaty ratified at home, the others agreed to accommodate Argentina's concern on this point.[15]

Article VI defines the "area" to which the treaty applies. It avoids defining the term *Antarctica,* which refers to the continent itself but which also can be construed to include the ice cap, ice shelves, and the fast and pack ice. This resulted from a controversy over an earlier version by the British, who had been asked to draft it because of their expertise regarding ice in the region. However, when they reported that "the Antarctic" had no fewer than fifteen different kinds of ice, agreement on what constituted "Antarctica" or "the Antarctic" became difficult. Subsequently, the Soviet representative suggested that the treaty area should extend to the Antarctic Convergence, the boundary between the Southern Ocean around Antarctica and the water masses of the other oceans. But because the Convergence fluctuates, this definition seemed too imprecise.[16] Thus, Article VI reads: "The provisions of the present Treaty shall apply to the area south of 60° South Latitude, including all ice shelves, but nothing in the present Treaty shall prejudice or in any way affect the rights, or exercise of the rights, of any State under international law with regard to the high seas within that area."

To enforce the arms control features, Article VII(3) allows each full party the right of unilateral inspection, at any time, of "all areas of Antarctica, including all stations, installations and equipment within those areas, and all ships and aircraft at points of discharging or embarking cargoes or personnel in Antarctica." To date, the Antarctic Treaty is the only international agreement in which the Soviet Union has agreed to unilateral on site inspection of its facilities. U.S. inspectors have made a point of visiting Soviet bases, as well as those of other nations, on their tours.

On-site inspection was one of President Eisenhower's general international goals, and his memoirs indicate he was pleased with the right to aerial inspection in the treaty. Indeed, Article VII sounds like the "Open Skies" proposal he had made to the Soviet Union in Geneva in 1955. There, the president had proposed that each side give each other a blueprint of its "military establishments, from beginning to end, from one end of our countries to the other.... Next, to provide within our countries facilities for aerial photography to the other country.... Likewise we will make more easily attainable a comprehensive and effective system of inspection and disarmament."[17] It is not surprising that this idea was added to the Antarctic Treaty at U.S. insistence. Ever since, U.S. arms-control negotiators have used the treaty's language on inspection as a model text in other negotiations.

[Richard E. Byrd made the first "aerial inspection" of another nation's base site in Antarctica. He and Siple accompanied Deepfreeze I to Antarctica in 1956–57. They flew to the South Pole to check conditions there for the airdrops bringing equipment for the planned U.S. base. Then, in a typical, flamboyant gesture, they veered off toward the Pole of Inaccessibility, where Soviet tractor parties soon would be trying to establish their South Pole base. Upon returning to McMurdo, Byrd sent the Soviets a message, ostensibly to inform them of conditions at "their" pole. (But the not-too-subtle, unstated, message was, "Aha! We got there first!"[18])]

Article VIII of the treaty takes up the messy problem of legal jurisdiction over people in Antarctica. Are they subject to the laws of the country that claims the land they visit? Or of the state that sponsors the expedition? Or of their own country, if it is not one of the above? Article VIII only addresses the status of "observers" carrying out arms-control inspections of Article VII and of "scientific personnel" on exchanges and of "member of the staffs accompanying any such persons," all of whom remain under the jurisdiction of the country sponsoring the expedition. Daniels subsequently has commented that the question of jurisdiction is among the important unfinished business in the treaty.[19] Or, as Finn Sollie, the Norwegian polar expert, has put it, "The question, for instance, of who shall have jurisdiction if a French tourist who arrives at the U.S. McMurdo station on board a Norwegian ship on a tour arranged by a British agency and who becomes subject to a serious crime committed by a person of unknown nationality during his stay in the New Zealand sector may not be any easier to solve than the question about the proper authority for regulation of mineral prospecting and mining."[20]

The jurisdiction question will have to be resolved before serious minerals exploration or exploitation begins or, for that matter, any other industry involving visitors who are neither scientists nor arms-control inspectors. Indeed, elsewhere in the world, offshore oil teams tend to include workers of many different nationalities.

Provisions for the Future

When the Antarctic Treaty was negotiated, the twelve nations sensed that their agreement was faulty and would have to be amended later. Article IX provides for the treaty to evolve through meetings held "at suitable intervals and places, for the purpose of exchanging information, consulting together on matters of common interest . . . and formulating and considering, and recommending to their governments" various measures. Article IX(1) lists six topics for such consultations, including the exercise of jurisdiction, and "preservation and conservation of living resources in Antarctica." The latter was suggested by Chile, which was concerned about fishing. Subsequently, the group has interpreted this phrase as a mandate to protect the region, and it is the basis of two treaty-based agreements dealing with living resource exploitation.

The group also discussed the possibility of a provision dealing with resources, but its exact content is not known. Several accounts indicate that because of dissension the subject was dropped.[21]

The system of administration remains unresolved. However, Article IX's requirement that they meet "at suitable intervals and places" resulted in the practice of meeting every two years in the capital of one of the treaty powers, taking turns by alphabetical order. At these "consultative meetings," as they came to be called, majority voting would not do, because individual claimants wanted the right to veto any measure that might undercut their claims. Voting, therefore, was by consensus. Consensus voting is often described as being the same as unanimous voting, but there are critical differences. In unanimous voting, *all* parties must vote affirmatively in order for a measure to pass; in consensus voting— even if one or more parties abstains—a measure passes so long as all the parties voting, vote "aye." Consensus voting in the consultative meetings has allowed room for many face-saving abstentions and maneuvers. To date, consensus voting has not been used excessively by one party trying to stymie the entire group.[22]

Article XII explains the term of the Antarctic Treaty and how it can be changed. The treaty runs indefinitely, and can be amended only by unanimous agreement—which makes it unlikely to be amended at all. But thirty years after it enters into force—that is, after June 23, 1991—any of the consultative parties can call a conference to review its operation. (The treaty assumes that this will happen once, because it speaks of "a" conference.) An amendment or modification to the treaty may be proposed at the conference if it is supported by a majority of all parties present, including a majority of the consultative parties. But to take effect, the proposed change must be ratified by all the consultative parties. Article XII(2c) states that if two years after the conference (1993 at the earliest) an amendment has not been ratified, then any of the parties may withdraw from the treaty within the following two years. So change, if it comes, would become effective at the earliest in 1995. The idea behind this provision was that if, after thirty years, a majority of the parties wanted to move in a certain direction but were blocked by a minority, the treaty could be said to be not working and due for change.[23] Today, much national effort is being spent preparing for 1991, as the original twelve consultative parties and more recent adherents stake their positions both down on the ice and politically back home. How the web of national interests is moving toward this deadline will be discussed in chapter 8.

The Treaty and Nonsignatory Nations

Most important, during the meetings held at the National Academy the group argued vigorously about their right to negotiate a treaty for Antarctica. In the early meetings, the Soviet Union argued that the group had no authority to negotiate an agreement; only a meeting of all nations did. New Zealand opposed inviting others but endorsed a principle of "fair treatment" for nonparties. Norway preferred an "open door" policy for other nations but wished to limit the conference to the twelve already there. Belgium said, "There was no question of monopoly of the Antarctic continent," and Australia agreed.[24]

The United States drafted an article on the rights of nonparticipants and tabled it in November 1958. It said:

> The administrative measures which become effective pursuant to Article VII of the present Treaty shall apply equally to all countries and shall be carried out in a uniform and non-discriminatory manner, with equal treatment being accorded to countries which are parties to the present Treaty and to countries not parties thereto, and to their respective nationals, so long as such countries and nationals respect the principles embodied in the present Treaty.

In explaining the article, Daniels told the others, "[I]t was inherent in the United States concept that fair treatment for all other countries should be provided in order to avoid the interpretation that the group was attempting to monopolize Antarctica for them-selves."[25]

In retrospect, this discussion has a hypocritical ring. Here were the historically active countries, saying that other countries would not be discriminated against in Antarctica, while at the same time they were exclusively and secretly deciding on a treaty. They drew up Article IX, which permits only those countries who go to the expense of conducting "substantial scientific research activity" in Antarctica to become consultative, or voting, parties. Doubtless this provision seemed only fair to the twelve in the board room whose representatives were then in the Antarctic conducting scientific research.

But to the rest of the world—especially the former colonies that at the time were joining the United Nations in record numbers—this test later might look discriminatory. Although Daniels contended that the twelve did not want to "monopolize" Antarctica, that is exactly what they were doing. Perhaps because it was obvious that this draft article, if included, might bring a storm of criticism from nonparticipants, the group excluded it from the treaty. And until now, it has never been published.

Nor did they anticipate in their talks the advent of a second tier of treaty adherents called acceding parties. These would include those countries who, because they did not conduct "substantial scientific research activity" in Antarctica, were not qualified to vote as consultative parties. However, acceding parties agreed to abide by the terms of the treaty and would ratify it. Soon after the treaty came into effect, NATO nations, as well as some Eastern Bloc countries, acceded. By the end of 1983, the acceding parties numbered thirteen, including the People's Republic of China.

Nonetheless, the acceding parties played no role whatever until the 1980s, when the consultative parties began serious talks on Antarctic minerals, and developing countries outside the treaty "club" took up the issue in the United Nations. In 1983, in a gesture of openness toward outsiders, the consultative parties invited the acceding parties to be observers at the twelfth consultative meeting in Canberra, Australia. However, when the acceding parties expressed their interest in attending the group's meetings on minerals, they were turned down.[26] To judge by the notes made in the Academy board room, and by the recollections of participants, the acceding parties' potential role as intermediaries between the treaty club and the outside world was not foreseen in 1958–59.

But the group's failure to address the problems of nonparticipants, of resources, or adequately to address criminal jurisdiction may be understandable under the circumstances. As the meetings dragged on—and for a long time it seemed that the Soviet representative had not been authorized by his government to conclude any agreement—the IGY was ending. Afterwards, rivalry would resume in Antarctica, and this time it was all the more serious because of the Soviet presence. Antarctica had been an expensive, time-consuming, politically hazardous problem for these countries for almost forty years, so the prospect of its resolution must have been alluring. This mood would explain the brevity and incompleteness of the treaty and its open-endedness about future arrangements. The spirit seems to have been one of do-what-you-can-and-finish-it-fast. After a document had been concluded in early 1959, the Washington Conference had been held, and the ratifications began to flow in, the negotiators, perhaps for the first time, realized that their hasty, vague deal was an act of global importance that would usher in a new era of peace.

In sum, the treaty is a mixed bag. Its articles on claims, demilitarization, and inspection erase the most difficult aspects of the claims dispute and of Soviet–Western rivalry down on the ice. These, and the Preamble's redefinition of the uses of Antarctica as peaceful and scientific, offer a creative formula for multinational coexistence there. Other articles dealing with information exchange, future meetings, and jurisdiction are inadequate. Other subjects, principally those of nonparticipants and resources, are not addressed at all. It is easy to see why so many international law scholars in subsequent years have criticized the six-page document for its brevity and incompleteness. But the counterargument is that it has worked, permitting historic rivals to cohabit Antarctica with minimal risks. To look at the treaty as international law is to miss the point: it is a profoundly political text that solves temporarily, and perhaps permanently, a profoundly political problem.

The Antarctic Treaty and the United States

The treaty allowed the United States to remain in Antarctica while removing most of the important risks of doing so. U.S. territorial rights would be preserved because Article IV of the treaty allows a country with the "basis of claims" made by itself or by its nationals to have free run of the continent. Also, other nations' claims or activities cannot be construed as infringing on U.S. rights. And most important, by elevating research in Antarctica from a fringe benefit of exploration to the chief activity of all nations in the region, the treaty gave the United States, with its growing Antarctic science community, something to do there.

President Eisenhower was pleased with his achievement. "There is one instance where our initiative for peace has recently been successful," he noted in his State of the Union message in January

1960. "A multilateral treaty signed last month provides for the exclusively peaceful use of Antarctica, assured by a system of inspection. It provides for free and cooperative scientific research in that continent, and prohibits nuclear explosions.... The treaty is a significant contribution toward peace, international cooperation, and the advancement of science."[27]

The usefulness of the treaty solution was clear in ratification hearings held before the Senate Foreign Relations Committee on June 14, 1960. Several witnesses—the heirs of Fanning, Reynolds, and Byrd—spoke up in favor of an outright U.S. claim and in opposition to the treaty. These included the American Legion, the Sons of the American Revolution, the American Coalition of Patriotic Societies, the Defenders of the American Constitution, and the National Sojourners, Inc. Sen. Clair Engle of California was alarmed. He claimed that the United States was giving away "something for nothing," that the treaty was the first step toward internationalization, and that the United States would never again enjoy its preeminent role there. Elizabeth A. Kendall, a private citizen who for years had urged the United States to claim Antarctic territory, said that the treaty was "an instrument of the peace offensive of the Soviet empire." And the treaty's muffling of the claims issue was criticized by Edouard A. Stackpole—the historian who had written that Americans were the first to set foot on the Antarctic continent. The treaty's fudge on claims, he said, flew in the face of history.[28]

But these comments were countered by Herman Phleger, who had represented the United States at the Washington Conference, by Philip C. Jessup of Columbia Law School, and by Laurence M. Gould, a member of the first Byrd expedition who had run the U.S. Antarctic effort for the IGY. They pointed out that under the treaty the United States would retain all its rights, a military presence, and freedom of access to all parts of the continent. Antarctic projects officer Rear Adm. David M. Tyree, who had succeeded Dufek, reassured the senators that contrary to rumor the U.S. Navy was not secretly opposed to the treaty. It seemed clear to most senators that the treaty made sense, and they did not seem to fear an imminent Soviet takeover of Antarctica. Because he sensed support for the treaty, Gould turned to another question—one that was of great interest to the junior senator from Minnesota, Hubert H. Humphrey—as to which was colder, Antarctica or Minneapolis.[29]

The treaty passed the Senate on August 10, 1960, by a vote of 66 to 21, and the United States formally ratified it on August 18. It was ratified by the other eleven nations too, with the most vocal claimants—Argentina, Chile, and Australia—ratifying last and simultaneously, on June 23, 1961.

The treaty not only was a political success, it was a *coup* for American science. Following Sputnik, federal support of science expanded exponentially. Scientists were being consulted as never before about the state of American defenses, education, and myriad other issues. Both Presidents Eisenhower and Kennedy relied on their scientist-advisers, and in the Kennedy administration, professors from Cambridge to California shuttled to Washington regularly.[30]

The U.S. Antarctic initiative rode the crest of this wave, and within a few years an entirely new constituency had formed for Antarctica in the United States. Unlike the explorers, who had been drawn to Antarctica by the memories of yapping sledge dogs and the hum of planes returning through the fog, these were well-traveled American academics who relished setting up laboratories in exotic places, pilgrimages across snowy landscapes for hard-won data, and the globe-trotting schedules that went along with doing science on its new, international scale.

Yet the treaty remains only a partial document, representing an unfinished diplomatic effort—as its negotiators well knew.[31] It solves the pressing political problem of the Soviet presence and continued national rivalries by giving competing nations in the region something constructive to do there instead—namely, scientific research. It allows the ambiguous U.S. role in Antarctica to continue, and does nothing to negate whatever legal weight the inchoate U.S. claim may have.

Nevertheless, it is silent on the other interests the United States had accumulated in Antarctica in the course of its 140 years of activities in the region. While encouraging continuation of the ambitious U.S. science program there, it provides no guidance as to what the nations should do about resources. What is the purpose of doing science in Antarctica besides keeping the peace? Specifically, everyone knew that resources were the ultimate stake in Antarctica, but was it a goal of the U.S. science program to find them?

In 1960 the scientists were confident that anything as exciting as Antarctic research would yield benefits, tangible and intangible. They shaded the resource question to imply that science would ultimately yield resource information. Gould told the Senate Foreign Relations Committee, in a formulation which would be repeated by other scientists over the years, "Even though the scientific data which come from Antarctica are not *immediately* translatable into oil or minerals, they are maybe fundamentally far more useful to us in understanding our planet" [italics added].[32] Thus the troubling question of resources—unresolved despite the great political and scientific gains that had been made—was kept at a safe distance.

A Permanent Antarctic Program

From 1959 on, the scientists gained on the older group of Antarcticans such as Dufek and Siple. They cemented their influence on Antarctic matters, in organization, programs, budgets, rationales, and personal loyalties, to assure that the United States would have an outstanding science program and be a model adherent to the treaty. The older group, seeing its control slip, was unhappy. In his book, *90° South,* Siple recalls that during planning for the IGY the "newcomers" disdained him. "I was amused to find that while I was accepted as a scientist [he had been chief biologist on the second Byrd expedition and had a doctorate in geography], I was not accepted by some as an Antarctic authority. Instead it seemed to be generally accepted that the 'old timers' knew less than untried newcomers. It was an eye-opening discovery to learn that though I had been on four Antarctic expeditions, I was considered by some to be full of old wives' tales unrelated to the modern scene."[33]

During the IGY, it had been clear to the scientists that a single eighteen-month period (which included one full austral summer) would only scratch the surface of the data lode to be gleaned from Antarctica. At a June 1957 meeting in Paris, the United States proposed that the IGY Antarctic work continue past December 31, 1958.[34]

Founded during the Lincoln administration to advise the government and the country, the National Academy of Sciences was unsuited to run an ongoing program. It had been given authority to run the U.S. effort for the IGY, and had received a one-time appropriation from the National Science Foundation for IGY activities (not counting military logistics support and the Department of Defense's IGY programs). But the Bureau of the Budget could not continue this arrangement permanently.[35]

With the government's growing stake in maintaining the Antarctic Treaty and the Antarctic region as a model of international cooperation, it became inappropriate for U.S. officials to discuss the American conquest of Antarctica and its long-standing territorial, national aims. Much of the material about U.S. claims—particularly the government's role in facilitating them—was classified. Indeed, most of those who wrote books and articles about Antarctica had security clearances. In time, the explorers' tradition would die, starved by the lack of newsmaking public figures, dockside sendoffs and tickertape parades, national Boy Scout contests, and corporate advertising of products selected for polar expeditions. Gradually, the U.S. presence there was defined exclusively—at least in public statements and pronouncements—as an interest in international cooperation, peace, and science.

The treaty was a precedent for other arms-control efforts; the State Department's Arms Control and Disarmament Agency (ACDA), which was concerned with compliance, was determined to make the United States a model adherent in the hopes of strengthening its negotiating position in other forums. U.S. inspection teams visited stations and made aerial overflights in 1964, 1967, 1971, 1975, 1977, and 1980, reporting that "the observed activities at each station were in compliance with the provisions and spirit of the Antarctic Treaty."[36]

Correspondingly, the U.S. military lost interest in Antarctica. In chapter 2, we saw that the military had found Antarctica useless for submarine travel with the rout of the *U.S.S. Sennet* from the pack ice during Operation Highjump. The military treated the IGY as another chance to compete with the Soviet's prowess in polar operations. Immediately following the IGY, for example, the Quartermaster Intelligence Agency of the U.S. Army, reporting on Soviet polar clothing, had said, "There is no doubt that the Soviets are striving for world supremacy in all aspects of living, working and fighting in polar climates."[37] In 1959 no defender of freedom could fail to rise to this challenge. But as the years went by, and the Soviets to all appearances adhered to the treaty, Antarctica hardly seemed first on the Soviet agenda for world conquest.

Gradually, the old Antarcticans dispersed. Siple, after his heroic tour managing the construction of the U.S. IGY base at the South Pole, returned to his army duties in Washington. In 1963 he became science attaché to the U.S. embassies in Australia and New Zealand. He died in 1968. In 1965 Mooney was transferred to the International Security Affairs Office of the Pentagon, and Dufek retired to run a naval museum. For a time, these Antarcticans held reunions and corresponded; but their sun set in the 1960s, as the public came to view Antarctica as internationalized and turned its attention to the next frontier, the moon.

The scientists of the sixties preferred to think that their work in Antarctica was for the benefit of all mankind—which, in many ways, it was. They tended to downplay the fact that the U.S. government was sending them there season after season for national, political reasons. They were understandably proud of the degree of international cooperation they had achieved there: Each winter a Soviet scientist wintered over with the Americans, and each austral summer season Soviets and Americans joined each other's expeditions.[38] It was easy for some scientists to make the semantic leap from international cooperation to the "internationalization" of Antarctica, which, they sometimes said, had already taken place. For instance, Albert P. Crary, chief scientist of the U.S. Antarctic Research Program and an astute IGY leader, wrote in

1962, "Under the provisions of the Antarctic Treaty, national claims to territory there have been at least temporarily abandoned. Antarctica can become a laboratory not only for the physical and life sciences but also for the development of international co-operation and understanding. There will be increasing opportunities for scientists of many nations to work together on this truly international continent to enlarge the heritage of all mankind."[39]

The Antarctic Treaty in Practice— Environmental Measures

Crary was right to praise the treaty; it facilitated the conduct of scientific research by dedicating Antarctica to that purpose, by providing for free exchange of data, and by permitting scientific expeditions access to others' territory with minimal problems of civil jurisdiction. Escudero's idea that a five-year period would provide an initial test was sound, and by 1966—indeed, even earlier—the new arrangement was working well. Scientific research and international cooperation flourished.

One way to see *how* this brief document of six pages evolved after five, ten, and twenty years into an administrative "system" is to look at the treaty's handling of environmental questions. The question of environmental protection was always a lively one in the consultative meetings—whether the discussion concerned the dumping of garbage around McMurdo station or the perennial problem of bird colonies being harassed by sledge-dogs which some nations still kept there. The group drew up some environmental measures long before such considerations became an international issue and before the 1972 UN Conference on the Human Environment. Nonetheless, the Stockholm conference spurred the group to pass more environmental measures and broaden the treaty's authority in this area. Later, the group used its environmental responsibilities as the basis for its authority to draw up agreements governing living resources and minerals. The evolution of these environmental measures shows how this brief, flawed document has come to govern one-fifteenth of the earth's surface.

Early environmental measures were aimed at common rules for expeditions operating in Antarctica. At a 1960 meeting in Cambridge, the Special Committee on Antarctic Research (SCAR) scientists drew up some general rules of conduct for expeditions: Flora and fauna not native to Antarctica should not be introduced on the continent; pilots should not fly helicopters too close to bird rookeries; firearms should not be detonated near colonies of seals;

dogs should not run free; and there should be no discharge of oil
in a manner that hurts plants and animals. The rules of the SCAR
were discussed when the twelve consultative parties met for the
first time in Canberra in 1961. There they passed Recommendation I-VIII, calling the "attention of all persons entering the area"
to the need to protect "living resources." This was the first of many
measures on resources deriving their authority from the treaty's
suggestion that the group consider "conservation of living resources" in their deliberations.[40]

Recommendation I-VIII said the group would consult on drawing up "internationally agreed measures" for preserving Antarctic
living resources. At the third consultative meeting, held in Brus-
sels in 1964, they adopted the Agreed Measures for the Conservation of Antarctic Flora and Fauna. "Agreed measures" are rules
that the treaty powers pass at their meetings. They are implemented unilaterally by each nation's adoption of administrative
rules or laws following the general lines agreed on by the group.
Each particpating government then becomes responsible for seeing that its own expeditions adhere to these rules while in Antarctica. Nonparticipants cannot become party to agreed measures
without first becoming parties to the Antarctic Treaty. Thus, by
1964, the method was established of separate but harmonious
national laws, which would become basic to the treaty "system." A
convenient by-product of this method was that it obviated the
need for a central administrative authority, or secretariat, to administer common laws (although the U.S. State Department carries out some central chores, as the United States is the depository
government for the treaty).[41]

These agreed measures designated the Antarctic Treaty area
south of 60° south latitude as a "Special Conservation Area."
Going beyond SCAR's 1960 rules, the measures designated places,
such as coasts having Emperor or Adélie penguin rookeries, as
"specially protected areas," where the killing, wounding, captur-
ing, or molesting of native mammals or birds was barred without a
permit issued by the applicant's host government. (Extreme emer-
gencies, such as imminent loss of human life, were excepted.)
Over time, the designated areas would change, but the system is
still in use, amplified by an Agreed Measure of 1975 creating Sites
of Special Scientific Interest. These can include areas of only bo-
tanical or zoological interest. The agreed measures broadened the
treaty's scope, moving its authority beyond the question of rules
for national expeditions to common rules implemented synchro-
nously by participating governments.[42]

The agreed measures protected many penguin and other bird
colonies, but seals were another matter. Many Antarctic seals do
not huddle in rookeries; some species live in twos and threes

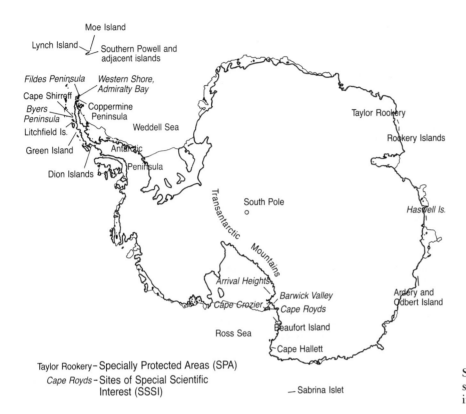

Taylor Rookery — Specially Protected Areas (SPA)
Cape Royds — Sites of Special Scientific
Interest (SSSI)

Specially protected areas and
sites of special scientific
interest.

throughout the vast extent of the pack ice. SCAR was concerned
about a revival of sealing; there were Soviet reports of 3,000 crabeaters clustered in one place, convenient for hunting. The rumor
spurred Norway to send a ship hunting for them—unsuccessfully.
The sealing threat never materialized; nonetheless, the treaty
group agreed to limit sealing at the fourth consultative meeting
held in Santiago, Chile, in 1966.[43] It had agreed at the third
consultative meeting to limit pelagic sealing voluntarily and, at the
next meeting, they agreed to common interim guidelines.[44] The
evolution of seal regulation, like that for specially protected areas,
encouraged the growth of voluntarism. Voluntarism was, after all,
in the claimants' and potential claimants' self-interest. If they
could govern Antarctica by harmonious action, they could avoid
creating a central secretriat that could threaten their positions on
claims.

The Convention for the Conservation of Antarctic Seals was
adopted at the seventh consultative meeting, held in Wellington in
1972. A draft of the convention had been tabled years before, but
the parties had voluntarily observed it in the interim. The convention prohibited killing of six species of Antarctic seals—the elephant, leopard, Weddell, crabeater, Ross, and southern fur seals—
unless the sealer obtained a permit from the government spon-

soring the expedition. An annex to the convention established a maximum sustainable yield principle for seal-hunting, and determined which seasons and geographic areas were open to hunters. In addition, SCAR was urged to study seal populations further.[45]

This convention extended the Antarctic Treaty beyond the agreed measures because it was a separate agreement negotiated by the consultative parties but independent of the treaty itself. Practically speaking, this format was necessary because anyone with a ship and a helicopter can hunt seals. The treaty parties—anticipating a growing sealing industry—wished to establish the right of any nation to become a signatory, regardless of whether they already were parties to the Antarctic Treaty. The group would follow this approach again in 1980, when it made a more ambitious agreement to regulate the marine living resources of the Southern Ocean, thereby reaffirming its authority to negotiate such agreements.

Successive consultative meetings refined these rules, changing designations of specially protected areas and addressing other environmental concerns, usually raised through SCAR. The 1972 Stockholm Declaration on the Human Environment broadened their environmental recommendations still further.[46] That year, at the seventh consultative meeting, the group agreed to voluntary guidelines, following a report of SCAR, to limit pollution around their Antarctic bases.[47] The eighth consultative meeting, held in Oslo in 1975, adopted a code of environmental conduct for both station activities and expeditions. Recommendation VIII-l3 introduced a new phrase extending the group's authority. It said: "Recognizing that prime responsibility for Antarctic matters, including protection of the Antarctic environment, lies with the States active in the area which are parties to the Antarctic Treaty...."[48] The ninth meeting—which also passed important recommendations on mineral and living resources—included a declaration of environmental principles which asserted, "The Consultative Parties recognise their prime responsibility for the protection of the Antarctic environment from all forms of harmful human interference.[49]

So it was that some unique biological spots on the planet came to be protected from the encroachments of hunters, tourists, and even scientists. The 1964 Agreed Measures for the Conservation of Antarctic Fauna and Flora, protected, among other sites, key rookeries for the Emperor penguin, including the largest one known at the foot of the Taylor Glacier in Mac. Robertson Land in East Antarctica, near Australia's Mawson station. The 1975 designation of sites of special scientific interest included Cape Crozier, on the edge of Ross Island near McMurdo station and

A pair of Emperor penguins. They will incubate their eggs during the long, dark winter and the chicks will hatch in early spring.

Ice-free coasts are vital to many forms of Antarctic life. Here, a crowded Adélie penguin rookery on the coast of Victoria Land near Cape Hallett exemplifies many such sites enjoying special protection under the Antarctic Treaty.

Antarctic life adapts to the long, bright summers and dark winters and months of twilight in between.

Scott's hut, from which the famous winter journey of Edward Wilson, "Birdie" Bowers, and Apsley Cherry-Garrard set out.

The Emperor penguins have been called by ornithologist Louis Halle "the most truly Antarctic of all birds."[50] They range farther south than other species, living on the coast and fast ice instead of the pack ice. Their colder habitat requires more volume-to-surface area, hence they are larger than their more northern-dwelling cousins, the Adélie, Gentoo, and Marconi. Because the Emperors' large size requires a longer incubation time for eggs (sixty to sixty-five days), they have evolved a unique life-style especially adapted to Antarctica's seasons, weather, and hardships.

Females lay their eggs in the Antarctic midwinter—when the wind howls, darkness reigns, and the flock huddles in sleepy masses along the coast. Cherry-Garrard, Wilson, and Bowers set out on their winter journey to capture the eggs before hatching to discover if the embryos showed evolutionary links between birds and reptiles. This theory was eventually discredited but their field work was remarkable. By surviving their own winter journey to Cape Crozier, they became the first humans to examine the Emperors' extraordinary winter life-style. The female Emperor, having laid her egg, passes it to her mate, who sits on them for months, losing one-third of his weight in the fast, while the mother sets off across the ice to the water's edge for food. In the spring the female returns; by then the chicks have hatched and are developed enough to start their own journey toward the ice edge, which is retreating toward them with the advance of spring. Thus, Nature shortens the Emperor chick's journey to manageable proportions—easing an otherwise hazardous existence. For the colony, the treacherous fractures of the spring ice and the giant fulmars sweeping down upon the young will cause the deaths of one-third of the chicks before they reach the safety of the sea.[51]

In the Antarctic not only well-adapted birds like the Emperors survive. Even the camp has survived that Wilson, Bowers, and Cherry-Garrard built in 1911 overlooking the rookery, and from which they slithered down bravely to collect the eggs. Antarctic writer Charles Neider visited Cape Crozier in the 1970s on a far easier journey—by helicopter—passing Mounts Erebus and Terror and the blue-crevassed mountainside of Ross Island where Cape Crozier lies. Below him was the igloo the trio had built during that gale in 1911, when their tent blew away.

We circled over Wilson's stone igloo and landed near it on a black knoll or spur about eight hundred feet above the ice shelf.... Some of the lower portions of the rock walls were still in place. One saw pieces of bamboo pole; parts of the skins of the Emperor penguins the three men had killed in the Crozier rookery and dragged here in order to flense them and use the blubber in the blubber stove and thus conserve oil; some blubber; pieces of pale Willesden canvas; a wooden box; either a khaki sock or a khaki balaclava; . . . a shirt; a ball of twine; some rusty bits of metal; pieces of rope and wire; and a bit of broken board.[52]

The protective measures engendered by the treaty have prevented scientists and tourists from disturbing the Emperor rookeries and special sites like Cape Crozier. Generally, most of the environmental measures the treaty powers adopted have been observed, although there have been complaints about garbage heaps near stations, helicopters "buzzing" the penguin rookeries, and other activities. We will see in chapter 6 how success in the environmental field spurred the group to pass still more ambitious recommendations regarding enviromental protection of resources—hence regulating future natural and minerals resource development.

Summary

Thus, by 1983, the treaty group had passed administrative rules on the environment, tourism, and the preservation of historic sites. In the cases of sealing and krill harvesting, they had drawn up separate conventions open for endorsement by nonsignatory nations. There were also agreed measures, to be implemented voluntarily, on telecommunications, rescue operations, weather, and even an agreement to refrain from minerals exploration and exploitation for so long as progress was being made in the negotiation of a regime.

Years later Daniels—the original U.S. negotiator who drafted many of the provisions tabled by the United States in 1958 in the Academy board room—would recall proudly what an impressive body of administrative law this had become.[53] It is true that the treaty powers' commitment to the environment had not undergone a substantial test; doubtless it is easier to rap the knuckles of a few scientific expeditions, or to speak quietly to an offending nation's diplomats about a dirty station or tanker discharge. Environmental groups were dissatisfied with the level of compliance, too. So the most serious test of the treaty powers' self-proclaimed commitment to the environment lay ahead. Nonetheless, the environmental issue had become a vehicle for the development of the original treaty's legal and political authority. The six-page document had, in effect, become much longer.

five

Resource Issues Emerge

The first sign that Antarctica was moving into a resource age occurred at a 1968 symposium held at Cambridge University under the auspices of the Scientific Committee on Antarctic Research, the nongovernmental group of scientists from many nations which had coordinated research in the Antarctic since the 1957–58 International Geophysical Year (IGY). This was the second symposium of the Special Committee on Antarctic Research (SCAR) Working Group on Biology, a committee that, despite its understated title, served as the fulcrum for international research in the field.[1]

From the quiet flow of the scientific work, there suddenly loomed an enormous applied research problem—the possibility that the Southern Ocean held a vast protein resource for mankind—one whose exploitation would be controversial because of its threat to the region's ecology, already unbalanced by whaling. It was obvious that from now on Antarctic biology would have to tangle with the often-political worlds of the fisherman and the ecologist. Antarctic science—at least research on the marine ecosystem—would no longer be a pure, unworldly, scientific activity.

The early drift of Antarctic biology had been reflected in the first symposium of the working group held in 1962. It was primarly concerned with classifying and describing the flora and fauna of Antarctica, including those of the dry valleys inland and the exposed shores.[2] What organisms lived there? How did they survive? This kind of science is called *autecology*, meaning the study of an individual organism. By 1968, however, enough was known about individual species to permit studies of entire populations and how species interact with one another.

The Antarctic ecosystem has relatively few lower species (zooplankton) on which higher species depend. Thus, many scientific papers discussed krill, the principal zooplankton animal of Antarctic waters. Krill are delicate, transparent, shrimplike animals, pink or ochre in color, and measuring 3 to 5 centimeters long, with knobby eyes and phosphorescent "lights" above their legs. They tend to swarm near the surface, by day appearing as great reddish patches; their nighttime swarms look like twinkling underwater galaxies. They were a familiar sight to whalers and explorers, and had been studied by British *Discovery* investigators; but otherwise Antarctic krill had not received much attention. It was known that there are six species of krill in Antarctic waters, of which one, *Euphausia superba,* is the most numerous and the principal food for whales, seals, and penguins and other birds.[3]

At the Cambridge symposium, several Soviet papers focused on krill. Everyone knew that Soviet whalers had found it increasingly difficult to find enough catch in the Antarctic. Whale populations were in the midst of a dramatic collapse, causing the Russians to harvest krill for processing and marketing back home. P. A. Moiseev of the Institute of Fisheries and Oceanography reported that the animals, who were almost 16 percent protein, swarmed in the upper 5 meters of the water column from December to March, and could be spotted visually by ordinary midwater trawlers or with echosounders. Catches, he reported, had reached 5 or 6 tons per haul.[4]

Moiseev plunged from the back waters of academic biology to the question of the world's future food supply, estimating that the total standing stock of Antarctic krill could reach 7,500 million metric tons. When the baleen whales were abundant, they had consumed more than 150 million metric tons of krill each year. Now this uneaten "surplus" was more than twice the world's total fish catch, which stood at approximately 60 million metric tons per year. In short, Antarctic krill represented the largest single fishery potentially available to man.[5]

Moiseev's suggestion was simple: since fish, whales, and other animals at the top of the marine food chain were decreasing, fishermen should turn to organisms at lower trophic levels, such as krill, to feed the world in the future. He concluded, "It seems well worth while for all countries to intensify their efforts to study the possibilities of establishing a commercial krill fishery in Antarctic waters. Success in investigations of this kind would probably allow us to double the present catch of aquatic organisms from the world ocean."[6]

Only the Beginning

Those attending the SCAR symposium had been prescient. Over the next fifteen years, more and more fishing vessels would come to the Antarctic in the austral summer until ten nations were plying the gray waters of the Southern Ocean, searching for the twinkling lights and reddish patches of the krill swarms. Research institutes and government laboratories began to study how to turn krill into edible protein, while scientific study of the marine ecosystem flourished. The ships' ease of access to the international waters around Antarctica would raise profound problems for Antarctic law, including whether the treaty encompassed resources—especially those of the Southern Ocean south of 60° south latitude—not merely the land. The 1968 SCAR symposium signaled the beginning of a sea-change in Antarctic science and politics; gradually the treaty powers—who for a decade had considered the region as a harmless, international scientific laboratory—began considering the Antarctic as a reservoir of potential food and mineral resources.

After the 1968 symposium, the krill discussion came to dominate Antarctic marine biology, and led SCAR to recommend conservation efforts, which in turn led to a convention to regulate the krill fishery. The Convention for the Conservation of Antarctic Marine Living Resources, as it is called, entered into force in April 1982. It is becoming an important test of the treaty powers' ability to deal with resource issues (its full text can be found in Appendix B).

The krill issue also prompted the treaty powers to take up the long-delayed question of Antarctic minerals. Part of the impetus for this was legal: the powers could not very well assert jurisdiction over the resources of the ocean south of 60° south latitude without—at least by implication—asserting their competence over the land resources too. But mostly, their motives were political and psychological. They had been spurred into action by the accidental discovery of trace hydrocarbons on the Ross Sea continental shelf in 1973. This discovery, followed by the OPEC oil price hikes, whetted international interest in Antarctica's oil potential.

But minerals were bound to be a more difficult issue than krill, although krill fishing was already under way and minerals development seemed unlikely in this century. Minerals raised the issue of rivalry over territorial claims and its extrapolation to Antarctica's continental shelves. As chapters 3 and 4 have shown, these claims had never been settled, only put on a back burner by the enactment of the treaty. For the United States and the Soviet Union, who have large potential claims there, the minerals issue would also raise the question of what each wants from Antarctica. Further, it served to awaken the developing world's interest in Antarctica: developing nations might well see Antarctic minerals as potential "riches" and dispute the treaty powers' control over them. Thus, the 1968 SCAR symposium had opened the door for major changes in Antarctic science, diplomacy, and law.

The Krill Problem

Although the numbers have changed somewhat, the principal conclusions of the 1968 SCAR symposium have been upheld in subsequent studies. Carbon can be fixed unusually fast in Antarctic waters, because the ocean upwells nutrient-rich waters to the cool, fresh surface waters, and because the twenty-four-hour daylight in the austral spring and summer makes for intense, continual photosynthesis. In 1968 George Knox of the University of Canterbury in New Zealand cited previous estimates that the Southern Ocean could be up to eight times more productive than parts of the North Atlantic, supporting a staggering biomass, or live weight, of animals.[7]

Furthermore, for all the scientists knew, the depletion of the whales had left a huge "surplus" of lower organisms, which had been formerly their food. This theory was elaborated at the SCAR meeting by N. A. Mackintosh, a *Discovery* scientist, who had made pioneering studies of whale populations from the whaling station at Grytviken, South Georgia. Mackintosh cut open dead animals, examined fetuses in the pregnant females, measured lengths, weights, the number of times females had calved, and growth rates. Thus he had compiled definitive data on whale biomass and abundance over the early decades of the whaling era. At the 1968 meeting, Mackintosh explained how the whale species, one by one, had decreased as a result of harvesting. First, in the 1930s, the blue whales—the pride of the whalers—declined. Then, in the 1950s, the fin whales decreased, and, by the 1960s, the sei whales had declined, although they were smaller and had been hunted only after the decline of the larger whales.

The whales fast eight months of the year in the rest of the world's oceans, but in the austral spring they swim southward, arriving in the Antarctic as the winter sea ice retreats, leaving a cold, freshwater surface layer that enhances food production. For 120 days the whales feed. Then, as the autumn ice reforms over their feeding ground, they disperse for another winter's fast— hiding, so to speak, from the whalers. Depending on assumptions about the average size and weight of each whale species, Mackintosh concluded it is possible that the "missing" whales that were hunted down ate 33 to 330 million metric tons of zooplankton each year. And most of this was krill.[8] Further, some other krill-eating species seem not to have increased, and others have a negligible effect on the "surplus" of krill, which he estimated to be 150 million metric tons.

Martin W. Holdgate, secretary of the working group, decided the krill issue was so important that he positioned the krill-related papers first in the published symposium volume. In his introduc-

Krill spilling from the opened stomach of a crab-eater seal. Many such higher species depend almost exclusively on krill for their food.

Crabeater seals in the ice pack. Antarctic crabeaters comprise one of the largest animal biomasses on earth, but they are impractical to hunt for food because they scatter in small groups across the pack.

Fur seals (*Arctocephalus gazella*) clustered near shore. Their habit of clustering made them easy prey to nineteenth-century sealers, who nearly extinguished them.

tion, Holdgate noted: "For many decades Antarctic exploration has been falsely represented as an expensive luxury yielding no return except heroism, obscure scientific data, and endearing pictures of penguins. Yet the first explorations...were made for economic reasons." Krill research, he wrote, prophetically, "is likely to be one of the most active growing points of Antarctic biology over the coming decade."[9]

Credit for defining the krill problem goes to Richard M. Laws, a British expert on Antarctic seals who in 1973 became director of the British Antarctic Survey, the government agency descended from the original *Discovery* committee to run the British Antarctic program.[10] After 1968, Laws tried to determine whether other species were increasing in response to the apparent surplus of krill food in the Antarctic. Which species had gained? Which had not? Who competed with whom for food? The answers to these questions would depend on a precise knowledge of various animal and bird populations.

But how to count, say, the crabeater seal, which lives singly or in pairs on the ice pack until the winter when it ranges more than 18.8 million square kilometers around Antarctica? This difficulty had been illustrated in 1964, after a Soviet researcher claimed he saw 3,000 crabeaters on the ice. The Norwegians followed up with an expedition to hunt them; but even using helicopters, they could hunt few. Although the treaty powers negotiated a convention to regulate future commercial sealing, the Norwegian expedition marked the end of modern "commercial" sealing in Antarctica.[11]

Laws and others found that several species had increased with the decline of the whales, suggesting that the Antarctic ecosystem was food-limited. J. Gilbert and A. Erickson estimated a total of 14.8 million crabeater seals with a total biomass of $2,868 \times 10^3$ tons. Laws examined the teeth of the crabeaters that had been killed for dog rations. The layered pattern in the cementum on the roots of their teeth indicated that they were maturing earlier than before. Animals are believed to mature sooner when they attain a larger body size early in life, something that occurs when they eat more food.[12]

Another researcher, M. R. Payne, counted the krill-eating, brown-pelted fur seal. This species barely survived extermination in the 1820s, after which it increased only to be hunted again in the late nineteenth century. The island of South Georgia, populated by only 100 fur seals in the 1930s, had a seal population of 15,000 by 1957. By 1976, the whales in the waters surrounding the island virtually had vanished, Payne noted, presumably leaving a krill surplus for the seals. That year, Payne estimated that there were more than 350,000 fur seals; more recently, however, their populations have been increasing quickly.[13] There also are significant increases in king, chinstrap, Adélie, and gentoo penguins, which feed mostly on krill. But many Antarctic birds eat other food besides krill, so that the krill-feeders have only a localized impact. Finally, the biomass of Antarctic birds is small despite their great numbers. Consequently, birds are deemed to have little overall impact on the availability of krill for other species.[14]

The best evidence that the Antarctic marine ecosystem is food-limited comes from the whales themselves. As other scientists continued Mackintosh's investigations of whale populations in the 1970s, they found that female whales were becoming pregnant at younger ages. Higher pregnancy rates also correlated with periods of intensive whaling, suggesting that greater food abundance causes the whales to attain full size earlier and to calve sooner. For example, a drop in whale pregnancies had occurred during World War II after whaling had stopped and there were more whales taking the available food.[15]

These increases, once documented, suggested that the Antarctic marine ecosystem was moving toward "a new equilibrium" in which the increased seal and penguin populations were taking new, larger shares of the krill surplus, as J. R. Beddington and R. M. May would write later.[16] Thus, krill fishing by man, es-

Chinstrap penguins are also major consumers of krill.

pecially in ocean areas with abundant krill swarms, possibly could hurt this new ecological balance.

Competition for food, moreover, was already intense on a geographic basis. When Laws and others mapped out where the different whale, seal, and bird species were (almost nothing was known about fish or squid), they found that the species seemed to have arranged themselves to minimize competition for food. The Weddell seals, who feed mostly on fish, live year round in the fast ice next to the shore. With their their remarkable teeth, they cut holes through the ice to breathe and feed. The crabeater seals live farther from shore and feed primarily on krill. The whales also distribute themselves in distinct regions. Every spring, the blue whales arrive first in the Southern Ocean and chase the pack ice as it retreats, feeding on the smaller krill found near the ice edge. Later on, the fin whales arrive to feed in the ocean area uncovered by the retreating pack ice. Humpback whales stay north of the maximum area of ice altogether, while sei and sperm whales remain even farther north, near the Antarctic Convergence.[17]

Laws's maps reveal critical areas where species overlap and apparently compete for food, especially in the South Atlantic. Robert J. Hofman of the U.S. Marine Mammal Commission, who reflects the view of some Antarctic biologists, concludes that these areas

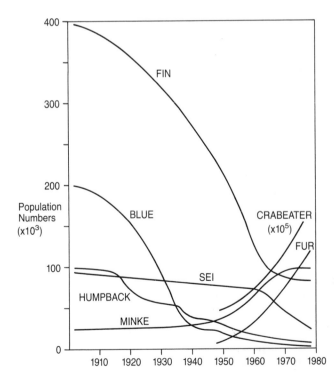

Massive declines of Antarctic whales led fishermen in the 1960s to harvest krill instead. The surplus of krill left uneaten by the whales may have helped the fur and crabeater seals and minke whale populations to recover.

will attract fishermen taking krill away from the whales and other species.

Wrote Hofman:

> ...[K]rill fishing will be concentrated in open-ocean (ice free) areas which are, or were, the major feeding grounds of krill-eating baleen whales...fishing will occur primarily in the summer months when whales are present on the feeding grounds...fishing effort will be selectively focused on the same kinds of high density krill swarms which are fed upon by baleen whales....Thus, levels of krill harvest that may have an immeasurable effect on overall krill density may result in significant reductions in the numbers or sizes of krill swarms and have a significant adverse impact on predators, such as baleen whales, that are adapted to feeding on swarms.
>
> [I]t further is possible that depleted populations of blue, fin, and humpback whales may be affected to a greater extent than more abundant populations of minke and sei whales.[18]

The question of whether the krill fishers would take food from other species and endanger the recovery of the whales became a key issue for environmentalists and Antarctic diplomats in the 1970s, as krill fishing grew and negotiations began for a convention governing the fishery.

Krill Biology

Besides their lack of knowledge of krill stocks, predator stocks, and competition for food, Antarctic biologists reckoned with their near-total ignorance of krill biology. J. W. S. Marr—who as a Boy Scout had gone on Ernest Shackleton's last expedition—had made an enormous contribution to the field. Marr's key work, *The Natural History and Geography of Antarctic Krill (Euphausia superba Dana)*, was published in 1962 before he died.[19]

Marr developed a theory of Antarctic krill development still largely regarded as accurate. The krill lay their eggs near the surface; the eggs sink for about ten days, dividing as they reach darker, deeper waters, from which they are carried south toward the edge of the continental landmass by Antarctic deepwater currents. There they hatch, perhaps when the water pressure is great enough, or for some other reason. As nauplii, they begin their long, struggling ascent to the surface, changing into calyptopes, and eventually, at the surface, into juvenile krill. This pattern explains the fact that near the pack ice edge, *E. superba* tend to be smaller, and larger krill tend to be found farther out.[20]

At the 1968 symposium Mackintosh and the Russian scientist R. R. Makarov, raised fundamental, and still unanswered, questions about how krill live. There may be different "races" of krill that remain in separate segments of the Southern Ocean. On the other hand, the krill swim against the current only with difficulty, and this suggests that they are carried by prevailing currents where they are evenly distributed around the continent.[21]

Almost every question dealing with krill biology has implications for the fishery. If the krill *are* evenly distributed around the continent, they are likely to replenish themselves in places where they have been depleted by fishing. If they live about two years, as Mackintosh predicts, then the stocks could replenish themselves quickly. If krill live longer than three years, as Nemoto suggests, they are more vulnerable to overfishing over one or two years. As often happens, the first time krill were studied under laboratory conditions, they produced many surprises, upsetting previous estimates of stock abundance, spawning, and larval development, as well as other aspects key to the fishery.

Throughout the 1977–78 and 1978–79 austral summers, two U.S. researchers—Mary Alice McWhinnie and Charlene Denys—succeeded in keeping krill alive in a flow-through seawater tank at the U.S. Palmer station. They observed that krill have a quality called "regression" which is found in some other crustaceans: they become smaller and less sexually mature in appearance after spawning. Thus, many of the animals previously counted as young krill in the stock estimates were instead older animals that had spawned already and would do so again. As McWhinnie wrote from her Antarctic laboratory: "If you must swim 24 hrs. a day to 'stay afloat!'—best you keep a 'tight fit' of your exoskeleton so your muscles have proper attachments.... Thus, when on reduced food as in winter, you shrink (many lower animals do!) and if you're a swimming krill you keep making a smaller 'house' to live in. Come the next season people call you younger because you're smaller—though indeed you're a year older!"[22]

Later work shed some further light on these observations. Other researchers, particularly an Australian team, became proficient at keeping krill alive in tanks to observe their biological properties close at hand. In addition, diving teams, videotaping the swarms, found them to be far denser than had been observed earlier. A pair of researchers from the University of California, Santa Barbara, found females to be twice as fecund as had been believed previously. But neither the later discoveries of denser swarms nor larger numbers of krill eggs pushed up the overall estimate of krill biomass. This was because the acoustic survey data from 1979–83 was hard to interpret since the instruments used in the survey were poorly calibrated. Although agreements were forthcoming on some points—that krill are cannibalistic, likely to feed over the winter, and live to be four years old—biologists had yet to fit these findings into a coherent ecosystem model capable of managing the fishery.[23] Antarctic marine biology had started on a new course, but it had a long way to go.

A New International Program

Even as scientific interest in the krill grew, so did krill fishing (table 5-1). The token harvests by Soviet vessels that had spurred the SCAR discussion in 1968 were followed by more intensive fishing by Japan, Poland, West Germany, Taiwan, East Germany, and Spain. By 1980–81, the annual catch had reached 0.57 million metric tons. This was hardly the doubling of the world's annual catch of fish that Moiseev had predicted in 1968, but it was not negligible—especially given how rapidly fisheries can expand and the difficulties encountered by many fishing nations as they fished in waters closer to home. As Hofman had noted, most fishing was done in areas where the krill's natural predators concentrate—thus increasing the chances that the fishermen were taking food away from the whales.

Scientists studying Antarctic marine ecology, as well as those from krill-fishing countries, agreed to study krill cooperatively through a new international program. In 1972 the SCAR Working Group on Biology established the Subcommittee on Marine Living Resources. By November of 1975, it had joined forces with the Scientific Committee on Oceanic Research (SCOR), SCAR's counterpart for coordinating international oceans studies. The SCOR–SCAR group became a working group in its own right. Its head was Sayed Z. El-Sayed, professor of biological oceanography at Texas A&M University and a prominent student of the biological productivity of the Southern Ocean. The subcommittee, at its first meeting in 1974, drew up a research plan called International

Table 5–1. Reported Catch of Fish and Krill in the Antarctic, 1972–73 Through 1980–81
(in millions of metric tons)

Species	1972–73	1973–74	1974–75	1975–76	1976–77	1977–78	1978–79	1979–80	1980–81
Finfish	13,500	106,100	25,300	57,100	268,700	257,939	125,091	115,427	121,357
Krill	n.a.	22,343	39,981	2,787	122,532	142,803	332,565	477,025	448,266
Squid	n.a.	n.a.	n.a.	n.a.	1	391	2	n.a.	n.a.
Total Antarctic	13,500	128,443	65,281	59,887	391,233	401,133	457,658	592,452	569,623
Total world[a]	62,824,400	66,597,100	66,135,600	69,590,200	68,677,700	70,399,400	71,314,200	72,376,800	74,760,400

Note: Nominal catches = live weight basis; n.a., data are not available.

Source: 1978, 1979, and 1981 *Yearbook of Fishery Statistics, Catches and Landings* (Rome, Food and Agricultural Organization of the United Nations) tab. C-48(A), C-58(A), C-88(A), and A-1(B).

[a] Figures are for end of the year. All other figures in the table are for a June 30 year end.

Biological Investigation of the Southern Ocean (IBISO). A fourteen-nation meeting was held at Woods Hole, Massachusetts, in August 1976, and at that time the research plan was revised and renamed BIOMASS. The acronym stood for Biological Investigations of Marine Antarctic Systems and Stocks, but it was word play on the basic question, How much biomass was in the Antarctic ecosystem, and how was it changing due to natural factors and to man?[24]

The BIOMASS objective was to combine basic marine biology with some of the emerging applied questions that both krill fishermen and conservationists were posing. Officially, it aimed "to gain a deeper understanding of the structure and dynamic functioning of the Antarctic marine ecosystem as a basis for the future management of potential living resources."[25]

BIOMASS included ambitious plans for a data center, a plankton-sorting center, setting common standards for identifying krill, and many other objectives. Its first major experiment, called the First International BIOMASS Experiment or FIBEX, was undertaken in the Southern Ocean in the austral spring of 1981. Twelve ships from ten nations participated, canvassing prearranged areas to correlate weather and ocean information with observations of animals and krill. The most spectacular find was the U.S. vessel *Melville's* sighting of the largest krill swarm so far recorded. Chief scientist, Osmund Holm-Hansen of the Scripps Institute of Oceanography, estimated the "superswarm" at 10 million metric tons. Word of the discovery spread, and soon there were some thirty-five Soviet fishing vessels trawling the waters around this unexpected marine boon.[26]

In addition, the scientists apprised the diplomats of the possibility that a sudden expansion in intensive localized krill fishing could severely damage whale recovery in the area. The Antarctic Treaty powers registered these concerns early in the 1970s and at the eighth consultative meeting, held in Oslo in 1975, member governments were urged by the group to "initiate or expand, insofar as is practicable within their Antarctic scientific programmes, detailed studies of the biology, distribution, bio-mass and population dynamics and the ecology of Antarctic marine living resources."[27]

Additionally, the group put Antarctic marine living resources on the agenda for the ninth consultative meeting. That session, held in London in 1977, passed a major resolution asserting that the treaty powers were capable of negotiating a regime governing the development of the fishery. Thus, the framework was established for explicitly asserting the treaty powers' authority over important resource questions.

The Minerals Issue Emerges

One reason the treaty powers launched discussions of Antarctic marine living resources by the mid-1970s was that the more difficult issue of minerals was looming on their horizon. The early explorers had believed that Antarctica would one day prove to be minerals-rich. According to one story, Adm. George Dufek, the leader of several U.S. Antarctic expeditions (who apparently knew more about ship operations than geology), used to brag that Antarctica would have gold, diamonds, oil—even brass (which is man-made). But, as no commercial deposits were actually found, the issue had been conveniently postponed. Indeed, as we saw in chapter 4, the group deliberately sidestepped the issue when they negotiated the treaty.

But in 1973 two things happened to raise the minerals issue: First came the accidental discovery in February of gaseous hydrocarbons in three of four holes drilled in the Ross Sea continental shelf by the *Glomar Challenger,* the U.S. scientific drilling ship; second came the sudden rise in the price of Middle Eastern oil.

Although for years the *Glomar Challenger* had taken core samples from the ocean floor, it was not permitted to drill in places with potential oil or gas deposits because it lacks a blowout preventer for plugging leaks. "Leg 28"—the cruise that included the Ross Sea—had been planned so the vessel would drill only in places deemed unlikely to have oil or gas. But gases—mostly methane mixed with ethane—were found in dominantly terrigenous sediments of Miocene age. In one hole, gas was found 45 meters down and at 150 and 265 meters in other holes. The crew quickly cemented each hole and abandoned the drilling effort.[28]

Since the appearance of hydrocarbon gases does not necessarily indicate the presence of commercial oil or gas deposits, the scientists on board the *Glomar Challenger,* Dennis E. Hayes and Lawrence A. Frakes, wrote, "It is premature to attach any economic significance to the Ross Sea hydrocarbons." Nonetheless, they were aware that their discovery would generate worldwide interest "in light of the fact that producing oil and gas fields are found beneath the shelves of southeast Australia and western New Zealand, areas that were contiguous with the Ross Shelf before continental drift."[29]

The hint that Antarctica might have oil and gas deposits—coming just when many capitals, including Washington, were reeling from the implications of costly oil from traditional sources—changed policymakers' views of Antarctica. The U.S. Geological Survey (USGS), which had made little systematic study of Antarctica since the IGY, was asked to study Antarctica's minerals potential. This study, coauthored by N. A. Wright and P. G. Williams, was published in 1974. The Wright–Williams report, which remained for years the only summary of the then-sketchy knowledge of Antarctica's minerals potential, was finally updated in 1983 by another USGS report edited by John C. Behrendt.[30]

While compiling the Wright–Williams study, the USGS made a confidential internal estimate of the potential oil and gas deposits in West Antarctica's continental shelves, that is, those located beneath the Ross, Weddell, and Bellingshausen seas. The estimate of 45 billion barrels of oil and 115 trillion cubic feet of natural gas was leaked to the press, causing a sensation. Later, the USGS explained that these were estimates of in-place deposits, for example, of the total resource in a single geologic formation. Because only one-third of an in-place deposit tends to be extracted, according to the USGS, the Antarctic continental shelves would yield only 15 billion barrels of oil, an amount comparable to the likely yield of 10 to 20 billion barrels from the U.S. Atlantic continental shelf and less than the 30 to 60 billion barrels that the USGS believed could be recovered from offshore Alaska. In fact, the Antarctic estimates were based on rough extrapolations from earlier productivity estimates of continental shelves located elsewhere in the Southern Hemisphere.[31]

Despite their hypothetical nature, Antarctica's hydrocarbon resources had political implications after 1973. The treaty powers believed they had to do something, both to assert jurisdiction over development—since mineral resources are not mentioned in the treaty—and to protect the Antarctic environment from any pell-mell rush toward development.

Until then, the issue had been simmering on a back burner. It had been discussed informally at the sixth consultative meeting, held in Tokyo in 1970; and in May and June of 1973, the Nansen Foundation had held a private meeting on Antarctic minerals at Polhogda, Norway. The group's report, published in 1974, makes clear that relationships among the treaty powers were strained. The eighth consultative meeting in Oslo in 1975 resolved to hold a special preparatory meeting on the sensitive minerals issue. SCAR was asked to assess the environmental impact of minerals exploration and exploitation, and the treaty powers urged national scientific programs, in coordination with one another, to obtain "fundamental scientific data on the geological structure of the Antarctic."[32]

Gondwana and Antarctica's Minerals Potential

As the Wright–Williams report explained, the argument for the existence of Antarctic mineral "riches" is that because the other six continents have rich, commercial-grade deposits, the seventh con-

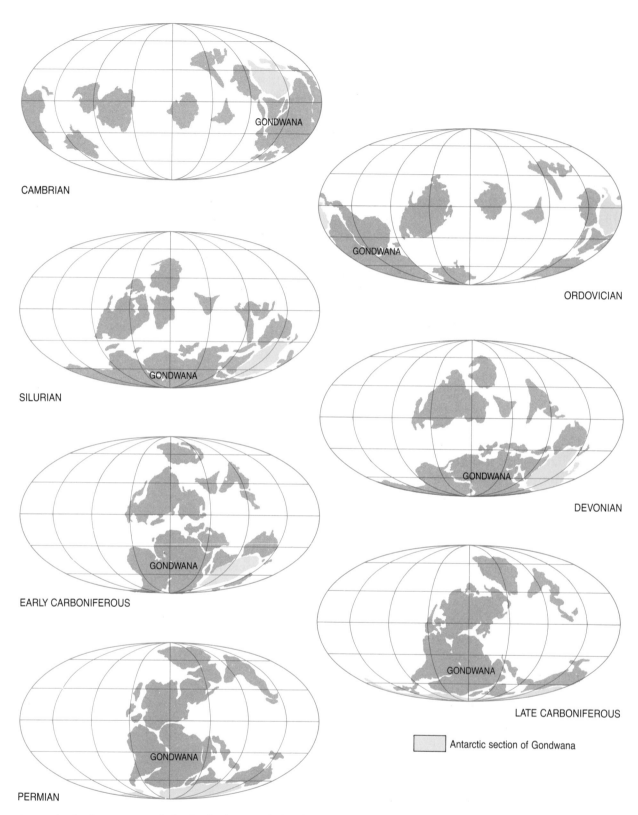

CAMBRIAN

GONDWANA

ORDOVICIAN

GONDWANA

SILURIAN

GONDWANA

DEVONIAN

GONDWANA

EARLY CARBONIFEROUS

GONDWANA

LATE CARBONIFEROUS

GONDWANA

Antarctic section of Gondwana

PERMIAN

GONDWANA

Antarctica lay in temperate latitudes for long periods of geologic time, as this reconstruction of continental drift by A. M. Ziegler shows. In the Cambrian period, Antarctica lay north of the Equator as a province of the super continent, Gondwana. Not until the mid-Permian, approximately 250 million years ago, did Antarctica start moving across the Pole.

tinent must have them also. To date, no deposits of economic value have been found in Antarctica, although extensive coal and iron deposits have been reported, as have traces of silver, gold, copper, manganese, and other valuable minerals.[33] One explanation is that only a small area has been surveyed. David H. Elliot of Ohio State University has estimated that, of the 2 percent of the continent not covered by ice, only 10 percent has been mapped in detail; and of the exposed area, less than 1 percent has been explored for minerals.[34] (This is analogous to prospecting in an area the size of Delaware for clues to the mineral wealth of the United States and Mexico.) So, until some method is found to explore beneath the ice cap—which hardly seems imminent—the continent's minerals are likely to remain hidden, in the explorers' phrase, by its "veil" of ice.

There are two arguments against Antarctica's having economic deposits of minerals. One is that the continents of the Southern Hemisphere, with the exception of southernmost Africa, have been somewhat less minerals-rich than those in the Northern Hemisphere. While this proposition may be debatable for minerals, it is nonetheless true that most giant hydrocarbon basins found to date have been in the Northern Hemisphere. A second

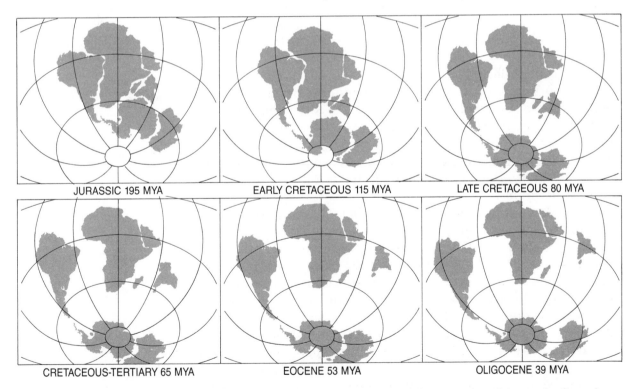

JURASSIC 195 MYA

EARLY CRETACEOUS 115 MYA

LATE CRETACEOUS 80 MYA

CRETACEOUS-TERTIARY 65 MYA

EOCENE 53 MYA

OLIGOCENE 39 MYA

The subsequent breakup of Gondwana, as depicted by Norton and Sclater. The separation of Antarctica from the other continents began 115 million years ago. The isolation of the continent at the bottom of the world caused it to become climatically locked within the circumantarctic current, and helped form the icecap.

argument was mentioned in the Wright–Williams report: if Antarctica's surface mineral deposits had been scraped up by the ice over millions of years, and carried, inch by inch, down to the sea, perhaps they are to be found offshore as pulverized ores.[35]

Indeed, the principal method for estimating Antarctica's minerals potential comes from reconstruction of Gondwana, the ancient continent which broke up into Antarctica, Australia, Africa, and South America. Gondwana itself has been the subject of a long-standing scientific controversy on continental drift in which Antarctic research has played a major role. So, as in the case of krill, knowledge of Antarctic mineral resources has emerged from fundamental scientific research. What remains to be accomplished are the detailed mapping, exploration, and baseline studies needed to fill out present skimpy information.[36]

The early explorers, picking away at Antarctic rock outcrops, often found coal, supporting the theory that the continent had once been temperate enough to support lush forests. Indeed, on Scott's last expedition, Dr. Edward A. Wilson collected some fossil-printed rocks, further evidence that the "dead" seventh continent had once been alive. In 1912 the German balloonist and Arctic explorer, Alfred Wegener, proposed that the continents had once been joined, but the scientists of his day mocked so ridiculous a notion.[37]

Thus, by the time of the IGY, a major scientific mystery was how ice-covered Antarctica could have once been warm. As George Doumani and William Long wrote in an issue of *Scientific American* devoted to Antarctica in 1962: "If the continent did not wander, where was the Pole when the continent was green? Where was the continent if the Pole did not wander? Ice and coal are not deposited in the same polar region under the same climatic conditions. Yet in Antarctica they coexist! This coexistence forcibly demands an explanation. The seventh continent holds a master key to the earth's ancient history."[38]

The IGY turned up more evidence that Antarctica resembled the other continents, supporting the still-disputed hypothesis of continental drift. In 1960 David Elliot, with an Ohio State team working in the Horlick Mountains, found a particular sequence of Permian age fossils on a bed of tillite and glacier-borne sand and mud. It was already known as the "Gondwana sequence" because it had been found in the Gondwana area of India and in South America and Australia.[39] Its existence in Antarctica suggests that Antarctica had been linked to these continents. Doumani and Long wrote excitedly, in 1962, that without the hypothesis of continental drift, "it is difficult otherwise to reconcile the geology and paleontology of Antarctica with the isolation of the continent on the bottom of the globe."[40]

Major ocean surveys undertaken by scientists during and after the IGY provided what Wegener had lacked: a plausible theory of how the continents might have moved. This theory, known as *plate tectonics*, opened up a new era in earth sciences. (Indeed, one reason for the *Glomar Challenger*'s presence in the Antarctic in 1973 was to compare its continental shelves to those of other continents.) The puzzle of how they all fitted together into Gondwana provides clues to the continent's geology and minerals potential. By studying the coasts of the other continents that once adjoined Antarctica, geologists can extrapolate what might lie under the ice. This was Wright's and Williams's approach in 1974, and it has been followed by others since.

Continental Shelf Hydrocarbons

Most experts think that if any Antarctic minerals are explored for seriously and developed before the end of the century, they will be the offshore hydrocarbons.

The breaking up of Gondwana and the sequence by which the continents separated is shown on page 127. Each break was preceded by millions of years in which one continent-sized province pulled away from the other, stretching and depressing the land in between, and forming basins filled with sedimentary rock. The Ross Sea and the Gippsland region of southern Australia probably drained both Antarctica and Australia over a long period. Like-

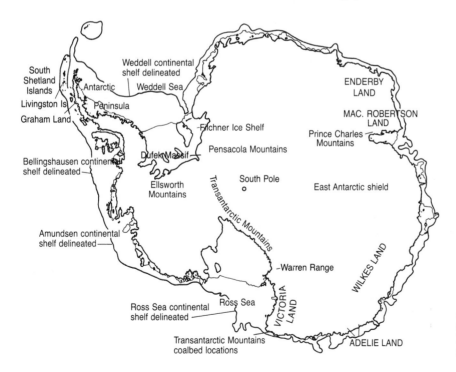

Antarctic continental shelves are narrow except for the large ones under the Weddell and Ross Seas.

wise, if the Weddell Sea and the Atlantic opened simultaneously, that side of the continent and possibly parts of southern South America would have been drained. The history of the Bellingshausen and Amundsen Sea shelves is unknown: They could be a sedimentary margin of East Antarctica that moved westward as the Weddell Sea opened; or they could be made of different pieces, plastered together by the movement of the plates.[41]

The Ross Sea continental shelf has an area of 772,000 square kilometers, the size of France and West Germany combined. Half the overlaying sea is covered by the ice shelf floating at its seaward edge. The floor seaward of the ice shelf is marked by scour marks, some made when the ice sheet reached its maximum extent thousands of years ago. Today, large icebergs can remain grounded for months on the sea floor, scouring new trenches.[42]

Following the *Glomar Challenger*'s discovery of trace hydrocarbons in the Ross Sea, a number of seismic surveys have been made in this area. West Germany's Federal Institute for Geosciences and Natural Resources made a 48-channel survey in 1980 over a course that the U.S. research ship *Eltanin* had charted with a single-channel survey in 1974–75. In 1981–82 the Institut Français du Petrol made a survey, and the Japanese visited the region in 1983 in the *Hakurei-Maru*. Behrendt's 1983 report said that neither the French nor the Japanese results were known, but that the German work supported estimates that the sedimentary rock measured at least 3 to 4 kilometers thick. One reason for the lively interest in the Ross Sea area is that it once adjoined the Gippsland basin of Australia which in 1974 had proved reserves of 2.5 billion barrels of oil and 220 billion cubic meters of gas.[43]

The Ross Sea probably adjoined New Zealand, although its precise fit is unclear, but the association is less promising in terms of oil potential. There are different reconstructions of the ancient history of both New Zealand and the subsea Campbell Plateau south of it. In any event, New Zealand's principal offshore find has been the Maui fields now supplying that country's gas market. Initial drilling on the Campbell Plateau was unproductive, but the government has plans to resume its search.[44]

Although the 1974 Wright–Williams report said the Ross Sea was the most promising place to look for oil in Antarctica, subsequent findings show that the Weddell Sea seems equally so. For five seasons, Soviet geophysicists made aeromagnetic surveys over the Filchner and Ronne ice shelves, finding sedimentary rock below that was estimated to be 12- to 15-kilometers thick. The British Antarctic Survey, which began aeromagnetic surveys as well, confirms the 14- to 15-kilometer thicknesses. Shipborne seismic surveys were carried out by Norway in 1976–77 (16-channel data), by the Federal Institute for Geosciences and Natural Resources in 1978 (48-channel data), by the Soviet Union in 1980–82 (12-channel data), and by Japan's *Hakurei-Maru* in 1981–82 (24-channel data). In 1980–81 the Argentines attempted a survey in the northwest portion, where sediments are believed to be thickest, but they were driven back by the ice. Consequently, they plan aerial surveys in the future. Although some of this new data was not yet available at the time of Behrendt's 1983 report, he speculates that thick layers of sedimentary rock, of Cretaceous and Tertiary age and much older, may extend under the West Antarctic ice cap as well.

The Weddell Sea floor is huge—some 901,000 square kilometers in area—and it is very old. The Atlantic edge dates prior to the late Jurassic or Cretaceous age, and it is from rocks of this period that most of the world's petroleum comes.[45] The floor has never been drilled, however, and it is unclear whether drilling would violate the treaty powers' policy of "voluntary restraint" on Antarctic minerals exploration and exploitation.

Little is known about the history or age of the Bellingshausen and Amundsen Sea shelves, which are 256,000 square kilometers and 210,000 square kilometers in area, respectively, and together almost the size of France. In 1981, as part of Japan National Oil Corporation's three-year survey of Antarctica's shelves, the *Hakurei-Maru* took 12-channel seismic profiles in the Bellingshausen Sea; it found sedimentary rock in one area that was 3 to 3.5 kilometers thick. Both this and the Amundsen Sea shelf are narrow, and ships have difficulty navigating over them because of the ice. The shelves are of geologic interest, however, because they resemble the "quiet" parts of the Pacific coast of South America that contain oil and gas.

Australia has shown particular interest in the continental shelf of East Antarctica, particularly that off the Amery Ice Shelf. Its Bureau of Mineral Resources made a closely spaced, 6-channel survey there in 1981–82. The Australian data is not yet available, although the 1983 Behrendt report called this region "the most favorable for petroleum resources" in East Antarctica.

How much oil is likely to be found in Antarctica? The 1973 USGS estimates were probably too low; both the Gippsland basin and the shelf of Argentina near Tierra del Fuego have proved to be more productive since. Oil has been found off the east coast of India, which once adjoined Antarctica. Much more is known about the Weddell Sea now—although its past remains something of a mystery. And if the current rash of survey data were made available and pooled, a new, far more accurate estimate, could be made. Behrendt's 1983 paper offers no quantitative estimates; but he does say that only giant or supergiant fields (defined as having a capacity of 0.5 billion and 5 to 10 billion barrels, respectively) in time could be economically exploitable in Antarctica, and only four to ten supergiant oilfields remain to be discovered worldwide.[46]

Parts of the Antarctic continent having subglacial basins of Cretaceous and Tertiary age might have oil, but because the practical technology does not exist for punching through several kilometers of ice, there has been no serious discussion of them. The continent does have minerals, though no economically viable deposits have been found. In the future, efforts to establish Antarctica's general geologic history and its geologic provinces may help assess its minerals.

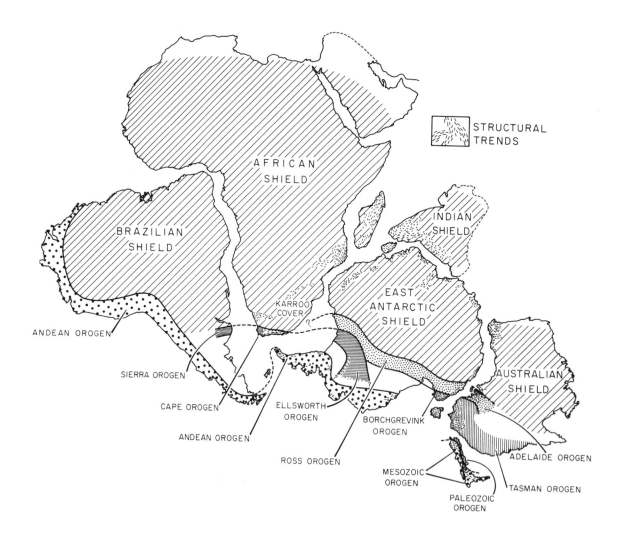

Geologic provinces of Antarctica and formerly adjoining continents, according to Craddock.

Continental Minerals

The "basement" or bottom rock layer in East Antarctica consists of Precambrian shield rocks similar to those underlying Africa, India, Australia, Brazil, and probably Patagonia in southern South America. Precambrian shields contain some of the world's most valuable mineral deposits, such as the Lake Superior banded iron formations, the great iron deposits of Australia, India, Africa, and South America; the gold of Witwatersrand, South Africa (which in 1973 produced 67 percent of the world's gold and much of its uranium). The Zambian copper belt and the lead-zinc deposits of Mount Isa and Broken Hill, Australia, probably are of this type.[47]

Typical shield-type minerals have been found in exposed areas of East Antarctica. Terre Adélie and Victoria Land have Precambrian gneisses and sheists, and gold, gold-copper, iron, beryllium, tantalum, tin, and tungsten. Sri Lanka offers a clue. Famous for its gemstones, Sri Lanka formerly lay somewhere along the edge of Enderby Land, Antarctica, where garnet has been found. Soviet geologists say that the Mahanadi Graben in India continues to the region at the foot of the Lambert glacier in Mac. Robertson Land, East Antarctica.[48]

An iron-formation subprovince extends from Enderby Land through Wilkes Land in East Antarctica, in which a large iron deposit found in the Prince Charles Mountains of Mac. Robertson Land was the subject of a long-term Soviet study. Their 1971–74 investigation established that in magnetite and silicate content the iron is comparable to that of Lake Superior, the Schefferville District in Canada, and the Krivi Rog of the Soviet Union. Aeromagnetic surveys showed that the concentrations may extend as much as 120 to 180 kilometers under the ice and that they range from 5- to 10-kilometers wide.[49]

Most natural diamonds come from a dark green basic rock of igneous origin known as kimberlite, which crystallizes under pressure at great depths in Precambrian rock. The diamond-bearing pipes of South Africa, Zaire, and Ghana shoot up through overlying, younger rock layers. As the pipes erode at the surface, the diamonds wash downstream, becoming alluvial diamonds. Such pipes formed the diamonds now being mined in one of the world's largest diamond mines on the beaches of Namibia. The Wright–Williams report said that there are probably diamond-bearing pipes in Antarctica too, but the chances of finding them by drilling through the ice cap are astronomically small. It instead suggested searching for diamonds along Antarctica's raised, exposed beaches that were uplifted by the retreating ice. The gemstones topaz, tourmaline, and garnet have been found, but their grade is unknown.[50]

Minerals from the Transantarctic Mountains and Victoria Land

The Transantarctic Mountains contain enormous beds of coal, perhaps the largest coalbed in the Southern Hemisphere, or even the world. In quality it ranges from low-volatile bituminous to semianthracite, and apparently extends from coast to coast under the ice mass. But the low price per unit volume of coal, and the likely high costs of shipping it in bulk to markets elsewhere preclude Antarctic coal from being an economic investment. If it ever were mined, it would be for fueling power plants in Antarctic settlements. Radioactive anomalies, which could indicate the presence of thorium or uranium, have been detected in the Transantarctic Mountains.[51]

The Transantarctic Mountains continue into Australia where geologists call them the Adelaide orogen. The Adelaide orogen has many deposits of lead-zinc-silver, gold, and copper-gold, barium, manganese, and antimony. A recent Soviet–Australian study suggests that the two belts formed as a single fold, 8,000-kilometers long, more than 400 to 450 million years ago.[52]

The northern tip of Victoria Land, near Cape Adare (where Carsten Borchgrevink's British expedition wintered over in 1898–99) represents still another geologic province, according to Campbell Craddock of the University of Wisconsin, an authority on Antarctica and Gondwana. The Borchgrevink orogen, as he calls

Sloping bare mountainsides in Wright Valley show sedimentary rock layers, indicating the continent once was temperate.

it, may have also upfolded Tasmania, which was then adjacent to Victoria Land. The Tasman belt has many economic minerals, including iron, gold, silver, copper-lead-zinc, tin, tungsten, and molybdenum. Joint teams from Great Britain, the United States, and West Germany have been investigating this region.[53]

Antarctica underwent a third mountain-building episode in the early Mesozoic period, about 100 million to 150 million years ago. It pushed up the Ellsworth Mountains in Antarctica, the Cape Fold Belt in southern South Africa, and the Sierra de la Ventana in South America. The Ellsworths are of interest because of their alignment. They may have formed along the present axis of the Transantarctic Mountains, rotating westward when the Weddell Sea opened. But neither the Cape Fold Belt, the Sierra de la Ventana, nor the Ellsworths have revealed significant mineral deposits.[54]

Another South Africa?

Some 600 kilometers from the northern edge of the Filchner Ice Shelf in the Pensacola Mountains lies the most interesting potential minerals site found to date—a layered intrusion called the Dufek Massif. A layered intrusion forms when magma collects in a very large underground chamber and cools slowly, so that the minerals separate into layers as it solidifies. Layered intrusions are found on all continents, but not all have large deposits of minerals. However, those that do have some of the most valuable sequences in the world. The richest is the Bushveld Complex of South Af-

Comparison of Antarctica and United States. To date, the area of Antarctica explored for minerals is comparable to exploring the states of Maine and Vermont for the minerals of the United States.

rica. There is also the one at Sudbury, Ontario, and the Stillwater complex in Montana. These are mined for platinum, nickel, copper, and chromium. Some layers of the Bushveld Complex have valuable tin and gold deposits. The Merensky Reef in the Bushveld, a single layer, is one of the world's most important sources of platinum. The Sudbury complex has the world's largest nickel mine.[55]

The Dufek layered intrusion was found by an IGY traverse party in 1957. It was revisited by USGS teams in 1965–66, 1976–77, and 1977–78. Soviet geologists investigated it from their base at Druzhnaya for several seasons beginning in 1975–76, but their findings have not been published. A 1978 airborne geophysical survey by the USGS and the Scott Polar Research Institute shows the deposit to be larger than earlier believed; it is now estimated to extend to 50,000 square kilometers with a thickness of 7 kilometers. The Bushveld is approximately 66,000 square kilometers and has a density of 8 kilometers. Each is more than ten times larger than those shown in table 5-2.[56]

There is much speculation on the potential of the Dufek layered intrusion. It is comparatively young—estimated as mid-Jurassic, or 160 million years in age—whereas the Bushveld, Sudbury, and Stillwater complexes were all formed in the Precambrian period, more than 600 million years ago. This suggests that the Dufek will not be as rich as these older formations. On the other hand, there are younger layered intrusions with economic deposits.[57] There also have been reports of other layered intrusions in Antarctica; one found by some New Zealanders in the Warren Range, and a second in the Ferrar group at Butcher Ridge reported by Edward Zeller of the University of Kansas. Little more has been said about them, however, and they probably have not been investigated further.[58]

Table 5–2. Comparison of Selected Layered Intrusions Worldwide

Intrusion	Area (km²)
Skaergaard intrusion (Greenland)	100
Stillwater complex (Montana)	200
Sudbury complex (Ontario)	1,340
Ingeli-Insizwa complex (South Africa)	1,800
Great Dyke (Zimbabwe)	3,260
Duluth complex (Minnesota)	4,710
Dufek intrusion (Antarctica)	50,000+
Bushveld complex (South Africa)	67,000

Source: Based on data from Arthur B. Ford, "The Dufek Intrusion of Antarctica and a Survey of Its Minor Metals and Possible Resources," in John C. Behrendt, *Petroleum and Mineral Resources of Antarctica*, Geological Survey Circular 909 (Alexandria, Va., USGS, 1983) tab. 1, p. 53.

Minerals of the Peninsula

Copper, molybdenum, and malachite occur at many points on the Antarctic Peninsula and nearby islands. Nonetheless, the old theory that this area is analagous to the mineral-rich Andes has not led those geologists exploring the peninsula to any El Dorado. Consequently, the peninsula's minerals potential remains unknown. The peninsula's mountain spine is assumed to have been formed during a fourth mountain-building episode that occurred when the Andes were formed. Henryk Arctowski, a Polish scientist on the *Belgica* expedition, suggested that the two have a common ancestry and proposed calling them the "Antarc-Andes." Early sealers named a bay in the South Shetland Islands Coppermine Cove after the presumed wealth there. In the early 1970s USGS geologist Peter D. Rowley named the Copper Nunataks in the peninsula because of their concentrations of that mineral. There are copper concentrations on King George Island, Livingston Island, and along the Danco and Northern Graham coasts, but none seem to be of economic interest. Nonetheless, in a 1983 summary of the Antarctic minerals situation, Rowley concludes that Antarctica's copper, as well as its coal and iron, could be considered "conditional resources," because they could be commercial if there were huge rises in unit prices.[59]

One of the most persistent students of the peninsula region, Ian Dalziel of Lamont Doherty Geological Observatory at Columbia University, continues to work the problem of the fit—or misfit—between the peninsula and southern South America in most Gondwana reconstructions. Major problems are posed by the fact that the South Georgia and South Orkney Islands are geologically distinct from the tip of the peninsula and southern Andes. If the Andes and the peninsular mountains were pushed up simultaneously and then bent into their present position, why are islands of different blocks along this line? One standard reconstruction points the peninsula into the gap between southern South America and Africa, but this does not account for the older Falkland Islands plateau that sits in the spot that the peninsula would have occupied. Another scenario explored by Dalziel is the chance that the peninsula originally lay alongside the East Antarctic craton and rotated to its present position with the opening of the Weddell Sea and the South Atlantic.

Most recently, Dalziel and Elliot support the idea of several West Antarctic microcontinents moving relative to each other and to East Antarctica. Originally, they speculate, the peninsula formed a straight line with southern South America; together they forged the Pacific margin of the ancient supercontinent. Meanwhile the

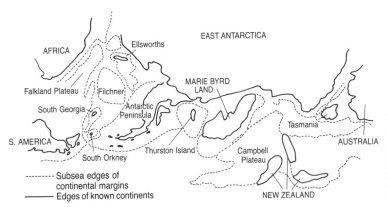

One possible early configuration of the microplates which now make up West Antarctica, showing locations of the Antarctic Peninsula, Marie Byrd Land, and "pieces" of New Zealand.

Falkland Islands plateau may have been far distant, connecting southern Africa with the East Antarctic craton. When Gondwana broke up, they write, the peninsula may have rotated its "shoulder" away from the East Antarctic craton some 90 degrees to its present position.

In this reconstruction, the Ellsworth mountains may have originally lain between the Cape Fold Belt and the Transantarctic Mountains. Riding on their own microplate, they could have rotated and moved south to their present position. Marie Byrd Land may be another microplate originally adjacent to the old East Antarctic craton which moved away as well, opening the large depression between the two called the Byrd Subglacial Basin.

Besides its significance for geology, Dalziel and Elliot note in passing that reconstructing the history of the West Antarctic microplates, including the peninsula, will have an impact on paleontology and paleoclimatology. For, as they moved away from the East Antarctic craton, the microcontinents may have opened sea passages between the early Atlantic and Pacific Oceans.

> [T]here may have been oceanic circulation between the southeast Pacific Ocean and the South Atlantic Ocean before Drake Passage opened about 30 [million years] ago.... This would in turn have influenced paleoclimate and hence affects views on possible causes of the onset of Antarctic glaciation and the Cenozoic ice age as a whole.... [D]istribution of fossil marsupials strongly suggests migration of primitive forms between South America and Australasia.[60]

Most experts on Antarctic geology agree that two conditions have to be met for an Antarctic minerals deposit to be worth exploiting. First, it will have to be inherently valuable, that is, commanding a strong price on world markets. Oil, for example, could be exploited if other reserves are scarce by the end of this century and adequate substitutes are not found. Coal, sand and gravel, and iron, which exist in abundance, do not have a high enough unit value to justify the cost of their extraction and transport. Second, deposits must be accessible, occurring in parts of the continental shelf that are ice-free in summer or in ice-free parts of the continent.

The chances of these conditions being met are slim; as Neal Potter wrote in his 1969 study for Resources for the Future, "The high cost of access to the continent and to most of the islands, as well as the great difficulties of operating in the offshore areas... will make it uneconomic to exploit any minerals but those which are high in unit value, and these only if they occur in large, rich ore bodies or fields. The likelihood of finding any minerals will be quite low because 95 percent or more of the land is covered by ice."[61]

Environmental Impacts of Minerals Extraction

The finding of trace hydrocarbons in 1973, and rumors of Antarctica's oil potential did not, however, set off a gold rush. The treaty powers felt that rules for development would be easier to negotiate so long as nothing of commercial value had been found. Suppose the *Glomar Challenger*'s drilling holes had gushed black oil, for example. If so, the resource would be in New Zealand's territory. Since New Zealand is dependent on Persian Gulf oil, its government well might be expected to assert rights to the resource. Meanwhile, the U.S. government, which could have the basis of a claim in the Ross Sea (see chapter 2), might feel obligated to assert discoverer's rights on behalf of its nationals. Positions could harden the interest of other oil-needy treaty powers, and developing nations could be piqued. Although not publicly stated, the treaty powers realized that a real find would open a diplomatic Pandora's Box.

Moreover, under the treaty, they are obligated not to take actions that could endanger the Antarctic environment without extensive prior study.[62] Thus, at the ninth consultative meeting in London in 1977, the treaty powers adopted, in Recommendation IX-1 on Antarctic minerals, a policy known as "voluntary restraint." They agreed to "urge their nationals and other States to refrain from all exploration and exploitation of Antarctic mineral resources while making progress towards the timely adoption of an agreed regime concerning Antarctic mineral resource activities."[63]

Serious exploration for economic minerals was renounced by the group for the time being; nonetheless, many nations were unable to resist the temptation to study more seriously and scientifically such areas of potential economic interest, particularly the Ross, Weddell, and Bellingshausen sea shelves. By the end of the decade, vessels of several nations were making scientific surveys of the hitherto little-studied Antarctic continental shelves.

At the same time, the treaty nations had no clear mandate for general environmental protection of the region. At the eighth consultative meeting in Oslo in 1975, the group asked SCAR to assess the environmental impact of exploration and exploitation. At SCAR's fourteenth meeting in Argentina the next year, a group for the Environmental Impact Assessment of Mineral Resource Exploration and Exploitation in Antarctica (EAMREA) was established. The EAMREA group was headed by James H. Zumberge, a geologist who worked in Antarctica during the IGY, who in 1975 was president of Southern Methodist University. The group produced a report in time for its 1977 meeting; it was later published in 1979.[64]

Meanwhile the U.S. Department of State contracted for a "framework" study of the likely environmental impact of minerals development in Antarctica. If the government negotiated such a regime, the State Department was required to assess such impacts in advance under a 1970 U.S. law. The study, carried out by Ohio State University's Institute of Polar Studies and led by its director, David H. Elliot, was delivered to the department in 1977.[65] The State Department released a final Environmental Impact Statement in 1982.[66]

But so delicate an issue as the environmental impact of minerals development was not to be left to SCAR or official government groups alone. In 1979 the Rockefeller Foundation sponsored a meeting of experts at its Bellagio, Italy, conference center. Their report was written by Martin W. Holdgate and Jon Tinker, a journalist. Holdgate had been secretary of the SCAR working group that held the 1968 symposium on krill, and Tinker was associated with the International Institute for Environment and Development, a London-based group with a continuing interest in Antarctic matters. The Bellagio report, although shorter and punchier than SCAR's or Ohio State's, reached similar conclusions. One indication of the skimpy state of this knowledge is the fact that all three reports are short; further, all call for extensive research to produce baseline knowledge of the Antarctic environment.[67]

The reports address minerals exploitation and all conclude that development of continental minerals is unlikely until the next century, but that offshore surveying for hydrocarbons and even exploratory drilling might begin earlier. It is recommended that such surveys should be far more detailed than the scientific scans currently undertaken. The rapid advance in polar hydrocarbon recovery technology, they note, should enhance offshore development. Indeed, historically, new technology that has revolutionized man's capabilities in Antarctica has been used there only after it had been tried in the Arctic.

Some of Antarctica's continental shelves are ice-free during the austral summer. However, offshore oil exploration and exploitation would take place under two very different environmental conditions—in the winter, when winter pack ice extends as far as 1,000 kilometers out from the continent, and wells on the ocean floor would have to be sealed; and in the summer, when constant daylight illuminates an ice-free sea, dotted with drifting tabular icebergs. Offshore wells close to the coast would face both fast ice and drifting pack ice. Farther out, in ice-free water, depths could reach 1,500 meters. Because the continent is depressed as much as 600 meters by the weight of its enormous ice cap, Antarctica's continental shelves are deeper than those of the other continents.[68]

Technologies now used in the Arctic could be adapted for use in developing an offshore Antarctic oil field. In order to explore, a drillship would have to remain over the hole but be able to disconnect and move away quickly from oncoming icebergs. This is the procedure followed now off the coast of Labrador, where enough icebergs drift by to justify calling one stretch "iceberg alley." It is also used in the Beaufort Sea, where, as fall approaches, pack ice closes in around the ship. In the Beaufort, a captain can disconnect from the riser and anchor lines and be safely away in ten minutes. Off Labrador, small support ships protect the drilling ships from the ice by "lassoing" the smaller icebergs and towing them away from the ship.[69]

Another hazard to oil-drilling activities in the north are the long pressure ridges of ice extending down from the sea ice. Like a long knife, they could scrape off a wellhead on the sea bottom, causing a large under-ice gusher that would be hard to plug, especially during the long austral winter. Dome Petroleum, working in the Beaufort Sea, has buried its wellheads in "glory holes" deeper than the deepest scoured-out trenches.[70] Until more is known about the history of ice movements and the contact between ice and the continental shelf in Antarctica oil drilling will not be safe.

New technologies are constantly under development. For example, instead of a rig extending to the water surface, it is possible to connect a group of wells with a single "subsea completion" unit on the sea floor. Already operational, the deepest unit proposed so far will operate in 250 meters of water off Tunisia. Such units may be serviced by divers or workmen in submersible craft. From underwater wells, semisubmersibles lay pipeline in water depths of 600 meters, and some North Sea underwater pipeline networks extend for 850 kilometers.

The promise of big Arctic discoveries has made industry engineers adapt these technologies for polar use. At one demonstration site, an underwater gas well located 1,900 kilometers from the geographic North Pole, Panarctic Oils Ltd. of Canada has demonstrated a pressurized, heated diver's suit that enables the diver to work for hours in freezing water and resurface quickly without getting the bends. Proposals have also been made for tanker submarines that would ship fuel directly from subsea wells over long distances under the icy surface.[71]

The greatest environmental risk from oil development is the leakage of oil in coastal, ice-free waters from tanker spills or undersea wells. Indeed, the effect of such an oil spill caught in coastal waters could persist for years. The lighter hydrocarbons in the oil would evaporate slowly while the heavier ones sank, possibly coating the underside of the sea and pack ice or the "anchor" ice on the sea bottom where many small organisms live.[72] However, the Ohio State report notes that offshore oil installations have been generally safe; they contribute only a fraction of the oil discharged into the world's oceans, most of which comes from tanker losses. Thus, the report concludes that repeated tanker losses en route to Antarctic ports and tank storage farms—and not

A diver in pressurized JIM suit is able to work in freezing Arctic waters and come up quickly without risk of the bends.

Artist's sketch of offshore oil rig proposed for the Arctic. The installation stores the oil until it can be removed by ice-strengthened tanker.

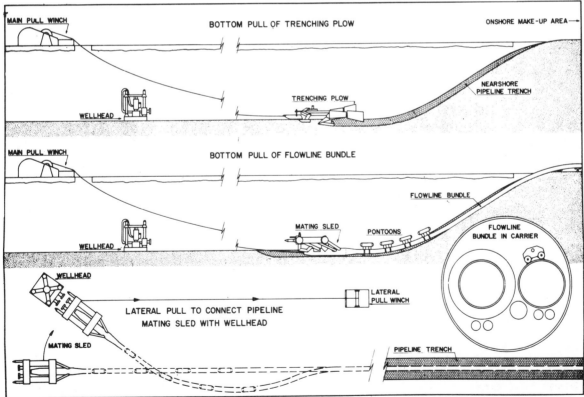

Offshore Arctic gas wells have been linked to shore via a trench dug from the land through which a pipe is threaded to the wellhead on the ocean floor. This system bypasses the hazardous ice conditions on the ocean surface.

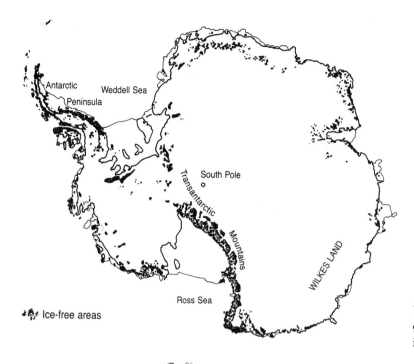

Antarctic
Peninsula

Weddell Sea

South Pole

Transantarctic
Mountains

WILKES LAND

Ross Sea

🐾🐾 Ice-free areas

A compilation published in
1983 showed Antarctica to be
97.6 percent covered with
ice, and 2.4 percent ice-free.

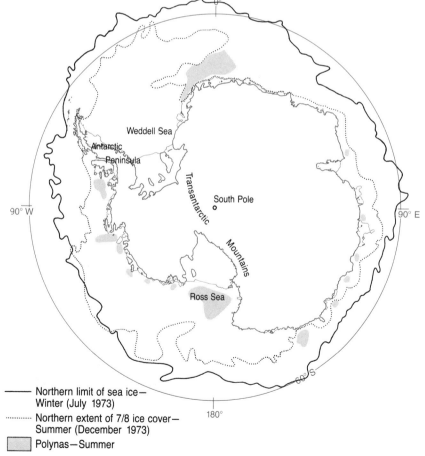

0°

Weddell Sea

Antarctic
Peninsula

Transantarctic

South Pole

90° W

90° E

Mountains

Ross Sea

60° S

180°

———— Northern limit of sea ice—
Winter (July 1973)

·········· Northern extent of 7/8 ice cover—
Summer (December 1973)

▓ Polynas—Summer

The extent of the winter and
summer sea ice. *Hatched areas*
show polynas, or areas of
open water in the austral
summer.

the offshore wells themselves—would be the most likely sources of oil pollution.[73]

Yet those delicate, ice-free coastal areas—which would also be likely sites for Antarctic ports—teem with life. The region's estimated 100 million birds must nest somewhere, even those that live on the winds and stay aloft for months. In nesting areas the birds are everywhere: "The rocks seemed covered with them as with a crust," Captain Cook wrote in 1776. In the South Atlantic islands, ships dim their lights at night to avoid attracting the birds that fly at them like moths. Even as gale-force winds and large waves pound the rocks, the islands echo with penguins' cries, elephant and fur seals huddle together for safety, and silent communities of fish, crustaceans, and simple marine organisms swarm into boiling waters. Once disturbed, such an intensity of life could be severely damaged.[74] As the Ohio State report notes, "There is no question that any resource exploitation will cause severe, and in many cases permanent, local impact on the environment because of the extremely slow rate of recovery that can be expected."[75] Life in the Antarctic moves slowly, the metabolism of organisms is slower there than in temperate climates; a footprint in the moss can last for decades. Although the impacts of a mine or offloading port might not be felt a few miles down the coast, the mosses, lichens, birds, and marine organisms at the site might not recover for decades.[76]

The Ohio State study details three kinds of proposed minerals activities and their environmental consequences. One is an underground, platinum-group metals mine in the Dufek Massif, capable of producing approximately 100,000 troy ounces of platinum-group metals per year, with workers and processing plants located underground and the output air-freighted to New Zealand. The second is an inland chromite mine capable of producing 100,000 tons of metallurgical grade chromite. Similar to the first underground mine, it would be located closer to the surface and use different processing facilities. Its output would be carried by tractor sleds 2,400 kilometers to a coastal port. The third case is a porphyry-copper deposit mine on the northern Antarctic peninsula, producing 100,000 tons of refined copper and molybdenum, gold, and silver. The copper would be concentrated and refined on site and shipped directly to market.[77]

Local impacts would be felt at the sites of the inland mines and along their overland routes, although none so severe as those occurring in closed basins and near rookeries and breeding grounds along the coast. Exploration would have little impact, generally; indeed, geologic and geophysical mapping would resemble present scientific activities. Likewise, while exploratory drilling on- and offshore would introduce drilling muds and circulating fluids, there would be minimal impact if procedures were used like those now followed in Antarctica's dry valleys. The Ohio State report recommends that exploration be avoided, if possible, in enclosed basins, the dry valleys, and in coastal rookeries and breeding grounds.[78]

The main environmental hazards of underground mines would be felt by the relatively unstudied and unmapped Antarctic permafrost. Most of the effects of an overland route to move the ores to port would be eliminated, eventually, by further snowfall. The report recommends using forms of transportation that have least contact with the ground, such as aircraft or air-cushioned vehicles.[79] Of course, local damage would occur where ports and tank farms were built.

The Ohio State report distinguishes, however, between localized impacts—some of which occur already in places of heavy activity such as McMurdo Station—and "maintaining the integrity of the Antarctic environment as a whole." (For example, if offshore oilfields, mines, and ports used an area as large as 132,090 square kilometers, they would utilize only 1 percent of the Antarctic continent's area. This would be analagous to developing an area the size of Maine and Vermont, while leaving the rest of the United States and Mexico intact.) Local oil spills, for example, would probably be carried away from the continent by powerful currents and dispersed into the Southern Ocean. On the other hand, the report warns that the sum of local impacts could be severe.[80]

Air pollution from open-pit mining or by particulates from refineries could pose a key problem for the Antarctic environment. It could alter the pattern of snowfall while the buildup of particulate matter in the atmosphere could alter the region's radiation balance affecting the rate of ice melt. The fact that the Antarctic interior is used to monitor global pollution levels also should be considered.[81]

As subsequent treaty recommendations and SCAR discussions have indicated, the minerals issue means a new agenda for Antarctic research. Zumberge wrote, "Even at the present level of effort, hundreds of man-years of scientific studies will be required before the components of the Antarctic environment and their interdependent relationships will be sufficiently understood to permit a reasonably confident assessment of the impact man's activities may have on the marine and terrestrial environment in Antarctica and the rest of the world."[82]

six

Resources Drive Diplomacy

As early as the mid-1970s, most of those in the international community who dealt with Antarctica realized that the *potential* of Antarctic resources would drive Antarctic diplomacy for years, perhaps decades, regardless of whether the Southern Ocean fishermen caught 1, 10, or 100 million tons per year, or whether offshore oil was systematically sought or found. Tensions were mounting over sovereign rights to minerals within the group of Antarctic Treaty nations, with at least one claimant state insisting that its ownership be recognized by the others before it would participate in talks on minerals. With comparable emotion, others outside the treaty circle called on the United Nations to sweep away what they deemed an obsolete and fragile treaty in favor of equal rights for all nations in Antarctica—much as the groaning, fissuring winter sea ice around the continent is forced to melt away each October with the arrival of the austral spring.

In 1977, in the town of Punta Arenas—the windswept port at the southern tip of Chile that services many Antarctic ships—Brian Roberts, the veteran British Antarctic diplomat, warned a meeting of fellow treaty nation representatives, "We are very rapidly being overtaken by events. The rest of the world will not wait while the treaty governments procrastinate."[1]

Roberts cautioned that an unnamed Antarctic claimant state should not be allowed to hold up constructive discussions about resources until its sovereignty was recognized by the others. And, showing a public candor rare for an Antarctic diplomat, he warned the treaty powers that even if they reached agreement on resource issues, "Anyone who seriously considers the future of the Antarctic Treaty Area must take into account the emergent hopes and wishes now finding expression in many nations which are not themselves active in the Antarctic."[2] Moreover, it was not obvious that the treaty nations could reach an agreement, let alone avert a showdown with the developing world at the United Nations.

The treaty powers were faced with two unsolved problems, one of resources and the other concerning nonsignatory nations. Both

issues, which had been avoided during the original treaty negotiations, were now of great concern. As the krill fishery grew and
interest in Antarctic oil intensified, the majority of the developing
countries—who over the years had remained signatories to the
treaty—had become a potent political force. Inspired by the success of the Organization of Petroleum Exporting Countries in
wrenching wealth from the industrial nations in exchange for oil,
the developing nations began to act *en bloc* in other UN forums to
obtain control of minerals and various other commodities. Their
economic theorists, calling for a new international economic order
to redress the imbalance between the industrial and developing
worlds, looked to ice-clad Antarctica, with its presumed mineral
riches, as a tempting target.

Chapter 8 will discuss more fully the developing nations who,
because many cannot afford to mount research programs in Antarctica, have difficulty becoming voting members under the existing treaty terms. Here, it is important to note how the developing
nations' interest in Antarctica served to spur the treaty powers into
action regarding resource regimes.

This chapter describes the long and difficult journey the treaty
powers undertook from 1970, when they superficially discussed
the minerals question in the corridors between sessions of the
sixth consultative meeting in Tokyo, through 1983. By then, the
group had concluded a wholly new regime governing the krill
fishery which, although untested, could serve as a model conservation regime for the protection of whales and other higher
organisms as well. Also, they were well on the way to concluding a
minerals regime—as is shown by the text of a treaty drafted by
Chris Beeby, a New Zealand diplomat, that was leaked to an environmental newsletter, *Eco*, in July 1983.

Their diplomatic journey proved to be as dangerous and delicate as any real trek across the Antarctic landscape. Developing
world interest was not their only challenge; in addition, they faced
the intransigence of some Antarctic claimant states, the USSR's
apparent unwillingness to conclude a krill convention until late in
the negotiations, and the Food and Agriculture Organization's
wish to launch its own Southern Ocean research program just as
BIOMASS (the treaty nations' effort in this field) was getting
under way. And, during the Nixon administration, the U.S. position changed. At a critical moment, when the other treaty powers
were close to approving a permanent moratorium on minerals
development in Antarctica, U.S. diplomats were instructed to
prevent it. The government's interest in assuring eventual access
to Antarctica for American oil companies may have been the deciding factor in preventing the group from zoning off Antarctica
from future minerals development.

By 1983, the outcome was still in doubt. The draft minerals regime under discussion gave no indication whether Antarctica would or should be opened to minerals development. It assigned no role or benefits to "mankind," that is, the developing world. It proposed no solutions for resolving the historic uncertainties about territorial ownership in Antarctica. Some developing nations took up Antarctica at the UN, while others sought to join the treaty. But although uncertainties remained, a decisive change *had* occurred: the treaty's scope had been broadened from science, arms control, and environmental protection to encompass all resources. The negotiators tried to solve the resource issue while, at the same time, they strengthened the treaty powers' claim to act as the legitimate government of Antarctica.

Antarctica and the Law of the Sea

As the group began serious discussion of resources, they felt threatened by the so-called Group of 77 developing nations (actually numbering some 112) organized at the time of the Third United Nations Conference on the Law of the Sea which began in 1974 in Caracas, Venezuela. From the Law of the Sea committee that addressed the disposition of manganese nodules on the deep ocean floor, sought after by industrial nations for unilateral development, the Group of 77 obtained the draft proposal to establish an International Sea-Bed Authority to govern all seabed mining.

According to the draft text, this would be a one-nation–one-vote body resembling the UN General Assembly and representing all the world's nations, a far cry from the system of unilateral government licenses sought by U.S. mining companies.[3] This particular struggle aligned Chile and Argentina, who were prominent in the Group of 77, against the United States, the Soviet Union, Great Britain, and other Antarctic Treaty powers. The deep-seabeds issue had been fought rancorously for years in the Law of the Sea Conference. Indeed, it so embroiled the otherwise successful final convention that the United States refused to sign. The struggle between the Group of 77 and the mining companies over control of these vast, unowned resources often appeared in the news after 1974. It was a constant reminder to the Antarctic Treaty powers as they met in their traditional way, behind closed doors, of what a UN takeover could be like. Indeed, whenever developing-country diplomats at the Law of the Sea Conference asked that Antarctica be included on the agenda, the treaty powers dissuaded them from doing so. Informally, the Group of 77 agreed not to include Antarctica in the Law of the Sea Conference on the understanding that it could be taken up when the conference was ended.[4] The promise would be kept.

The question of UN jurisdiction was not only a rhetorical one. Its jurisdiction over the manganese nodules—found in mid-ocean, on the flat ocean plain—had been established in 1970 by a resolution passed unanimously by the General Assembly. It declared that "the sea-bed and ocean floor, and the subsoil thereof, beyond the limits of national jurisdiction. . . . The resources of the area, are the common heritage of mankind."[5]

This language could conceivably be read to mean that the continent of Antarctica was "seabed . . . beyond the limits of national jurisdiction," and therefore part of the world's common heritage. Or it could be interpreted to mean that the unclaimed sector of Antarctica, which includes Marie Byrd Land, was "beyond the limits of national jurisdiction" and common heritage. This interpretation would be geologically wrong, as deep seabeds and continents have different geologic origins and an entirely different composition. In addition, it would stretch the widely accepted interpretation that "common heritage" applies only to the deep ocean floors, including those south of 60° south latitude (although there is some dispute about whether it applies to the continental slope and rise which connect the sea floor to continental shelf). Nonetheless, this interpretation allowed the Group of 77 to argue that Antarctica was within the purview of the Law of the Sea conference. And if Antarctica were incorporated into the Law of the Sea Convention, its minerals development would be likely to be controlled by the International Sea-Bed Authority designed to run deep-sea mining. Although far-fetched, this argument seemed a likely way for developing nations to attack the treaty.[6]

The treaty powers began serious resource negotiations largely to preempt such assaults. It is no accident that the group passed key, substantive recommendations on living and mineral resources at their ninth consultative meeting in 1977, when the Group of 77's power and militancy were at their height. The treaty group reached a second peak of resource diplomacy in 1982, just as the Law of the Sea Conference ended, and India, Brazil, and the People's Republic of China indicated serious interest in becoming treaty powers. A speech on Antarctica given at a meeting of nonaligned nations by the prime minister of Malaysia also spurred them on.[7]

The treaty powers resisted any suggestion that the United Nations should act on its own with respect to Antarctica; even UN-sponsored research was discouraged. In 1976 the UN Food and Agriculture Organization (whose Southern Ocean specialist, John Gulland, had taken part in the original 1968 SCAR symposium that focused on the krill) considered launching a $45 million, ten-season program of research on krill exploitation and use, possibly using Soviet vessels. But some claimant states objected:

Chile and Argentina reportedly did not want Soviet vessels working near their coasts, even on UN business. Instead, in January 1976, the FAO began a more modest effort with the $202,500 Southern Ocean Fisheries Survey Program, whose financing came from the UN Development Programme. Although it produced three key reports summarizing what is known about krill fishing and processing technology,[8] quiet words from treaty power representatives apparently cut back the program.

The New Krill Convention

Whatever political motives brought it into being, the Convention on the Conservation of Antarctic Marine Living Resources, as the new krill convention is called, represents a bold new effort by the treaty powers to regulate the fishery.[9]

Concluded in May 1980, and in effect since April 1982, it is in many ways a breakthrough in marine conservation law. Recommendation IX-2 adopted in 1977, which outlines the key principles of the regime, notes that it "should provide for the effective conservation of the marine living resources of the Antarctic ecosystem as a whole."[10]

This was new. Scientifically, it meant that the standard for overfishing used in most fishing agreements, called the "maximum sustainable yield," would be inappropriate for the krill fishery. Since the maximum sustainable yield measures only the health of the stock being fished, the SCAR scientists argued that some broader standard was needed for the Antarctic in order to protect not only the krill stocks but the predator species that depend on krill. The U.S. Marine Mammal Commission drafted details of an "ecosystem standard" proposed by the U.S. delegation early in the negotiations. In essence, it was included in the final convention. James N. Barnes, a public interest adviser to the U.S. delegation and chief spokesman for environmentalists concerned about Antarctica, later wrote that P. A. Moiseev, the Soviet scientist who had raised the krill issue back in 1968, convinced the Soviet and Japanese delegations—who otherwise sought maximum leeway for fishing—to accept the standard. In general, Barnes notes, other scientists, including Richard Laws, the British seal expert who had delineated aspects of the ecosystem, were instrumental in the adoption of the ecosystem standard.[11]

The negotiators expanded the convention beyond the Antarctic Treaty area south of 60° south latitude out to the natural oceanic boundary of the Antarctic Convergence. As J. W. S. Marr had found earlier, the krill swarm as far as the Convergence. The convention's area of application, defined by longitude and latitude in

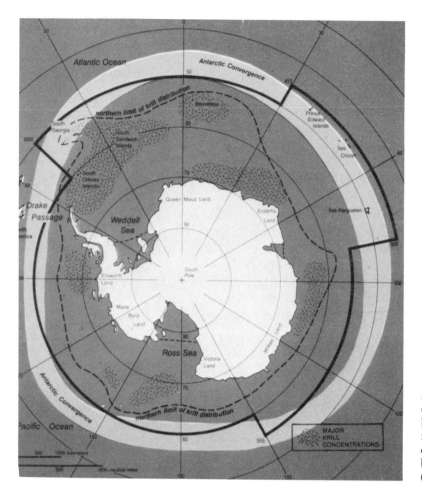

Antarctic krill are found
south of the Antarctic
Convergence oceanic
boundary (*dashed line*). The
new convention extends to
established, fishing-zone
boundaries to the north
(*solid line*).

its text, coincides with boundaries used by the FAO. In places, this
northern boundary extends to 45° south latitude.[12]

Conservationists won an important victory with this match
between an organism's natural habitat and the jurisdiction of
the law governing it. It was seen as a useful precedent for other
marine resource agreements. But on another matter—interim
measures—they lost. Such measures would have bound the parties
to voluntarily regulate their fishing in the period prior to the time
that the new convention entered into force. Recommendation
IX-2 called for interim measures, but they were not included in
the final convention text.[13] Such measures would have been critical if the fishery had expanded suddenly in 1978 and 1979, or if
the final convention had taken years to be effectively ratified. As it
happened, the lack of such measures did not matter because fishing expanded slowly and steadily, and because the new convention, signed in May 1980, went into effect twenty-three months
later.

Because krill also are found close to the Antarctic coast and some key fishing grounds lie off coasts claimed by Chile, Argentina, and Great Britain, serious negotiating problems have arisen. Claimant states wish to protect their "rights" to jurisdiction over their "exclusive economic zones" offshore. Moreover, Chile and Argentina have actively asserted their jurisdiction out to 200 miles offshore from "their" Antarctic claims. The issue of offshore jurisdiction is complicated by the fact that within the expanded area of the convention are several islands whose ownership is *not* disputed, such as Bouvetoya (which belongs to Norway), Prince Edward Island (belonging to South Africa), the Heard and McDonald islands (belonging to Australia), and Kerguelen and Crozet (belonging to France). All the parties are sensitive on the offshore jurisdiction issue, influenced no doubt by an urge to protect their ultimate rights to offshore hydrocarbons. The final convention handles the issue by using bifocal language that could be interpreted as supporting the positions of all claimants in the region—both those whose claims are disputed, and those whose ownership is not.[14] Indeed, the final text of the convention, just to be safe, repeats in *its* Article IV the entire text of the treaty's Article IV.

So, although the new convention is a strong conservation agreement in theory, it may prove to be weak in practice, for the region's territorial disputes may interfere with enforcement of any future fishing limitations. Elsewhere, most nations restrict fishing within 320 kilometers of their shores by imposing catch quotas on foreign (and sometimes their own) fishermen. But the territorial dispute in Antarctica means that no agreement has been reached as to which nation(s) will perform this function. Thus, lack of agreed offshore jurisdiction prevents the adoption of traditional national quotas. It will be left to the new commission created under the new convention to manage the fishery, to find a way around this problem. Probably, it will have to assign catch limits directly to fishing states, even though such states are generally lax about enforcing strict limits on their own vessels. Furthermore, the provision for consensus voting on matters of substance in the new convention may make it difficult for the commission to take strong enforcement action.

The convention tries to come to grips with the interest of "outsiders" to the Antarctic Treaty club in the Antarctic fishery. Maritime nations, such as South Korea, that fish in the Antarctic are neither acceding nor consultative parties to the treaty and had at the time shown no interest in joining it.

First, the treaty group decided to make their krill agreement a "convention," that is, a free-standing treaty in its own right, which any state that qualifies under its terms can join. This makes it stronger than the alternative, which was to adopt "agreed measures" under the treaty. Nevertheless, only parties to the treaty are bound by agreed measures, and their enforcement depends upon voluntary self-regulation. The agreed-measures approach could neither include nor regulate South Korea, for example. Instead, parties to the treaty followed the approach used in the Convention for the Conservation of Antarctic Seals concluded in 1972, that of opening it to accession by any state. In the new krill convention, therefore, any state may accede to the convention. Any acceding state may vote on the new commission if it is "engaged in research or harvesting activities in relation to the marine living resources" of the Antarctic.[15]

But the treaty powers also give themselves special status as the original governors of Antarctica. All treaty consultative parties automatically become voting members of the new commission, regardless of whether they engage in harvesting or research on Antarctic marine living resources. (In addition, the European Economic Community, to which Great Britain, France, and Belgium have delegated their fishing authority, sits on the commission despite longstanding protests by the Soviet Union.[16])

Furthermore, parties to the krill convention who are not also parties to the Antarctic Treaty must "acknowledge the special obligations and responsibilities" of the latter, and must agree to abide by the main provisions of the Antarctic Treaty, including, of course, Article IV on claims.[17]

The treaty powers also retain their authority through the voting scheme of the new convention. Despite the fact that the United States, in its effort to make the new commission more workable, advocated a two-thirds majority vote, the final text requires consensus voting on all substantive matters. Although the scheme advocated by the Americans would have enabled the commission to take more controversial positions, it also would have prevented any single party (such as a claimant state which felt its offshore "rights" threatened) from having a veto.

Not surprisingly, Antarctic claimant states objected to the U.S. proposal. Thus, the final text requires voting to be done as it is in Antarctic Treaty meetings, by consensus on matters of substance, and by a simple majority on matters of procedure. If there is doubt as to whether a question is one of procedure or of substance, it is automatically considered to be a question of substance.[18] This and several other features in the new convention in effect form a legal umbilical cord, tying the new convention to the Antarctic Treaty. The minerals regime, the group concludes, may well follow similar lines.

The umbilical cord approach seems to have worked. West Germany was not a consultative party to the Antarctic Treaty at the time the convention was concluded, although it became one in March of 1981. Nonetheless, the treaty group kept it and other interested nations informed of the negotiations, and both West and East Germany signed the final convention in Canberra in May 1980. The Netherlands, which had followed the negotiations, also seemed likely to sign. These signatures gave the treaty group the first outside recognition of their special responsibilities in Antarctica and endorsed their approach to Antarctic marine living resources. These ratifications were a milestone, then, for the group of nations that had operated in the Antarctic for two decades on the assumption of legal authority which, although never disputed, had never been acknowledged by those outside the club.

Although they had won recognition of their authority to negotiate resource regimes for the Antarctic, their journey was far from finished. Yet to be determined was whether the new regime would work, or whether, like so many other internationally endorsed fishing conventions, it looked good only on paper. In effect, by pushing for a conservation regime, the SCAR scientists had set themselves another, larger task. The new convention's success depended on whether the ecosystem could be understood and modeled well enough to detect overfishing. They needed an advanced model of the ecosystem—and one which worked—to boot.

Early Workings of the Convention

The first meeting of the new commission took place in Hobart, Australia, from May 25 to June 11, 1982. It was decided that the chairmanship would rotate among the countries in alphabetical order, and if the chairman were from a non-fishing state, the vice chairman would be from a fishing state, and vice versa. As it had during the negotiations, the Soviet Union continued its objections to the role of the European Economic Community.

The significance of the meeting was not the business it transacted—but that it took place at all. Inside the meeting room in Hobart, delegates from Great Britain and Argentina faced one another amicably across the table, seemingly oblivious to the fact that soldiers of both sides were shivering on ships in the cold South Atlantic as the two countries waged war over the Falkland Islands. Thus, the first meeting proved a successful test of Antarctica's quasi insulation from conflicts elsewhere on the globe.[19]

The key to the convention's effectiveness would be the scientific committee created to advise the commission on whether krill (or other marine living resources within the area) are being overfished. The committee chairman, Dietrich Sahrhage of the Federal Research Institute for Fisheries of West Germany, earlier had achieved prominence in the BIOMASS research program. Laws, Robert Hofman of the U.S. Marine Mammal Commission, and others studying the krill fishery also were appointed, and the FAO's John Gulland served as an observer. The committee's initial discussions centered on an assessment of existing research on the marine ecosystem, and how to standardize reporting from fishing vessels.[20]

By 1983, the scientific committee had a huge agenda—and virtually no budget to undertake the research. In order to measure local changes in the ecosystem, the committee will need detailed information on annual fish and krill catches—a highly sensitive subject for the fishing states. The yearly catch data supplied to the FAO by national fisheries institutes for all regions in the world (which is the basis for table 6-1) is very general. It does not list the efforts made in specific localities where krill fishermen are likely to interfere with the feeding of whales. Moreover, there is no way of knowing what part of the catch and what efforts are *not* being reported to the FAO. Ultimately, the commission will have to draw up detailed standards for obtaining data from the fishing states.

But it has been clear from the first meeting that the knowledge base of the scientific committee would depend heavily on the BIOMASS program and on the individual national research programs. The committee agreed to follow certain efforts closely, such as the SIBEX surveys of localized areas planned for late 1983 to early 1985. Article XV of the krill convention also empowers the committee to "formulate proposals for the conduct of international and national programs of research." For example, the committee may decide to encourage a study of squid—for the wide variation in krill stock estimates, shown in table 6-1, cannot be reconciled without some knowledge of how many squid there are, and whether, as Moiseev contends, squid consume millions of metric tons of krill each year. Data centers and good local models

Table 6-1. Reported Catch of Fish and Krill in the Antarctic, 1972–73 to 1982–83
(metric tons)

Catch	1972/73	1973/74	1974/75	1975/76	1976/77	1977/78	1978/79	1979/80	1980/81	1981/82	1982/83
Finfish	13,500	106,100	25,300	57,100	268,700	353,045	125,091	115,704	123,278	124,395	197,522
Krill	—	22,343	39,981	2,787	122,532	128,340	333,634	478,526	448,252	528,287	225,138
Squid	—	—	—	—	1	391	2	—	—	—	—
Total Antarctic	13,500	128,443	65,281	59,887	391,233	481,776	458,727	594,230	571,530	652,682	422,660
Total World[a]	62,824,400	66,597,100	66,135,600	69,590,200	68,224,600	70,154,700	71,060,400	72,008,300	74,777,100	76,470,200	76,470,600[b]

Note: Nominal catches = live weight basis.

Source: 1978, 1979, and 1983 Yearbook of Fishery Statistics, Catches and Landings (Rome, Food and Agricultural Organization of the United Nations), tab. C-48(A), C-58(A), C-88(A), and A-1(B).

[a] Figures are for end of year. All other figures on this chart are for June 30 year end.

[b] Figures are for 1983.

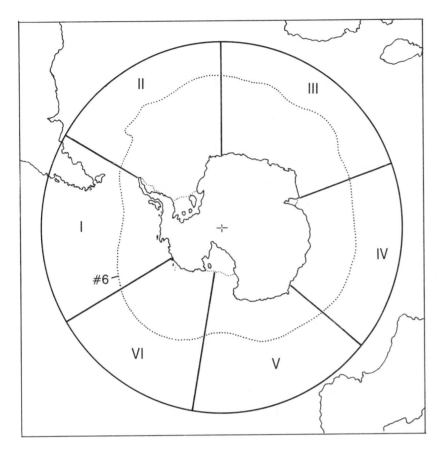

To test the effect of krill fishing on the Antarctic marine ecosystem, one scientist proposes allowing fishing in alternative International Whaling Commission zones (II, IV, VI) while leaving the ecosystem intact in the zones between.

of the marine ecosystem are needed for the committee to do its work effectively.[21]

Eventually, the scientific committee's job would be to implement, through the commission, an overall management scheme for the fishery. One plan, proposed by Hofman, would be based on the International Whaling Commission fishing areas for the Southern Ocean. Data collected in these areas by the commission already are widely recognized by fishermen and fisheries institutes.

Hofman also proposes that some fishing be permitted in International Whaling Commission areas II, IV, and VI, that is, in the Weddell Sea region, off North Victoria Land, and off Marie Byrd Land on either side of the Ross Sea. The remaining areas in between, that is, I, III, and V, would serve as controls; fishing there would be allowed only to "identify and monitor" krill stocks.[22]

By 1983, more detailed schemes and models of the ecosystem had been devised. These efforts, as well as those of the scientific committee, emphasize the need to gather good, basic data, much as the *Discovery* investigations assembled whale data for the previous half-century. Furthermore, the scientists were excited by the prospect that the relatively simple Antarctic marine ecosystem, if it could be both modeled and managed, would shed light on how to manage other, more complex fisheries that were being overfished or were suffering from fishing of lower trophic levels. As May and Beddington conclude in their *Scientific American* article, "The Southern Ocean provides a testing ground where one can begin to understand the effects of exploiting several species that occupy different levels in the food chain, but much more work is needed. The task of managing other multispecies fisheries, where the relations between stock and recruitment are intrinsically less predictable and where changes can take place faster, is correspondingly even more formidable."[23]

Political Implications of Antarctic Minerals

If the krill convention represented a major summit in the treaty powers' trek, a more challenging peak loomed ahead—how to reconcile their old disputes and hold off the Group of 77 while drawing up a regime for Antarctic minerals. Though difficult, the negotiations on living resources were considered by the diplomats as a dry run for solving the minerals problem.

The *Glomar Challenger's* find of trace hydrocarbons in the Ross Sea, industry demands for exploration licenses, and the imminent completion of the krill convention spurred the treaty powers to negotiate on Antarctic minerals in the late 1970s. Predictably, the subject had produced enormous strains. In 1973 the Fridtjof Nansen Foundation invited representatives of the treaty powers to Polhogda, Norway, for an informal meeting on Antarctic minerals. The report of this session, which was published in 1974 after the advent of the world oil crisis, offers a rare public glimpse of these differences. It also shows that fear of an Antarctic gold rush was creating pressure for a moratorium on minerals development.[24]

In view of the existing legal literature about Antarctica—much of which argues that the powers should pool their claims or renounce them altogether—the Polhogda report is useful in that it shows the group's rationale in rejecting a condominium, a "common heritage," and other radical solutions.[25] Four points made in the report and summarized here explain how the group arrived at the particular consensus.

They wanted to arrange for exploitation for Antarctic minerals before they determined what was there.   Noting the oil companies' inquiries, the report said that "it is realistic to expect this interest to increase within five or ten years...[by which time] governments should have decided how they propose to deal with the problem; not at a much later stage when commercial exploitation may become practical."[26] Later, they would formalize their wish to remain deliberately ignorant—on paper, anyway—by adopting a policy of voluntary restraint on Antarctic minerals exploration and exploitation so long as progress was being made in negotiating a regime.

They considered the possibility that the treaty does not apply to Antarctic minerals. Because they are not mentioned in the text of the treaty, one could assume that this allows any nation to explore and exploit Antarctic minerals unilaterally. Or it could permit a completely different organization (the Law of the Sea's International Sea-Bed Authority, for example) to assert jurisdiction over Antarctic minerals. At Polhogda and later, the notion of a minerals accord outside the treaty gained little ground. It would lead to too much disagreement. If one country objected to another's actions, Antarctica would then be, by definition, the "scene or object of international discord" prohibited in the treaty's preamble, and the treaty powers therefore would have to intervene in the dispute, in accordance with their treaty obligations.

They considered the problem of a nonsignatory to the treaty exploring for or exploiting Antarctic minerals. At Polhogda, the group felt that together they could urge the nation in question to join them.[27] As a practical matter, should a nonsignatory nation get in trouble "down on the ice," the treaty parties always could hint that they would be unable to divert resources to rescue the expedition. A nonsignatory nation would, in all likelihood, find minerals exploration too risky, without receiving assurance of help from the treaty parties should their expedition require rescue facilities.

Finally, they discussed territorial claims. Could they settle the ancient dispute? One participant (probably Finn Sollie, director of the Nansen Foundation, who had made the suggestion elsewhere) suggested that they form a condominium, but this fell on deaf ears. Critics argued that it would upset too many established positions or, what is worse, would require reopening the text of Article IV of the treaty. Moreover, the historic political balance between claimants and nonclaimants and that between the rival superpowers and the array of smaller nations who feel an equal stake in Antarctica would be threatened by a condominium.

The Polhogda meeting marked the beginning of the treaty parties' consensus approach to the difficult minerals question. It involved delaying systematic exploration—perhaps permanently through a moratorium—in order to forestall a gold rush. It rejected a condominium and retained Article IV, that is, it agreed *not* to settle the sovereignty dispute. By implication, therefore, it involved keeping the treaty intact and negotiating a satellite agreement. Such a consensus would seem to reject both a UN-based minerals regime, such as the International Sea-Bed Authority, as well as any convention severed from the treaty. Therefore, the parties would have to reconcile the practical and legal requirements of a minerals regime with the ambiguities and contradictions inherent in Article IV.

But in 1974, many extreme positions obscured this consensus. Several legal articles called for a common heritage approach to Antarctica and the sweeping away of the treaty. At the eighth consultative meeting in 1975, New Zealand's labor government proposed making Antarctica a world park—a move vigorously advocated by environmentalists. British Antarctic diplomat Brian Roberts later revealed that, at the time, at least one Antarctic claimant state was requesting recognition of its claim before it would allow any serious minerals discussions to go forward. Disarray still threatened, and Roberts warned that it would lead to the paralysis of the group and the much-feared UN takeover—an eventuality, he added coolly, that meant "administrative chaos."[28]

Near-Miss on a Moratorium

Antarctica might have been zoned against all minerals development forever if the permanent moratorium, frequently discussed in 1973–75, had been adopted by the group. By the eighth consultative meeting, held in Oslo in 1975, a majority of the treaty powers favored one. The Soviet Union was a strong advocate, apparently hoping to catch up with Norway, Great Britain, and the United States in offshore oil technology. The United States was almost alone in opposing a moratorium, according to later testimony by R. Tucker Scully of the U.S. State Department.[29]

The U.S. refusal to go along with a moratorium has been discussed in chapter 4. The result, however, was that the moratorium question was left unresolved at Oslo. All that passed was a weaker compromise agreement on a policy of "voluntary restraint," which was reaffirmed at the 1977 meeting.

By 1973–74, the Nixon administration was anxious to ease the way for U.S. oil companies to explore in Antarctica. It did not believe Antarctic exploration was imminent, but the world oil crisis had produced acute concern about future sources of supply. Officials of the energy agencies who favored maximum leeway for development were fighting with those at the State Department and at the National Science Foundation where it was feared that a sudden, pro-development posture by the United States would upset the treaty group's deliberations.[30]

An argument ensued over which agency representatives would attend the Oslo meeting at which a permanent moratorium would be considered, and what their instructions would be. If mishandled, U.S. behavior could be interpreted as a willingness to defeat the treaty, if need be, to gain access to Antarctic oil. Finally, the U.S. delegation was instructed to oppose a permanent moratorium, to show strong support for the treaty, and to advance the discussion on minerals. Later, this would develop into strong U.S. support for the policy of voluntary restraint.

Four principles were sketched by SCAR in 1976:

(i)   the Consultative Parties should continue to play an active and responsible role in dealing with the question of Antarctic mineral resources;

(ii) the Antarctic Treaty must be maintained in its entirety;

(iii) protection of the unique Antarctic environment and of its dependent ecosystems should be a basic consideration;

(iv) the Consultative Parties, in dealing with the question of mineral resources in Antarctica, should not prejudice the interests of all mankind in Antarctica.[31]

At the ninth consultative meeting held in London in 1977, the parties implemented SCAR's suggestion in Recommendation IX-1, on minerals. Agreeing to deal with the issue within the treaty framework, they asserted their "special responsibilities" to "ensure that any activities" in Antarctica—including minerals exploration and exploitation—should not harm the environment. They made a stab at reading the question of nonsignatory nations, too: "The Consultative Parties, in dealing with the question of mineral resources in Antarctica, should not prejudice the interests of all mankind in Antarctica."[32]

In Recommendation IX-1, "voluntary restraint" was adopted in which the parties

> Urge their nationals and other States to refrain from all exploration and exploitation of Antarctic mineral resources while making progress towards the timely adoption of an agreed regime concerning Antarctic mineral resource activities. They will thus endeavor to ensure that, pending the timely adoption of agreed solutions pertaining to exploration and exploitation of mineral resources, no activity shall be conducted to explore or exploit such resources.[33]

The idea that the group would "not prejudice the interests of all mankind in Antarctica" harked back to the draft article on non-signatory nations that had been considered and rejected in the treaty negotiations (see chapter 4). But such a vague phrase hardly clarified what the treaty powers were prepared to offer. And while it was rumored that some benefit—such as a royalty payment from oil revenues going to the International Sea-Bed Authority—is being considered by the treaty parties, by the end of 1983 the treaty parties had hardly enlightened "mankind" on the point.

In effect, the tenth consultative meeting, held in Washington in 1979, added a fifth principle. Recommendation X-1 said that a future minerals regime would be responsible for "determining whether mineral resource activities will be acceptable." In other words, the regime would not presume that any mineral activities would take place—the parties would make that decision within its framework not before.[34]

It was also important, as the Law of the Sea Conference continued its acrimonious course, for the treaty powers to deny it jurisdiction over Antarctica. At their eleventh consultative meeting they adopted a statement in Recommendation XI-1 that allowed the International Sea-Bed Authority jurisdiction over the deep ocean floors around Antarctica that lie beneath what the treaty acknowledges to be the high seas. The statement also supported their position that the Law of the Sea Conference had no jurisdiction over the Antarctic continental shelf. The recommendation noted the "unity between the continent of Antarctica and its adjacent offshore areas." While "mindful" of the Law of the Sea negotiations, it added that a future Antarctic minerals regime shall "apply to all mineral resource activities taking place on the Antarctic Continent and its adjacent offshore areas but without encroachment on the deep seabed. The precise limit of the area of application would be determined in the elaboration of the regime."[35]

Within these principles, several minerals regimes are possible, and a number of ingenious suggestions have been made. Since both the treaty and Article IV must be kept intact, the regime could use the treaty's bifocal language on territorial claims. The krill convention's Article IV maintaining all the competing positions regarding ownership of offshore Antarctica could be included. Thus, the regime would permit title to the land in which the minerals are found to remain disputed—yet arrange to give title to the minerals themselves once a developer has met certain conditions.

Roberts, in his Punta Arenas talk, said he saw no particular obstacle to such an arrangement. At another unofficial meeting in 1982 (this one held at Chile's Teniente Marsh station on the

Antarctic Peninsula), a staunch defender of Australia's ownership of "its" territory and "its" oil in Antarctica, Ambassador Keith G. Brennan, took a similar view, saying: "If there is too much respect for the legal position of the States concerned, solutions will evade us. . . . In the search for pragmatism, States will have to avoid legalism. . . . Sovereignty will have to find an alias."[36]

Luckily for Antarctic pragmatists, there are cases in international law where a resource that might have competing owners while in the ground (such as the North Sea oilfields, or a hydrocarbon basin that straddles disputed boundary) can be handed over with clear title to a developer meeting requirements endorsed by its rival owners.[37]

The minerals regime would also avoid setting up permanent institutions to run Antarctic minerals development in much the same way as the International Sea-Bed Authority plans to control seabed mining.[38] In the past, claimants and would-be claimants have not created a permanent secretariat because doing so would imply that they lacked sovereignty. If such an international body proves necessary, it probably would meet only occasionally and have limited powers.

According to Philip W. Quigg, one proposal "thought out in considerable detail" in the negotiations would divide Antarctica into four quadrants, starting at 30° east longitude and drawn in such a way that at least three of the four quadrants have "promising areas for exploration." Quigg continues,

> Each region would have a panel composed of parties to the regime to serve as resource manager in accordance with the standards set forth in the regime. No party . . . could be a member of more than two panels and no panel would include more than 50 percent of the parties to the regime. A decision to open an area for exploration and development would be made by the regional panel by consensus, but it could be reviewed at the next full meeting of the parties.[39]

Another ingenious proposal, put forward by Jonathan I. Charney of the Vanderbilt University School of Law, would have the group form an umbrella organization committed to upholding principles of free access, uniform standards, environmental protection, and the sharing of some financial benefits. Under Charney's plan, implementation and enforcement could be left to the countries who would be appointed "managers" of specific "zones" in the Antarctic. The managers could also be claimants of those sectors. Nonclaimants would have a stake in assuring that the umbrella organization was viable. Thus, Charney notes, the regime would have a built-in political balance between claimants and nonclaimants.[40]

The Beeby Draft

The zonal manager idea has been incorporated into a draft minerals regime prepared by Chris Beeby, the New Zealander who
chaired an informal meeting of the treaty powers on Antarctic
minerals. Held in Wellington in January 1983, it is one of a series
of drafts that supposedly will lead to a final regime. The Beeby
draft was obtained by the editors of *Eco*, an occasional publication
of environmental groups (in this case, Greenpeace, the Antarctic
and Southern Ocean Coalition, and Environment and Conservation Organizations of New Zealand, Inc.). "Beeby's Slick Solution,"
the *Eco* headline cried, in reference to the editor's view that the
draft would inadequately protect the Antarctic environment. But
the text also could be called "slick" in another sense, in the way that
it circumvents myriad legal and practical problems.[41]

Beeby's draft would create a commission much like that set up
by the new krill convention. Made up of the Antarctic Treaty consultative parties and any state that sponsors a group wanting to
develop Antarctic minerals, the commission would have ultimate
decision-making authority over whether an area of Antarctica
would be open for exploration, whether a proposed "management
scheme" for an area would adequately protect the environment,
and whether a development permit would be issued. Thus, it
would be the body that guards the treaty powers' ultimate interests. But, in keeping with the group's wish for a decentralized
administration, the commission will operate only at the request of
a treaty party or of a nonsignatory nation wishing to sponsor an
expedition. Once the commission meets to consider the request, it
can decide, among other things, "upon the necessity for further
meetings." However, the commission can have a secretariat, and a
permanent headquarters site is not ruled out.[42]

Under the Beeby draft, decisions on Antarctic minerals will be
made on an *ad hoc* basis, considering the individual proposal, the
area and the willingness of claimant states, the superpowers, and
the sponsoring state to cut a deal over royalties and rights. Other
than its veto authority, the commission has few deliberative powers, however. It would actually serve as a broker between a powerful scientific, technical, and environmental advisory committee
and regional committees specifically created to manage Antarctic
development.

The procedure would start with notification by a sponsoring
state that it, or a private company operating under its jurisdiction,
wishes to prospect a given area of Antarctica. The proposal would
describe the general area to be explored, how long prospecting
would last, the resources being sought, and the methods to be
used. ("Prospecting" is not defined in Beeby's draft, but it would

seem likely to be limited to minerals prospecting on land or to seismic surveys at sea and not include exploratory offshore drilling.)

A sponsoring state wishing to explore—a phase that probably includes exploratory offshore drilling for oil—submits a request to the commission that the area be opened for exploration. The request, including an environmental impact assessment, in turn can be forwarded by the commission to its advisory committee, made up of one expert from each state that is a member of the commission. The advisory committee evaluates the request and recommends a decision.

Whenever an exploration application is received by the commission, a key mechanism central to the Beeby draft comes into play. This is the regulatory committee, which would oversee the exploration and exploitation of the area in question and remain in existence so long as development continues. The regulatory committee is an advanced version of the "zonal manager" proposed by Charney. But instead of giving a single claimant the job, regulation is carried out by a committee of up to eight states reflecting Antarctica's web of political interests. The eight include the two superpowers (which have competing "bases of claims" in nearly all of Antarctica, as noted in chapter 2); two states, including the sponsoring state and another chosen by it; and up to four additional states, including those chosen by the one or more states that claim the area outright as well as the claimants themselves. The *Eco* editors note that the superpowers are guaranteed a seat, while the claimant(s) and the sponsoring state must vote in the majority on key committee decisions.[43]

The regulatory committee is responsible for adopting a so-called management scheme containing the key contract between the would-be developer of Antarctic minerals, the state(s) that claim the area under exploration, and the superpowers. The management scheme which avoids the issuance of blanket rules for Antarctic development, instead presumes that these can be developed *ad hoc* for individual situations and regions.

The management scheme ought to address key issues: what criminal or civil law applies to the operator in the Antarctic (the old problem postponed when the treaty was drawn up), licensing arrangements, inspection and enforcement, taxes and royalties, technical and safety specifications, monitoring all aspects of the operation emergency plans, collection and reporting of data, liability and insurance, and, finally, the circumstances under which an exploration permit may be suspended, either because of noncompliance with the management scheme "or in the event of unforeseen risks." Operators are guaranteed secure tenure and the sole right to apply for a development permit for the block they explore. The regulatory committee submits this scheme to the commission, which can approve or veto but cannot modify it. But the key decision—the deal between the claimant state, the sponsoring state, and the superpowers—already would have been struck.[44]

Environmentalists had some objections to parts of the Beeby plan. They noted that, whereas approval for exploration requires setting up a regulatory committee and drawing up a management scheme for review and approval by the commission and the advisory committee, going to development was far simpler. Under Beeby's proposal, when the operator wishes to begin development—that is, actual oil recovery or commercial mining—he or she must submit an application to the commission's secretariat, which in turn refers the application to the advisory committee. If the advisory committee finds that there has been either a change in the development activities envisaged in the original application, or that there are previously unforeseen environmental hazards, it may require modification of the application. The commission, upon receiving the original development application (or one modified by the regulatory committee at the advisory committee's request) must approve it without more ado.[45]

The Beeby draft is just that. It does not address the many subjects a more complete agreement should. For example, it makes no provision for mankind to benefit from Antarctic minerals development—that is, the majority of nations represented by the United Nations or the International Sea-Bed Authority—as is implied in Recommendation IX-1. It offers no role to interested outsiders in the treaty club: the all-important commission, which decides whether Antarctica is to be explored and passes on requests for minerals exploration, is made up of only the Antarctic Treaty consultative parties and states sponsoring minerals operators. No role is offered to those nations only interested in protecting the Antarctic environment unless such a state gains a seat on the advisory committee by acceding to the Antarctic Treaty and to the minerals regime, and by conducting "scientific research relevant to Antarctic mineral resource activities."[46]

"We are alarmed at the extent to which it facilitates mining at the expense of adequate control," wrote Eco's editors in an accompanying commentary. "The document gives the lie to the negotiators...who assert that concern for the pristine Antarctic environment is their first consideration. It reveals that providing a political solution to the Treaty partners' internal conflicts is an objective that has overridden protection of that environment. Political expediency has triumphed over sound management."[47]

Finally, problems could arise because of the extent to which the Beeby draft sidesteps such key considerations as title, royalties,

and civil and criminal jurisdiction by relegating them to negotiation between claimants, mining sponsors, and superpowers, in
the regulatory committee. This key feature seems to be an *ad hoc*
arrangement for Antarctic governance and is somewhat typical of
the treaty system. But it also puts an enormous burden on the
regulatory committee to solve all the critical problems that are
likely to plague any attempt at Antarctic minerals development.

It was a race—this scramble of the treaty powers to conclude a
minerals regime across a wilderness of historically unsolved issues,
ambiguous language, and divergent national ends. The race was
with time and with the emergent aspirations of the developing
world. By 1983, these same countries were articulating their interests in Antarctica at the United Nations. Moreover, the race was
also with the treaty itself, whose operation could be reviewed in
1991, on its thirtieth birthday. The treaty nations had hoped to fill
in the gaps left in Antarctic law, especially the lack of a minerals
provision, but it was not clear if they would succeed.

In the meantime, there was nothing more to do but to "go
forward and do our best," as Robert Falcon Scott remarked dryly
upon learning that Amundsen was racing him to the Pole.[48] For
the treaty powers now, as it had been for Scott, the goal itself was
less important than the fact that the race was on.

Resources Change National Antarctic Programs

Historically, any period of diplomatic ferment regarding Antarctica induces interested nations to signal their interests with expeditions, stations, and logistical feats using airplanes and ships. As
we have seen, this occurred in the 1930s as various nations announced their territorial claims and, again, in the IGY and during
the treaty negotiations that followed. It resurfaced in the late
1970s when, for example, the Soviet Union, Argentina, and Australia all planned or completed runways for wheeled aircraft to
enable them, as well as the United States, to fly in to Antarctica.

Both the krill fishery and preliminary geophysical data suggest
that the Weddell Sea region—where ice had gobbled up Shackleton's *Endurance* and another ship had sunk in 1975—was of potentially great resource interest. By the early 1980s, several new bases
were under construction along its fringe. The Argentines established the Belgrano II and III stations; the Soviets, Druzhnaya II;
the British reopened their older base, Halley; and, in 1981, West
Germany attempted an installation on the Filchner Ice Shelf, the
scene of Wilhelm Filchner's short-lived camp in 1911–12. Delayed
by the forming ice, however, the expedition was forced to establish
its Georg von Neumayer station 1,200 kilometers short of its goal.

Increasingly, expeditions regularly visited the ice-free part of the Weddell Sea to do research.[49]

Generally, the voluntary restraint policy prevented nations from declaring that they were undertaking minerals exploration in their geologic work and offshore surveys. Indeed, most of the expeditions volunteered to make all their work public so it would count as scientific research, not proprietary information. (An exception was the Soviet Union, whose geology papers and Tass announcements declared that the search for Antarctic minerals was under way. Over the years, however, other geologists active in Antarctica came to regard such statements as political necessities rather than as accurate descriptions of Soviet research.[50]) Indeed, at this point it would be difficult for any of the scientific programs to undertake serious minerals exploration in Antarctica. So general is the geologic knowledge, and so skimpy are the maps, that almost any geologic study counts as a contribution to basic science.

Nonetheless, some nations' behavior in Antarctica has been ambiguous. Numerous countries undertook seismic surveys of the Antarctic continental shelf, but the grid lines were so far apart as to produce only the most general information.

The *Glomar Challenger* did not return to Antarctica after drilling its holes in the Ross Sea Shelf. The U.S. *R. V. Eltanin* explored there as well, permitting Dennis E. Hayes and F. J. Davey to make a small-scale profile of some rock layers of the Ross Sea.[51] Vessels from other nations explored there too; however, unable to drill as deep as the *Challenger*, they had to be content with taking piston cores. By the 1983–84 season, however, the USGS's ice-strengthened vessel, *S. P. Lee*, ventured there.

The Norwegian Antarctic Research Expedition collected multichannel and other seismic data in the Weddell Sea during 1976–77. The West Germans' *Explora* gathered data from the Weddell Sea in 1978 and from the Ross Sea in 1980; and the Institut Français du Petrol sponsored *Explora*'s journey to the continental shelf of East Antarctica in 1981–82. The Soviets, British, and Argentines also planned seismic studies in the Weddell Sea.

In 1979 the Japan National Oil Corporation announced a three-year program of offshore surveying for oil in Antarctica, and for several seasons, the Weddell, Ross, and Bellingshausen seas were surveyed by the Japanese ship *Hakurei-Maru*.[52]

Although the data from all these expeditions were to be published, as of 1983 a remarkable amount of it had not been. Analyzing seismic data is expensive and time-consuming, and it is easier to store the data on tapes than to analyze and publish it. The Norwegians indicated that anyone interested in theirs could come to Norway and look at it. But some observers—among them geologists—wondered whether delays indicated that the treaty powers' agreement for the free exchange was being deliberately ignored by countries seeking "inside" knowledge of Antarctica's oil potential.[53]

Shipbuilding increased during this period. The Japanese built a new polar icebreaker, *Shirase,* a larger and more modern one than their present vessel. West Germany invested DM 185 million on a commodious polar research vessel, *Polarstern.* Brazil acquired an older Danish vessel, *Thala Dan,* while Great Britain, which had planned to sell one of its polar ships, decided to keep it after the Falkland Islands war.[54]

West Germany founded a new polar research institute, the Alfred Wegener Institute for Polar Research in Bremerhaven, which would be directed by Gotthilf Hempel, a biologist specializing in fisheries research. Its Antarctic program clearly reflects a decision by several German institutions and by the government that Antarctica was worth a major investment. For example, West Germany had sent krill research expeditions, led by Hempel, to the Antarctic as early as 1975–76, and sustained them over a period of years. German glaciologists participated in the Ross Ice Shelf project managed by the U.S. National Science Foundation. The Federal Institute of Geosciences and Natural Resources sponsored the *Explora*'s work; and a second ship, *Polarsirkel,* carried out ecological investigations in 1979–80. A third ship, *R/V Meteor,* participated in FIBEX, while a fourth, the *Walter Herwig,* carried out several seasons of work on krill. A crowning effort was the German Antarctic North Victoria Land Expedition (GANOVEX), that was to extend over three seasons in conjunction with scientists from New Zealand, the United States, and other nations. GANOVEX was carried out despite the sinking of *R/V Gotland II* off North Victoria Land on December 18, 1981.[55]

One explanation for West Germany's effort was its desire to qualify for full consultative status under the Antarctic Treaty. This was achieved in March 1982, but its scientific activities had exceeded by far the minimum requirements for consultative status. (Poland, for example, had done much less before being admitted in 1977.) Another explanation was that because West Germany considered Antarctic science and the emerging minerals and living resource issues important, it wanted to have a strong voice in future developments there.

Smaller nations, traditionally less active in the region, also were beefing up their programs on the resource side. Argentina's Antarctic programs had consisted primarily of meteorological work from its stations on the peninsula and surrounding islands. But during the 1980–81 season the icebreaker, *ARA Almirante Irizar,* performed seismic reflection studies and took krill samples in the Weddell Sea. At the Almirante Brown station, biologists were looking for evidence of pollution from natural petroleum seepage. The Argentines participated in attempts to find and measure krill though radio echosoundings from aboard *FRS Dr. Eduardo Holmberg.* A new Argentine polar ship, *ARA Bahia Paraiso,* was scheduled to begin operation, and Argentine aircraft were planning to make airborne magnetic surveys of the Weddell Sea basin. Australia likewise planned seismic surveys of the deep, narrow, continental shelf around the coast of the Australian Antarctic Territory.[56]

In sum, the 1968 SCAR meeting in Cambridge had borne fruit. As SCAR advised the diplomats and members of national scientific programs what resource and environmental research was needed, many nations, upon hearing the message, responded. Although this preliminary work lacked orchestration and much of it was rumored to be of poor quality, it represented a new, international awareness of Antarctica's potential resources.

seven

The United States in Antarctica
Present and Future

By 1983, the Antarctic Treaty group's deliberations on fishing and minerals were giving concrete meaning to the age-old dream of Antarctic riches. Krill fishing expanded season by season, and although no decision had been made to explore for Antarctic minerals, let alone to exploit them, the environmental impact of minerals development was a matter of intense debate. Thus, indirectly, the new resource diplomacy called attention to the beauty and uniqueness of the vast white continent and its value as an environmental preserve. The negotiations were having practical side effects: They burdened the underfunded, nongovernmental Scientific Committee on Antarctic Research (SCAR), which the treaty group leaned on increasingly for expert advice; they transformed the programs of several nations by incorporating resource-related work; and finally, they brought Antarctica to the attention of the developing world—a political shift whose consequences were still unknown. Regardless of whether the dream of Antarctic riches ever came true, a new era in Antarctic diplomacy had begun.

In addition, the treaty group had negotiated a new convention in order to prevent overfishing that could endanger the recovery of Antarctic whale stocks. They also negotiated a minerals regime under the framework of the treaty and moved to accommodate interested states who were not consultative parties to the Antarctic Treaty.[1] For example, they invited the thirteen acceding parties to the treaty (heretofore ignored) to the twelfth consultative meeting held in Canberra, Australia, in September 1982.[2]

The United States, which played a key role in the new Antarctic diplomacy, urged developing nations with an interest in Antarctica, such as Brazil and India, to accede to the treaty. The fact that they became full consultative parties was a victory for both U.S. policy and for the treaty system. In addition, the United States urged the treaty powers to conclude practical resource regimes in a timely manner and argued in favor of a conservationist position during negotiations for a krill convention.[3]

This role flowed naturally from the U.S. historic role in Antarctica since 1948, when it had begun brokering a regional peace—a quest that culminated in the 1959 Antarctic Treaty. The U.S. role as peacemaker and power balancer had become more important when the Soviet Union sent an expedition to the Antarctic for the 1957–58 International Geophysical Year (IGY). Thereafter, U.S. allies would look to the United States to counter the Soviet influence. And in the krill negotiations, when the Soviets took adamantly pro-fishing stands that nearly prevented concluding an agreement, the United States helped push one through. The need for continued U.S. influence seems greater than ever with the increasing interest of the United Nations and the approach of 1991—the date when any of the treaty powers may call a conference to review its operation.

But while U.S. diplomacy has adapted to changing circumstances, the same cannot be said for the two other elements in the U.S. Antarctic posture—its logistics effort and its scientific program. In 1983 both were run on much the same premises as in 1963. Both paid little more than lip service to the fact that things were changing in Antarctica itself and in the world of policy and diplomacy that governs it. Gone, by and large, were the scientists and explorers who were responsible for the U.S. program of the IGY; missing was the flexibility the United States had shown in deftly adapting its plans to suit the IGY's shifting scientific and political needs.

What seems to have happened is that, as the U.S. program became a permanent one under the direction of the National Science Foundation (NSF), and after Antarctic logistics became a permanent mission of the U.S. Navy, both became captives of their parent bureaucracies. Logistics changed little over the years. The Amundsen–Scott South Pole station had been maintained without interruption, and in most years a second inland station was maintained in West Antarctica, the region explored most by Americans. Many of the coastal stations the United States had built for the IGY were turned over to other nations—except for McMurdo on the Ross Sea. McMurdo remained the continent's "brain," its chief airport and switching center, just as the geographic South Pole was its heart. Air rather than sea power remained the dominant feature of the U.S. presence; this pattern continued without change for twenty years, except for the refurbishing of stations; the buying and retooling of aircraft; the relinquishing of the program's only ocean-going, ice-strengthened research ship, the *R/V Eltanin*, in 1972; and the 1965–70 expansion of Palmer station on the Antarctic Peninsula.

The U.S. scientific program had become similarly institutionalized. The IGY had used an uneasy coalition of older explorers and younger scientists, many of the latter beginning careers in Antarctic research. As the organizational base of the program shifted to the NSF, the program office consolidated its control; in time it funded not only the university scientists going to Antarctica each year, but also the U.S. Navy's logistics and the private contractor who had been hired in order to save money.

The funding for the U.S. Antarctic Program (USARP)—most of which went for planes, fuel, support staff, and supplies—made a peculiar bulge in the budget of the NSF. For example, for fiscal 1984, the USARP request was $102.1 million out of a total foundation request for $1.292 billion.[4] Nonetheless, the NSF put its stamp on the USARP. The NSF determined the USARP's procedures and priorities: its orientation toward basic university research on individual grants, its lack of an in-house scientific staff, and its habit of responding to random proposals sent in by scientists rather than planning ahead all mimed the institutional style of NSF. However laudatory this style was for the support of small-scale basic science, it impeded the kind of research that Antarctica was demanding more and more. Thus, the natural addition of work on applied resource and environmental questions that was occurring in other nations' programs (see chapter 6) was precluded in the U.S. program. Protests to the contrary, by 1983 the United States was falling behind in the race for knowledge of the Antarctic that dealt, not with the mysteries of past climates, meteorites, or cosmic rays, but literally with how many fish—and krill—are in the sea.

As the new treaties evolved regarding krill and minerals, and as SCAR requested ever-greater levels of expertise on resource and environmental matters in its members' national programs, it seemed unclear how U.S. leadership in the treaty group would weather the changing tide.

Chapter 2 described the explorers' failure to successfully annex the southern continent and integrate it into the U.S. sphere; this chapter relates another U.S. disappointment. In the 1970s the United States seemed to have little clearer idea about what it was doing in that distant part of the world than it had in the days of the early explorers. "But what good is it?" the young George Dufek had asked Byrd decades earlier as they stared at the great ice mass before them. The question still held.

What did the United States want to do with Antarctica? Exploit it—even though U.S. mining and oil companies were lukewarm, for the moment, about the prospect? Or was the Antarctic adventure one of raw power, in which the United States had called the shots because it flew more airplanes, sent more people, and *knew* more about Antarctica than did the other nations? As for President Eisenhower's decision in January 1956 to downplay the U.S.

A Buckminister Fuller geodesic dome houses trailer buildings at the U.S. South Pole station.

interest in territory and play up its interest in science and peace, was this policy shift a profound change or a temporary expedient? Were science and peace the critical U.S. goals in Antarctica? Or were they only the tip of a broader underlying interest that the explorers, despite their mock imperialism, had understood? The story of the United States in Antarctica in the 1970s and early 1980s is one of ambivalence—the vague and somewhat defensive reaction of the scientific establishment to the threat that some other priority might invade its Antarctic turf. One ironic result was that the USARP found itself opposing certain kinds of international cooperative research. It also resisted, for a time, having other agencies sponsor Antarctic research—and on one occasion it even opposed acquiring a much-needed polar ship—merely to protect the primacy of basic science and of the NSF. Sometimes it seemed that certain managers of U.S. Antarctic science were so taken with their Antarctic laboratory that they were blind to other national policy needs.

Meanwhile, down on the ice, the U.S. presence showed signs of its quarter-century tenure. The South Pole, now a tourist attraction, sported a new geodesic dome, dedicated in 1975 at a ceremony graced by Paul Siple's widow, Ruth. (The first women had landed at the Pole in 1970; showing more equanimity than either Amundsen or Scott, they jumped, all at once, from the plane.[5]) The dome housed the station buildings and a little contemplation

McMurdo, the main U.S. station on McMurdo Sound in the Ross Sea, is often compared to a mining town.

room called Sky Lab. It was carpeted, equipped with arm chairs, a stereo tape deck, and books. There, of an austral twilight, one could retreat from the crowded, brightly lit galley and bunkrooms and gaze at that historic plain where Scott's and his companions' party had manhauled their sledges past Amundsen's dire black flags. The American symbol remained: the barber pole surrounded by the flags of the treaty nations. Lost in this blazonry was the real South Pole, marked by a wispy bamboo stick that could be moved each year. One could also see the original station built by Siple in 1956, now so drifted over with snow that officials declared it unsafe.[6]

For twenty-five unbroken years, Americans had volunteered to occupy this spot, in the bright clear austral summer and in winter which runs from February when the last plane leaves until the following November when the first one usually lands. In 1980, for example, seventeen people (sixteen men and one woman) wintered over (as would five Americans at Siple in West Antarctica, eleven at Palmer on the peninsula, and seventy-eight at McMurdo).[7] Professors, naval officers, and government officials generally went home during the austral winter; and the group remaining were mostly young people, technicians and junior scientists, all lured by the white continent and the elation and isolation that all wintering parties had experienced since the *Belgica* expedition of 1897.

In contrast to the tinseled beauty of the Pole, the main U.S. station at McMurdo had become a sprawling city. Visiting journalists often wrote that it resembled a mining town—which was ironic since mining was the last thing in the minds of the NSF officials who ran it. With the years it had grown willy-nilly; build-

ings were constructed as needed without design or plan. Katherine Bouton visited Antarctica in 1979 for *The New Yorker* and
found that McMurdo had no fewer than four bars in keeping with
navy rules against mixing among ranks. She was escorted to the
officers' club and the chiefs' club; she saw the enlisted men's club.
She was warned away from the Erebus club (where, she wrote,
some airplane mechanics with shaved heads apparently appeared
naked except for their balaclavas and mukluks). "It was not a place
for a lady," her guide had cautioned—echoing a sentiment that
many Antarctic old-timers shared regarding the introduction of
women into Antarctica.[8]

McMurdo's disorderly appearance reflected the tensions between the three groups of Americans that cohabited this bizarre
capital. There was the navy, which flew planes and ran communications and medical services, but was unable to run McMurdo like
a regular naval base because it was subordinate to the NSF. There
was the private engineering firm (for many years, Holmes and
Narver, Inc., and later ITT Antarctic Services, Inc.), which built
and maintained buildings and serviced vehicles, but had little experience of command. Finally, there were the scientists, who in
theory were in charge of everything, but whose presence at Mc-
Murdo, in their trim dormitories and remote laboratories, was less
conspicuous. The lines of command were intricate: each station
had two chiefs, a leader in charge of the scientists and a project
manager in charge of contract employees. The navy commanded
only its own personnel at McMurdo; but at the same time it controlled everyone, because it controlled logistics.[9] Visitors sometimes wondered what would happen in a real emergency.

All of which begged the question of what the United States was
doing at the bottom of the world anyway, using USAF Lockheed
C-141 Starlifters part-time, six Lockheed C-130 Hercules planes,
four UH-1N Iroquois helicopters, and shipping in millions of
gallons of fuel each year. Edward P. Todd, who became director of
the NSF's Division of Polar Programs in the 1970s, told the Senate
in 1979 that the NSF had "responsibility for planning, managing,
and funding the total U.S. Antarctic program." The science program run by the NSF is, he said, the "principal expression" of U.S.
interests there. But its critics—among them a House committee—
berated the NSF for not having a serious program to study krill.
The published committee report concluded that "the U.S. position is rapidly eroding" with respect to other countries in Antarctica, and that the U.S. effort on living resource questions was
"virtually nonexistent."[10]

This chapter picks up the story of the United States in Antarctica with the ratification of the treaty and the arrangements for a
permanent U.S. presence there. The United States faces serious

problems there because its scientific and logistics programs have become steadily more expensive, leaving little money or momentum for anything new. It also discusses the oft-ignored question of U.S. logistics in Antarctica, and those of the Soviet Union, which now has a ring of bases around the continent's coastline. The Soviet presence could expand by 1991; yet, as far as can be determined, the United States has no strategy for dealing with this contingency. Finally, what does the United States want Antarctica to become? A wilderness preserve? An internationally managed polar oilfield—a Prudhoe Bay of the south? Should the United States insist that Antarctica be devoted exclusively to basic research? Or is there some blend of basic science and more directly useful resource that it should encourage, more suited to a resource-conscious era in Antarctic diplomacy?

This chapter argues that U.S. institutional arrangements for Antarctica are outmoded, especially in light of the rapidly shifting diplomatic and resource scene there and the question of 1991. Institutions, far more than individuals, seem to have held the USARP frozen in its IGY posture. During the 1970s, three directors of the USARP strove to move the program in new directions—but could not. Thus, a new institutional arrangement may be needed to serve the broader range of U.S. interests and to deal with many more contingencies that the present one cannot handle. The difficulties of change could be eased by roughly doubling the size of the U.S. Antarctic budget to accommodate new resource studies, new international initiatives, and a better logistics posture.

The Establishment of a Permanent Program

As chapter 4 explained, the Antarctic Treaty solved several historic U.S. dilemmas in Antarctica. It mooted the long-standing problem of whether the United States should assert a territorial claim, which could conflict with the claims of its allies, and it made scientific research—already the principal U.S. activity—the centerpiece of other national programs as well. By demilitarizing the region and providing for unilateral inspection, it calmed fears of the Soviet presence. Finally, by making the region a model of international cooperation, the treaty was a useful spur for other diplomatic and arms-control efforts.

To underpin these triumphs, Eisenhower needed permanent institutional arrangements. Therefore, at a meeting of the National Security Council held on April 2, 1959, he directed the Bureau of the Budget to examine the post-IGY management of U.S. activities in Antarctica. In July, after talks with the Departments of State, Interior, Defense, and Commerce, and with the CIA and the NSF, the bureau recommended an "integrated" program run by one agency. It concluded the "most logical agency" was the NSF.[11]

The suggestion met resistance both from older "Antarcticans," who had influenced Antarctic policy through the military, and the Antarctic Working Group—which the bureau had recommended be abolished, along with the post of the U.S. Antarctic Projects Office then held by Rear Adm. David M. Tyree. Some of the objections were self-serving, but others would prove relevant later. For example, Rear Adm. George Dufek, when asked by the secretary of defense for advice on the matter, wrote: "Since the activities in Antarctica represent varied and diverse activities, of fourteen departments and agencies, not all of which are scientific, . . . we question the advisability of vesting in any one of the components, the responsibility for development and justification of the overall Antarctic program."[12]

Henry B. Dater, the thoughtful historian of the Antarctic Projects Office, noted that while NSF's founding act did not prohibit it from "management of a large-scale research operation . . . with so many implications outside its usual field of endeavor. . . . [Nevertheless] such an activity runs counter to the spirit of the legislation which repeatedly emphasizes the awarding of contracts and grants to individuals."[13]

Their objections were not without effect. Both the Antarctic Working Group and the post of Antarctic Projects Officer survived, although the Working Group died when President John F. Kennedy abolished the Operations Coordinating Board in February of 1961,[14] and the project officer's job was eliminated in 1965. But, in designating the NSF as the leading agency, Eisenhower had implied that the assignment would be temporary. His letter to NSF Director Alan T. Waterman said: *"Inasmuch as the principal activity to be carried on in that region over the foreseeable future will be scientific in character,* I am assigning to the National Science Foundation responsibility for leadership in planning and coordinating a national program for activities in the Antarctic regions" [italics added].[15]

These wishes were expressed in organizational terms by Bureau of the Budget Circular A-51, which gave the NSF "the principal coordinating and management role" in carrying out the U.S. program for Antarctica, while the Defense Department would carry out operations in support of "scientific or other programs."[16] In 1961 President Kennedy abolished the Antarctic Working Group and gave the assistant secretary of state for international organizations the job of coordinating Antarctic policy, but this effort failed. Whereas Circular A-51 was implemented quickly because it carried forward the scientists' momentum on Antarctic matters, the State Department failed to exert the leadership Kennedy sought. Admiral Tyree commented in 1965, "No statement of national policy and objectives, no plans for the totality of Antarctic activities, no adequate overall guidance for a total program has been issued under this arrangement."[17]

This structure remains to the present day, with three noteworthy changes. The first came as a response to the State Department's failure to coordinate the work of the agencies under the 1961 arrangement, and to several bills introduced in Congress to establish a U.S. Antarctic Commission, conceived as a sinecure for the "Antarcticans." In 1965 President Johnson replaced the State Department arrangement with the Antarctic Policy Group (APG), a committee made up of representatives of the secretary of state (who would be its chairman), the director of the National Science Foundation, and the secretary of defense. Other agencies could be invited to participate on an *ad hoc* basis. At first the group was active, but later it would merely air interagency quarrels and ratify positions for U.S. delegations to consultative meetings.[18]

The second change came in 1970, as a result of continued quarrels between the navy and the NSF over accounting for Antarctic logistics. An APG review concluded, and the president agreed, that funding and management of all U.S. activities including logistics in Antarctica would henceforth be under the NSF. The NSF was also permitted to hire a private contractor for some services previously performed by the navy.[19]

A third change followed another APG review and a policy directive signed by President Ronald Reagan, and issued February 5, 1982. By then, the NSF clearly had no room for resource studies in its scientific program, but such studies were of interest to other government agencies, notably the National Oceanic and Atmospheric Administration (NOAA) in the Department of Commerce and the U.S. Geological Survey (USGS) in the Department of the Interior. The new directive said that the NSF would remain in charge, but other federal agencies could "fund and undertake directed short-term programs of scientific activity related to Antarctica upon the recommendation of the Antarctic Policy Group" so long as they coordinated with the NSF's logistics.[20]

The directive seemed unlikely to usher in an era of organized, coordinated, basic and applied research in the region, however. The budgets of NOAA and the USGS were already tight, and despite the enthusiasm of some staff Antarctic experts, neither agency had much of a mandate to work in foreign areas away from the United States, nor did either have the logistics capability to make more than an occasional foray. However, the directive encouraged the USGS to send its ice-strengthened vessel, the *S. P.*

Lee, to the Antarctic in the 1983–84 season, and to sponsor a symposium in October of 1983. NOAA, by contrast, did not mount an organized effort—despite the new krill convention giving marine research a critical role which the United States had been instrumental in negotiating. Although its intention was to improve flexibility, the 1982 directive also limited any other agency seeking to work in the Antarctic. Under the directive, the agency had to request permission from the Antarctic Policy Group to mount a project. Its proposal had to meet the requirements of the NSF's logistics plan and be only for "short term" projects.

The Balanced Program

Key to the story of the U.S. Antarctic program in the 1970s and early 1980s was the attitude of its parent agency, the NSF. In his illuminating history of the NSF in the 1950s, J. Merton England shows how the fledgling agency evolved to serve the needs of university scientists in traditional disciplines. Where a conflict existed with other agencies—such as the National Institutes of Health—which were invariably more powerful, the NSF would fund only the most basic, nonapplied work. The NSF evolved a universitylike disciplinary structure and the practice of funding on a project-by-project basis those proposals originating with scientists that had passed peer review. One result of this procedure was the agency's belief that it existed only to fund the best "basic" research and not to direct it in any way.[21]

These attitudes profoundly influenced the Antarctic program. Officially, the Antarctic effort of the IGY included only ionospheric physics, glaciology, meterology, and oceanography. To this agenda, the NSF added work in biology, geology, and cartography. This longer list meshed neatly with the structure of the university departments where the Antarctic scientists were located. And in time it became formalized as the "balanced program," a policy of giving each of these disciplines, each year, its "fair" share of the scientific funds of the USARP. Table 7-1 shows how it worked in selected years.

The balanced program gave the U.S. scientific effort in Antarctica a strength, scope, and quality that many other national Antarctic science programs lacked, and it also encouraged fields of study in Antarctica that otherwise might have been ignored. And it worked well so long as science was, as Eisenhower had written, "the principal activity to be carried out in that region."[22] But in the 1970s and early 1980s, as the costs of scientific work mounted and as fuel and logistics became startlingly expensive, the effect of the balanced program was to stifle major shifts in program emphasis.

Table 7–1. Allocations Within U.S. Antarctic Research for Selected Years
(in millions of $)

Discipline	FY 1973	FY 1979	FY 1980
Atmospheric sciences (meteorological and upper atmospheric physics)	.890	1.691	2.0
Biological sciences	.598	1.403	2.2
Oceanography	.256[a]	1.019	1.0
Earth sciences	.694	1.394	1.4
Glaciology	.372	.854	.9
Information	.602	.651	.5
Total	$3.413	$7.232	$8.0

Source: National Science Foundation, Division of Polar Programs.
[a] Does not include operating costs of *R/V Eltanin*.

It became axiomatic that no one discipline could be cut because it had been unproductive, or because another was on the verge of major breakthroughs or needed expensive new equipment such as a ship. After all, it was not the NSF's business—so the agency believed—to make such tradeoffs. This environment and belief structure hindered top-down development of large-scale programs, and as a result, the USARP was only able to mount relatively few basic science projects, such as the plan to drill through the Ross Ice Shelf. Thus, by the 1970s, it became commonplace to reject proposals submitted for resource-related research or environmental impact surveys; they seemed "applied" rather than "basic" in content; they required expensive logistics; and, in effect, they threatened the NSF Antarctic program. Table 7-2 shows the funding history of the USARP, including the extraordinary increases in logistics costs that tended to squeeze out such initiatives.

A second theme was the persistent dichotomy between official U.S. policy for Antarctica—as enunciated in periodic reviews by the APG and the National Security Council which were signed by the president—and what the agencies actually accomplished in Antarctica. It was far easier to include in a policy statement written in Washington that the United States should assess Antarctic resources than it was for the NSF and the navy to undertake such a program down on the ice. Nearly every White House review of U.S. policy (there were approximately nine such reviews from 1948 to 1970, and several more in the 1970s) concluded that resources were a key aim. Several called for assessments of Antarctic resources, and the 1965 guidelines expressed the hope that "these great projects of peaceful cooperation in Antarctica will yield resources which every nation needs and every nation can use."[23]

Yet carrying out these plans required capabilities that the U.S. arrangement lacked—interagency coordination was extremely

Table 7–2. Annual Obligations for the U.S. Antarctic Program for FY 1953–84

(in millions of $)

Fiscal year	Science (NSF)	Support (Department of Defense, Coast Guard contractor)	Total	Major capital items
1953	0.002			
1954	0.004			
1955	0.073			
1956	2.2	17.8	20.0	
1957	0.7	30.0	30.7	
1958	1.9	17.3	19.2	
1959	2.3	26.7	29.0	
1960	6.2	16.3	22.5	
1961	5.5	25.0	30.5	Overhaul of *Eltanin*
1962	7.2	20.9	28.1	Byrd and McMurdo stations rebuilt
1963	6.4	21.5	27.9	
1964	7.2	21.6	28.8	
1965	7.6	21.0	28.6	*Hero* built ($1.06 million)
1966	8.4	22.6	31.0	
1967	7.6	26.9	34.5	
1968	7.8	28.2	36.0	
1969	6.9	25.6	32.5	
1970	7.4	24.8	32.2	
1971	7.8	19.0	26.8	
1972	7.0	21.0	28.0	
1973	5.3	39.8	45.1	Three new LC-130s ($19 million)
1974	4.8	20.0	24.8	Pole Station rebuilt ($6 million, FY 71–75)
1975	4.7	21.4	26.1	
1976	4.3	44.4	48.7	Two new LC-130s ($18 million)
1977	6.3	39.0	45.3	
1978	7.0	41.4	48.4	
1979	7.3	43.8	51.1	
1980	7.3	48.5	55.8	Diesel fuel Arctic (from $0.61 to $1.29 per gal.)
1981	9.0	58.4	67.4	Diesel fuel Arctic (from $1.29 to $1.37 per gal.)
1982	8.5	60.0	68.5	
1983	9.0	74.2	83.2	NSF assumes full costs of Coast Guard icebreakers
1984	10.0	92.1	102.1	

Source: National Science Foundation, Division of Polar Programs.

difficult, and top-down management telling scientists what to do was a role the NSF shunned. Further, the constrained funding of the late 1970s required resource assessments to be allotted from funds that would otherwise go to basic science, something else the NSF was unwilling to do. Early in the 1970s, the NSF Antarctic program tried to accommodate the new resource concerns, but by the late 1970s, support for organized resource and environmental research was not forthcoming. Scientists, concerned that the United States was failing to address the new issues, expressed amazement at the hostility of program officials toward some international resource-related programs.

Although the NSF was not well equipped to carry out the resource assessments that its official policy called for, it habitually claimed its scientific research was resource-related. In its public statements, when seeking money, it often advertised its work as bearing on important, emerging resource questions. The claim was true, because almost any bit of research done in Antarctica *bears* on resource or environmental questions in one way or another. But it also veiled the fact that NSF's program was refusing to *address* resource and environmental issues in an organized manner that would directly support the new diplomacy.

The USARP and the Krill Problem

After 1968, when the krill fishery problem emerged in international scientific circles, the USARP initially gave assurances that U.S. research would provide national expertise and leadership. U.S. scientists, such as Sayed Z. El-Sayed of Texas A&M University and Don Siniff of the University of Minnesota, played an important role in defining the issue internationally. Likewise, the USARP Biology Program Manager, George A. Llano, who later became acting chief scientist, encouraged several colleagues to study the problem. Llano concurred with Richard Laws, director of the British Antarctic Survey, that whether whales and other higher organisms would be hurt by krill fishing was a far-reaching concern for Antarctic biologists. Thus, the NSF supported the international meeting held at Woods Hole in August 1976, at which the BIOMASS plan was developed.[24]

Llano also encouraged Mary Alice McWhinnie, a biologist specializing in crustacean biology, to study krill. McWhinnie of DePaul University in Chicago, who had worked on several *Eltanin* cruises from 1962 to 1972, was something of a feminist heroine in Antarctic circles. Besides being one of the few prominent women scientists working there, she was the first female station scientific leader. That the NSF encouraged her and other scientists to work in the field reflected the aim of U.S. diplomacy at the time, that of seeking a living resources agreement within the framework of the treaty.[25]

But the NSF's role had been ambiguous from the start. Was the USARP orchestrating diverse U.S. interests there? Or was it merely "expressing," as NSF officials said, that particular interest which happened to be science?

At the time, the head of the Office of Polar Programs, which included the USARP, was Robert Rutford, a geologist who had been on several Antarctic expeditions in the early 1960s. He viewed the NSF as the manager of the entire range of U.S. interests there. "Because the Foundation has national management responsibility," he wrote in a widely circulated memorandum in 1975, it would coordinate the government and private sector in developing a national position on the krill fishery.[26] Such a plan, according to Rutford, should be the basis for diplomacy, to guide other agencies' actions and the work founded by the NSF. He suggested holding a conference of interested parties to develop such a position. This, in turn, would be followed by an international meeting for coordinating research with other nations. He also expressed the hope that the United States would follow through on some of the ideas being developed by SCAR. And, since the krill issue would be around for a long time, the United States should "generously invest in developing a cadre of young marine biologists/fishery biologists to become the leading experts in the krill and its fishery. Support for outstanding young biologists will increase the potential U.S. contribution to the management of this important world resource...unless we have a presence in krill fishery, it would be difficult for the U.S. to have a strong voice in its management."[27]

But the NSF's "national management responsibility" proved elusive. The USARP office hardly could compel other agencies to take action on krill or anything else, and other agencies were mostly silent in response to Rutford's memo. Circular A-51, which guided U.S. Antarctic activities, seemed to leave policy guidance to the Antarctic Policy Group (APG). But State Department leadership through the APG was rare; the APG sometimes went for years without meeting. In 1977 Rutford left the NSF for the University of Nebraska, and his replacement was Edward P. Todd, a long-time NSF official. Todd's primary concern had been the fact that individuals on the National Science Board had criticized the quality of the science done in Antarctica;[28] his tenure would be marked by an effort to have the work pass muster as good, basic research. Rutford's plan was not carried out; by 1980, the United States still lacked a coordinated plan for addressing the emerging krill issue.

Meanwhile, in Antarctica, McWhinnie's work on krill biology had moved to the laboratory at the neat little station at Palmer, on the west coast of the peninsula. Charles Neider, a writer who portrays Antarctica from a humanist's point of view, visited Palmer on sheltered Arthur Harbor. With its mountain backdrop, glacier snout, and still waters, its green grass and orange lichens, its cycles of day and night, Palmer seemed like a resort to Neider, compared to the dryness and eerie brightness of the "true" Ant-arctic around McMurdo.[29] In the Palmer station lab, McWhinnie kept krill alive for a time in a flow-through seawater tank (al-though eventually they were killed by a reveler pouring liquor into the tank). Later efforts were more successful; however the krill survived in the seawater tank all winter and showed interesting behavioral and biological features. As word of McWhinnie's work spread among scientists of other nations working in the penin-sular area, she received requests from Chilean, Polish, and West German scientists to work with her at Palmer. Many of them had failed to keep krill alive in their own labs.[30]

El-Sayed and McWhinnie pressed to develop a U.S. plan for BIOMASS, which had become steadily more controversial. Sup-porters argued that it was a portent of the future and linked U.S. Antarctic science to an important new rationale, but its critics, including Todd, said it was poorly designed and unworthy of U.S. science.[31]

Whatever the merits of BIOMASS, it unmasked the weakness of the NSF arrangement. The NSF chose only the best proposals from those that came in essentially at random from university scientists. Program officers did not want to direct the scientists, although they often gave informal encouragement to some. Since year-to-year funding for the overall program was uncertain, multiyear plans were difficult to carry out. The USARP tended to fund one-year proposals, meaning that a scientist who competed successfully to go to the Antarctic for one austral season had little certainty that he or she could return. The higher councils of government debated perennially whether the NSF should do more applied research, why it rarely did it well, and why it had few ties to industry. In the mid-1970s this debate proceeded in the abstract, the participants apparently unaware that this very issue was being fought out in the USARP.

Any major international research program was incompatible with this style of management. For the FIBEX observations, for instance, scientists from different countries worked in prear-ranged sectors of the Southern Ocean, from South America around to Australia, during certain weeks in 1981. The USARP could not participate because its program officials did not sit on SCAR committees and it had no suitable ship of its own (in 1972 the *Eltanin* had been dry-docked in a sudden cost-cutting move). Nor could the USARP control the schedules of the twenty-seven ships belonging to the U.S. oceanographic fleet, which are run by independent university centers. This problem resurfaced repeat-edly in discussions of possible research efforts during the 1970s and early 1980s.

Todd felt that the pro-BIOMASS faction was getting too strong.

In an interview, he told the author that the NSF was not going to "count krill"—a phrase he used repeatedly when talking about BIOMASS. He also added that the USARP would not be "taking orders" from the international group.[32] When McWhinnie submitted a proposal in 1978 to extend her krill biology work for five years, it underwent peer review and was declined.

Todd told the author that McWhinnie had not published her Palmer station findings in peer-reviewed scientific journals and was working in too many issues concerning krill. The rejection of her proposal was meant to start a dialogue with her about a more modest plan, he said. In 1980 El-Sayed sought funds to work on Antarctic phytoplankton in relation to krill, but his proposal was rejected as well. Todd told the author that El-Sayed had not published enough in peer-reviewed journals, although El-Sayed protested that his reviews were favorable and that the NSF had deemed his record good for the last twenty years.[33]

As an international figure, El-Sayed found other support and was able to do his work on foreign vessels until he later received NSF support. McWhinnie, who submitted a two-year proposal, set about rebutting her reviewers for the earlier one but showed signs of overwork and exhaustion. She suffered a stroke in September 1979 and later died. The USARP gave her colleague, Charlene Denys, who had run the tank at Palmer during the 1979-80 season, a small grant to get the work published. In the meantime, the USARP awarded Osmund Holm-Hansen of Scripps Institution of Oceanography a large grant for two months of work in 1980–81 in the Scotia and Weddell seas aboard the Scripps ship, *R/V Melville*.[34] (The USARP had planned to sponsor a cruise every two years, and several major institutions submitted proposals for marine ecosystem work for 1982–83; in the end, the USARP lacked funds for a ship and turned the proposals down.)

At the same time USARP officials told Congress that its work was relevant to the krill fishery. During his 1979 testimony to the Senate, Todd said that the research "is expanding our understanding of the food needs and species diversity of the Antarctic marine ecosystem. However, we have much more to learn before we will be in a position to recommend with any surety those steps that will be necessary to manage responsibly the harvesting of antarctic marine living resources."[35]

A NSF *Program Report* for September 1979 describes BIOMASS as "an international program, a large and expensive one" and "an ambitious proposal."[36] The USARP, it said, "is already funding programs related to the BIOMASS objectives, and this funding is increasing," and it listed McWhinnie's and Deny's findings in the tank at Palmer station as evidence of the USARP's achievements.[37]

The claim was accurate. The USARP had been prescient in directing some fieldwork in the early 1970s, and funding was increasing because of the award to Scripps. Moreover, almost everything in Antarctic marine biology "related" to BIOMASS. Left unsaid was the fact that, to be truly supportive, the USARP needed a major effort of targeted research paid for by cutting another program or by fighting for extra funds. Also ignored was the relatively greater commitment of other nations to Antarctic marine biology. Nor was anything said about the need for a ship, since it was overall NSF policy not to seek funds for a polar ship. Finally, not mentioned was the fact that, with the exception of a few individuals, U.S. scientists knew less about the fishery than did those of Japan, Great Britain, West Germany, and some other maritime nations.

Finally, in the spring of 1981, more than a decade after the subject first arose in SCAR, the NSF, the Marine Mammal Commission, and NOAA agreed that an *ad hoc* committee of the Polar Research Board and the Ocean Sciences Board of the National Academy of Sciences should study U.S. needs for research on the Antarctic marine ecosystem. The committee's charge gingerly avoided references to planning. Instead, it was to "evaluate" national and international research plans, to recommend "targets" for U.S. research, to "examine" the national and international structure for BIOMASS, and to make "suggestions" to U.S. agencies. Most important, an outsider to the dispute, John H. Steele, a distinguished oceanographer and director of the Woods Hole Oceanographic Institution, was brought in to chair the study. The Steele Report concluded that the Southern Ocean offers a "great range of interesting and important ecological problems."[38] However, it noted, "The general aim of understanding the Antarctic marine ecosystem is necessarily long term, possibly Utopian."[39]

The Steele report recommended two "coherent projects" be undertaken by the United States. One was a study of the physical, chemical, and behavioral processes underlying krill swarms. The other was a study of processes at the ice edge, when the pack ice forms in the fall and retreats in the spring. Both could pass muster as basic research as well as filling in gaps in other nations' research activities.

To Todd's concern that quality of basic science be the basis for selecting research projects, the Steele Report added two others: "relevance to resource and conservation aspects," and international commitment and cooperation. It could not avoid institutional issues, stating that "The present NSF system, typified by unsolicited proposals from individual investigators, will probably not be adequate for the resource questions....We wish to avoid a dichotomy between basic and directed research, but we realize that an exact balance in interests between basic science and resource-oriented studies is not practical or possible...."[40]

NOAA, it said, was best suited to carry out some of the studies, but more important was the need to integrate field and laboratory work. International cooperation was necessary and it was recommended that the United States set up a committee to relate to BIOMASS. The Steele Report estimated that such a program would cost $3 to $6 million per year, not counting a ship.[41]

The Steele Report and the 1982 recommendations in due course moved the USARP toward greater tolerance toward marine ecosystem research. A greater effort was made to find a ship; the *R/V Melville* made a voyage of eight weeks in 1980–81 and returned again in 1983–84. Most U.S. scientists wishing to work in the Southern Ocean, however, went aboard Soviet or other nations' ships. The krill study at Palmer station was resumed in the 1981–82 and 1982–83 seasons by a group from the University of California at Los Angeles, led by William M. Hamner, who observed krill behavior, feeding habits, schooling, and molting. Researchers from the University of California at Santa Barbara, aboard the *Hero*, Robin M. Ross and Langdon B. Quetin, observed the spawning frequency of krill in the seas off Elephant Island and the South Shetland Islands and found them to be five to ten times more fecund than previously had been estimated.

In order to mount the "coherent" program envisioned by the Steele Report, it was necessary to change procedures. How could an individual scientist writing a proposal know what aspects of the ice edge might "fit" into a coherent plan? An *ad hoc* committee formed under Donald B. Siniff, the University of Minnesota seal expert who had worked on the report's ice edge suggestion. The committee members urged individual scientists to submit proposals to the NSF. When the submission was sent out for review, the *ad hoc* committee enclosed an explanation of the overall study. Although some recommended studies did not survive the review process, the USARP did fund eleven, to be carried out from the *Melville* and the icebreaker, *Westwind*. In all, the package was arguably "coherent."[42]

These projects, along with the krill studies mentioned above, constituted the implementation of the Steele Report. By the end of 1983, it was unclear what would follow. The projects had no specific relationship to the BIOMASS study, nor were they specifically designed to help the new scientific committee established by the krill convention. An institutional experiment had been carried out. But it had had to be tailored to fit within the confines of the NSF's basic research mission; it had occurred only at the Academy's instruction. Thus, this U.S. venture into krill research stayed a safe distance from the political and applied science questions that had been looming for years.

Meanwhile, NOAA encountered predictable problems in trying to carry out the 1982 presidential policy. Eager staffers indicated that the new krill convention would give NOAA a role in supporting U.S. diplomacy. But the agency—and its parent Department of Commerce—refused to support a specific program. A NOAA official testified that its budget for the Antarctic ranged from $200,000 to $300,000 per year and had been taken from other accounts, including the administrator's discretionary fund. But that agency's National Marine Fisheries Service, which was the logical branch to carry out research in support of the krill convention, had no coherent program although it housed some scientists expert in the Antarctic. NOAA awaited an outside push—this time from Congress, where friends of the agency, concerned about the inability of the executive board to anticipate or carry out major new research responsibilities arising from the krill convention, prepared legislation mandating a NOAA role.[43]

At about the same time, the Polar Research Board attempted an overall review of the Antarctic science program. Todd explicitly had ruled out a study that would link the USARP-sponsored research to the "plethora" of difficult new resource issues. In February 1982, he responded to the board's query about the scope of their study, saying it should look only at fundamental scientific questions and determine the relative emphasis among disciplines within the balanced program. It might have to measure the weight accorded to the different disciplines—the "most complex and difficult problem" before it. But, Todd continued:

I do not think it desirable, at this time, that the Board consider either the costs of alternative logistic systems, or the structure and needs of management regimes and the plethora of resource issues that these management regimes must resolve. Instead, the Board should focus on science as a means of expanding fundamental knowledge about Antarctica and the processes or phenomena there that may contribute to the advancement of a scientific discipline in its broader context.[44]

The board recommended—albeit politely—that the USARP change the way it did business. Because most of the problems studied in the Antarctic were "essentially interdisciplinary," it felt that the USARP should give priority to a few, large interdisciplinary programs, requiring long-range planning and advance notice to possible participants, including those from abroad. As for dealing with fundamental science instead of the Pandora's box of resource questions, the committee suggested that a proposal's "relevance to resource and environmental issues" as a key criterion in judging projects. It was hardly surprising then that, among the six "highest" and "very high" priority projects the board recommended, the second was study of the "structure and function of marine biological communities" (which would require a ship) and the fourth was the geologic history of the Antarctic continental margin (which would require specially equipped aircraft, as well as a ship). Still others related to krill and to environmental processes.[45]

The USARP and Antarctic Minerals

The story was the same concerning U.S. research on Antarctica's mineral resources. In the early 1970s, the USARP office redirected its geological research toward minerals of economic interest. A USGS team returned to the Dufek Massif, and other minerals studies were encouraged. Deducing the broad geology of an entire continent is a massive task, and although these studies were unquestionably basic research, they also were necessary to assess Antarctica's minerals potential.

Joseph O. Fletcher, head of the NSF's Office of Polar Programs (forerunner of the Division of Polar Programs) in the early 1970s, went further, asking the USGS to submit a plan for a geophysical study of Antarctica's continental shelves. Such a project could shed light on the original configuration of Gondwana and furnish basic data for an evaluation of Antarctica's hydrocarbon potential. Fletcher hoped to pay for it with funds from the $100 million that President Richard Nixon had set aside for energy independence. The plan the USGS submitted sought $12 million over five years for a survey of the accessible parts of the continental shelf, and included the refitting of the *Eltanin,* the addition of seismic gear to the *Hero,* and another Antarctic survey by the *Glomar Challenger.* The extra funding was needed, the proposal said, because the area is "not likely to be comparably investigated along these lines for several more decades unless a project such as this one is implemented...."[46]

Either the NSF did not forward the proposal to those handling the apportionment of the Energy Independence money, or the project was advocated unsuccessfully; in any event, the plan was neither funded nor encouraged. (Another proposal from the Office of the Polar Programs, a leading agency in Arctic research, for an environmental assessment of offshore Alaska also was rejected.[47]) In short, Fletcher was constrained by his parent institutions: while the NSF did not forbid the Office of Polar Programs from making geophysical surveys or environmental assessments, it did nothing to encourage such projects, either.

Likewise, working with industry was not the NSF's style, although it would be logical to work out a framework for U.S. oil or mining companies' involvement in Antarctic minerals exploration. In 1975 the Aquatic Exploration Company of Dallas proposed an Antarctic seismic survey in a brief letter to the NSF, and Texas Geophysical Instruments, Inc., made inquiries about Antarctic work to the State Department.[48] And in late 1978 Joel S. Watkins, representing the Gulf Research and Development Company, made persistent inquiries with the USARP office, offering what the scientists most needed—a ship. He proposed the use of one of the most advanced seismic exploration ships in the world, The *R/V Hollis Hedberg,* for a multiyear survey of the Antarctic continental shelf beginning in the Ross Sea.

With its reinforced hull and 92-channel seismic array (the university ships have 24-channel arrays), the vessel—with a well-designed program—could yield better results than those achieved previously by other nations (see chapter 5). Gulf proposed joint sponsorship between the NSF and a consortium of domestic agencies, foreign governments, and corporations. Under the proposal, the NSF would support U.S. academic participation, while the consortium would underwrite the cost per field season, which was estimated to be $2.5 million. U.S. or foreign corporations and governments could participate in the consortium by putting up $250,000 apiece. The NSF would appoint a planning committee to establish track lines and priorities, and, in cooperation with Gulf, would select participants and look after the scientific review and publication of an initial report. Gulf alone would be responsible for the conduct of technical and field operations. According to Watkins, a report would be sent to the printer within fourteen months of the field research, and the data would reach the government's geophysical data center in Boulder, Colorado, within twenty-four months for interpretation by jointly appointed panels.[49]

Todd encouraged the proposal's submission and told Watkins what he should do to have the plan considered by the NSF. In an interview he told the author that he was delighted at the prospect of having the ship. Most of its reviewers were also enthusiastic about the brief proposal, especially from the standpoint of basic science. The only point of criticism was the fourteen-month period that Gulf would hold the data before publishing it. Ultimately, the proposal was dropped. "We were very busy that season," Todd told the author, "the season was under way in Antarctica and a lot was going on."[50] The extensive file for the record, kept by NSF officials, does not indicate that the obstacle represented by the fourteen-month holding period was ever discussed with Gulf however.[51]

Clearly, the minerals work that the USARP preferred to do in Antarctica was the kind it did best: putting a few university scientists in the field to do basic science that also related to minerals. An example of this approach is the uranium "survey" conducted for successive seasons by two scientists of the University of Kansas, Edward Zeller and Gisela Dreschhoff. Each year they flew by helicopter to areas of exposed rock in the Transantarctic Mountains, in the Ellsworth Mountains, in South Victoria Land, and elsewhere. Like a giant dragonfly, the chopper would sweep back and forth over the mountainsides—often over places humans had never walked—while the gamma-ray spectrometer measured radiation levels of the rock.

The survey accomplished two things. First, by measuring radioactivity, it identified different rock types, hence the broad geology below; and second, it showed deposits of uranium, thorium, and potassium in the rock. Although such findings are used as preliminary guides to uranium exploration, Zeller and Dreschhoff were not "exploring" for minerals. They did find thorium concentrations in the vicinity of the Darwin Glacier in the Transantarctic Mountains, but they did not drill as a surveyor might have done. Instead, a contemporary account noted, "At the end of the first three years of their survey, Gisela Dreschhoff and Edward Zeller estimated that they had measured radiation levels over about one-tenth of 1 percent of all the exposed rock in Antarctica. The work had just begun."[52]

The NSF often cited the minerals information the study would collect as the reason for funding the project. On the other hand, it was also clearly basic research. And since the exposed rock area of Antarctica is enormous compared to what they could survey in a few weeks per season, the odds of finding a large surface deposit remained comfortably small. It was not obvious that the work, in the words of the 1965 guidance, "would yield resources which every nation needs." More to the point, during this same period other nations began planning seismic surveys of the Antarctic continental shelf and other resource-related geologic work. By restricting its own work thus, the United States risked losing its edge in key information about Antarctica which traditionally had cemented its influence in Antarctic diplomacy worldwide.

No Polar Research Ship

The issue of devoting more of USARP's effort to emerging resource and environmental issues would have been easier had a ship been available, but, by 1982–83, the USARP had less sea-going capability in the Antarctic than at almost any time previously.

The red-sailed little *Hero*, which services the U.S. Palmer station, sails along the Antarctic Peninsula.

The *Northwind*, a U.S. Coast Guard icebreaker, has a rounded hull for riding through the ice.

The old U.S. polar research vessel *Eltanin* was built to be squat and stable, so research could be carried out even in heavy seas.

The only ship the program operated directly in Antarctica during the 1970s was the red-sailed, 125-foot *R/V Hero*: it had a wooden hull with metal plating around its bow to cut through ice. The *Hero* plied the coastal waters of the peninsula and serviced Palmer station in the summer; in the winter it worked along the coasts of southern South America. Neider called it "frail, innocent, anachronistic and romantic."[53] It had been launched in 1968 to service Palmer and do coastal marine research—hence the sails, for stability in stormy seas and to avoid engine noise that might disturb the fish.

The *R/V Eltanin,* by contrast, was an older ship that had been converted for research, serving in the Antarctic since 1962. It was 266 feet long and capable of pushing through pack ice. It displaced 3,886 tons and could squat in stormy seas as scientists on board went about their work.

In December 1972, when the Nixon administration was imposing across-the-board cuts, the USARP was instructed to slash several million dollars from its budget. Several research grants were canceled, and the *Eltanin,* then plowing through the Southern Ocean swells, was ordered to return home. Subsequently, an arrangement was worked out with Argentina whereby the vessel would be operated by that country (and thus it was renamed the *Islas Orcadas*). U.S. scientists, in return for a contribution from the USARP, could work on board during half of each season. In 1979 the agreement expired and, in October 1981, the ship was tied up because the NSF had no money to refit or mothball it.

During the 1980–81 season, the *R/V Melville* worked in the Scotia Sea, the Weddell Sea, and the Drake Passage and was to return during 1983–84. But since the *Melville* was scheduled and operated by Scripps Institution of Oceanography, the USARP could not control its schedule. For the USARP to send shipborne scientists to the Antarctic—even for a few weeks—it needed to know what proposals would be submitted, reviewed, and funded *before* the oceanographic institutions had scheduled their ships.[54]

Thus, only two coast guard icebreakers sent to the Antarctic each austral summer could be regarded as certain research platforms. Neider sailed on one, the *Burton Island,* from McMurdo to Palmer around the West Antarctic coast, retracing Byrd's route and crossing the route which Captain Cook had followed in January 1774. Built in the 1940s, the *Burton Island's* gunwales were frighteningly close to the water, and the nearly keelless ship rolled incessantly, sometimes taking rolls of thirty-five and forty degrees. Tired of watching the old movies that had been provided as entertainment, Neider took to his hammock-bed, which swung wildly day and night. While lying there, he wondered if the rumor were true that the ship could do a complete flip in the water. Obviously, such a ship was hardly ideal for extended scientific laboratory work in those seas.[55]

The USARP estimated that the cost of building its own research ship would range from $20 to $22 million, and that equipping it would cost another $4 million. Yearly operation would come to $2.75 million, and the ship would have a lifetime of twenty-five to thirty-five years. On the other hand, to extend the life of the *Eltanin* another ten years would cost only $9 million, but the annual operating costs were estimated to be $5 million. The USARP could hardly command such sums, especially when the NSF's far more influential University–National Laboratories Oceanographic System sought upwards of $50 million to modernize its twenty-seven-vessel academic fleet, in addition to the fleet's yearly operating costs of $25 million.[56] The polar office, which needed a new ship for both the Arctic and the Antarctic, was unsuccessful in persuading the NSF to go to the Office of Management and Budget and the Congress for the extra funds.[57]

With no prospect of a polar ship, the Steele committee in 1981 suggested chartering a Norwegian ice-strengthened ship for studying Antarctic marine biology. This proposal was rejected because the USARP hoped to use a ship in the academic research fleet for a few weeks every other season.

The lack of a full-time research vessel underscored the fact that the USARP was not oriented to hardware. Indeed, the Steele Report said that advances in Antarctic marine research were limited by technology. For instance, the echo-sounders used by most nations to detect krill swarms had error rates of 50 percent. Only with more accurate instruments could scientists estimate swarms precisely enough to detect stock changes. Similarly, although satellites could be very useful in recording sea surface temperature and color, and could convey much about the physics and biology of the Southern Ocean, rather little use was made of them. Drifting buoys now used for meterological work could be modified to collect oceanographic and biological data as well; it would cost something but would be cheaper than other methods of collecting the same information. Nonetheless, the USARP felt its purpose was to put individuals on the Antarctic continent and to fly airplanes and helicopters to signal the U.S. "presence." Studying Antarctica with satellites or drifting buoys would not accomplish this. Moreover, the atmosphere of constant worry over funds that permeated the USARP office precluded long-term investments in new technology. Indeed, the USARP's major technical initiative—a drill designed to penetrate the ice shelf, sample the water below, and take a core of the continental shelf—failed miserably. It is not surprising then that the Polar Research Board's 1983 report recommended that in the future more heed should be paid to advances in technology.[58]

Similar problems precluded the United States from undertaking offshore seismic surveys of the kind that other nations were undertaking. As we saw in chapter 5, the use of the *R/V Hollis Hedberg* was ruled out when the NSF failed to continue the dialogue with Gulf.

Scientists within the USGS continued to propose a study of Antarctica's continental shelves, using NSF money and the *S. P. Lee,* an ice-strengthened geophysical ship with a 24-channel seismic array. Only later, when the USGS itself came up with the money for a voyage, did the *Lee* sail in Antarctic waters. One reason for the *Lee's* voyage during the 1983–84 austral season was to obtain data that the USGS could barter with other nations, who had failed to publish data gathered during their own seismic studies. Just before the *Lee* reached the Antarctic, the USGS worked out an arrangement with some West German scientists under which their country's seismic data would be made more readily available. A USGS conference was also held on Antarctic (and Arctic) geology in October 1983. Nevertheless, the USGS was limited to doing only "short-term" work in Antarctica by the 1982 guidelines, so that an ongoing U.S. program of offshore seismic work still seemed unlikely.[59]

As for U.S. ships capable of drilling the Antarctic continental shelf to study the age and composition of the rock there, only the *Glomar Challenger* was equipped for the task. However, her captain decided that the Antarctic seas and icebergs were too hazardous for the ship. Indeed, the ship is not adapted to work in the ice, and it had exited the Antarctic surfing down waves 80 to 90 feet high. In addition, the Deep Sea Drilling Project, which had pioneered marine geophysics and made key contributions to the study of continental drift, along with its parent agency, the NSF, currently were preoccupied with trying to replace the old *Glomar Challenger,* which was scheduled to retire at the end of fiscal 1983. Thus, it paid little attention to the USARP's faint cry for a polar research ship.[60]

"Frozen" U.S. Logistics

Historically, a country's influence in Antarctic diplomacy has been in proportion to its expertise down on the ice. For example, the series of U.S. expeditions (described in chapter 2), and the knowledge gained during the IGY formed the basis for U.S. influence during the Antarctic Treaty negotiations and set the tone for subsequent international activities. And even a small country—namely, Great Britain—proved to have enormous influence in the krill convention negotiations because of its expertise on the marine ecosystem of the Antarctic.

Logistics in Antarctica, therefore, serves purposes other than simply moving people and supplies around. It also signals political intentions. For example, the U.S. South Pole station symbolizes U.S. nonrecognition of the six sector claims that converge there. This signaling through logistics continues today: in 1981 a West German expedition was sent to establish a year-round station on the Filchner Ice Shelf on the Weddell Sea. Delayed by ice conditions, the expedition built the station at 70° 37′ south latitude and 8° 22′ west longitude, far from the intended location, in order to have it in place before the treaty powers met in Buenos Aires to vote on admitting West Germany as the fourteenth consultative party.[61]

By the early 1980s, a number of countries had signaled their interest in the krill fishery and in offshore hydrocarbons by sending marine seismic surveys to the Ross, Weddell, and Bellingshausen seas. New bases were being built by West Germany and Argentina in the Weddell Sea region, where a number of cruises had been undertaken because of the region's resource potential.

The year-round station established by the United States at the South Pole also symbolizes U.S. determination to keep the Soviet Union from occupying the "heart" of the great continent. Both inland U.S. bases served as depots and fuel caches, making it possible for U.S. aircraft to fly almost anywhere on the continent. This unique capability, useful for research and for rescue, also symbolizes the U.S. political wish to maintain access to all parts of Antarctica.

Yet, the United States is too constrained by its continental logistics program to signal its interest in living and mineral resources offshore. If, for example, another nation were to find a major hydrocarbon basin in the ice-clogged Weddell Sea, the U.S. program—which requires use of its icebreakers full-time to service McMurdo—would be hard pressed to deploy a vessel to the site to assert U.S. interests and to monitor what happens there. Should the scientific committee of the new commission regulating the fishery suspect that the Soviets and Japanese are overfishing for krill in the Scotia Sea, the United States would have difficulty visiting the area in question for a season or two to investigate the issue independently. Without a capability to roam freely in offshore areas, the United States could have difficulty influencing the evolution of the fishery.

Should a mineral deposit of economic value be found on some virgin Antarctic coast, it would be very hard for the United States to investigate without disrupting the rest of its scientific program and giving the impression it suddenly was shifting gears. The present U.S. logistics effort of four year-round stations, with air and icebreaker support, costs $60 million per year. Another $19.6 million goes for construction and procurement,[62] so that most of the program's costs are fixed, leaving little money or flexibility to put down a station in a new area of interest. How, for example, would the United States signal its disapproval to a nonsignatory nation which went to the Antarctic and engaged in some taboo activity there? It might well be limited to lamenting the newcomer's behavior at an international meeting, or resorting to unrelated diplomatic efforts, such as denying the newcomer the sale of arms. Most areas of resource interest are offshore waters or along the coast, whereas U.S. logistics are organized around its continental air presence.

The U.S. presence is centered, logistically and financially, at the South Pole station. From the staging area at Christchurch, New Zealand, supplies are flown into the U.S. station at McMurdo, from where they are flown to the Pole. Even after twenty-four years of practice, the navy's C-141 flight in from New Zealand is hardly certain. After a 10,400-kilometer flight from California to New Zealand, the traveler can spend days in Christchurch waiting for the weather to clear for the 4,000-kilometer flight to McMurdo. The only country with major air access to the Antarctic continent, the United States made twenty-two turnaround flights from Christchurch to McMurdo in the 1980 season. Less-sophisticated air access is afforded the Argentines who have a runway for wheeled aircraft at Marambio; by the Chileans at Teniente Marsh; by the Australians at Casey; and by the Soviets at Molodezhnaya.[63] However, few locations are suitable for landing fields in Antarctica, even on the Peninsula where the shores are steep and the weather windy and treacherous.

The U.S. South Pole station serves as a fuel cache for transcontinental flights by ski-equipped C-130s. During the 1978–79 season, a Soviet plane crashed at Molodezhnaya on the coast of Enderby Land in East Antarctica. The pilot, copilot, and a passenger were killed, and eleven others were injured. Delayed perhaps by waiting for permission from Moscow, the Molodezhnaya staff requested emergency aid from McMurdo. There, a naval flight surgeon and two orderlies stepped aboard a LC-130 to begin one of the longest transcontinental flights in Antarctic history—a journey of over 8,960 kilometers. They flew to the South Pole, where they refueled before flying to Molodezhnaya to pick up the injured—a journey of 2,880 kilometers in twelve hours. Returning to McMurdo after another refueling at the Pole, the injured were

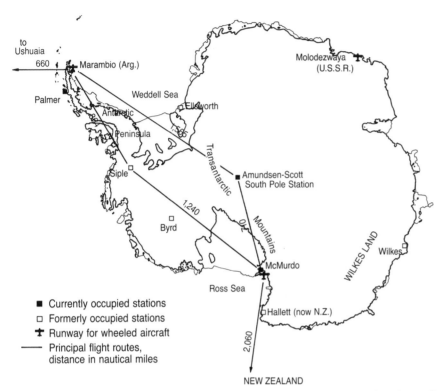

- ■ Currently occupied stations
- □ Formerly occupied stations
- ✝ Runway for wheeled aircraft
- —— Principal flight routes, distance in nautical miles

U.S. logistics based at McMurdo include the capability to fly anywhere on the continent. A major activity is the maintenance of the year-round South Pole station.

evacuated by C-141 to New Zealand, where they received medical care.[64]

The U.S. program is inevitably shaped by such logistics. Some of the U.S. scientific projects still depend on old-fashioned daring and mountaineering skill. One such is vulcanologist Philip Kyle's attempts to climb down the inner crater of Mount Erebus, a hot lava volcano, to collect gases. Usually successful, he is often driven back by lava bombs and hot steam. However, most projects move people around by air, such as Zeller's and Dreschhoff's uranium survey, the work in the dry valleys, or the field searches for meteorites near the Darwin Glacier. Temporary stations have been built in the East Antarctic interior as well.[65]

But such outposts are vulnerable to rising fuel costs. For example, transporting teams of scientists away from McMurdo by helicopter cost $7,500 in fuel alone in 1979; a comparable effort in 1981 cost $15,900. For years, White House reviews concluded that the United States should maintain two inland stations (one at the South Pole and the ionospheric research station at Siple) and two coastal ones (at McMurdo and Palmer). But because of rising costs, particularly of fuel, serious thought was given to keeping only three stations and only opening Siple occasionally.[66]

However, except for McMurdo, no U.S. station is near an area of current resource interest. Palmer, on the west coast of the Peninsula, is away from the main krill-fishing grounds located north and northeast of the peninsula. There is no U.S. station on the Weddell Sea, now thought to have the thickest sediments and some hydrocarbon potential. Inland from the Weddell Sea lies the Dufek intrusion, which has been compared with the Bushveld complex in South Africa. The Soviets, Argentines, and British are located together along the coast of the Weddell Sea, and the West Germans are nearby. Although no country has erected a station on the Amundsen or Bellingshausen Sea coasts, the Soviets have a small one on the coast of Marie Byrd Land. There is a Soviet station, but no American one, near the Amery Ice Shelf, also thought to be promising for hydrocarbons.

The Soviet Presence

Prior to the IGY, the Soviet Union applied to Australia and New Zealand for permission to stage its Antarctic operations but was turned down. Nonetheless, the Soviets have a strong desire to fly into Antarctica, for this capability permits a country to reach its wintering-over stations sooner in the austral spring and to leave them later in the fall, and enhances its ability to move people in and out more efficiently. The alternative—supplying stations by ship—is hazardous: Indeed, two icebreakers were trapped by ice in the Weddell Sea in 1975; and, in 1981, a German ship was caught in the ice off Victoria Land and sank. With this in mind, the Soviets, in the first months of 1980, completed a snow-ice runway at Molodezhnaya and flew in to it from an airfield in Madagascar. The flight of the Ilyushin-18D originated in Moscow and touched down at Odessa, Cairo, Aden, and Mabutu before flying almost due south to Molodezhnaya. Further demonstration flights were made during the next season, and subsequently the Soviet representative to a SCAR meeting explained that this had been a major step toward having an air presence like that of the Americans in Antarctica.[67]

In recent years, the number of Soviet stations, like Soviet air power, has grown. In fact, during the IGY, it was feared that the burgeoning Soviet stations would lead that country to a sector claim in the territory previously claimed by Australia. This fear was one factor that prompted negotiation of the Antarctic Treaty. But after the IGY, the Soviet interest in Antarctica declined. The number of wintering-over personnel decreased, and several stations were abandoned, including Vostok, the principal inland one.

But, in the late 1960s, Soviet interest in Antarctica seems to have revived. In addition to Novolazarevskaya in the Norwegian sector

(Queen Maud Land), the Soviets reestablished Vostok and Mirny, built Bellingshausen station on King George Island in the South Shetlands, and built Leningradskaya on the border of the Australian and New Zealand sectors near Victoria Land. The Soviets also built Molodezhnaya in the Australian sector of Enderby Land and Druzhnaya on the Filchner Ice Shelf. Druzhnaya, Tass announced, would be the staging area for minerals exploration, presumably of the Dufek intrusion. After several unsuccessful attempts, the Soviets managed to establish a small year-round station, Russkaya, on the coast of Marie Byrd Land.[68]

Many have visited the bare Soviet stations, where holidays, birthdays, and visitors' arrivals and departures are celebrated with banquets, gifts, and many vodka toasts. The American who winters over with the Soviets each year usually has his own hut for privacy. Generally, it has been a positive experience. For example, Frank Sechrist, a meteorologist from the University of Wisconsin, was able to improve his Russian language skills by helping out in the kitchen.

Elsewhere in Antarctica, dinner conversation often turns to whether the Soviets would occupy the South Pole were the United States ever to evacuate—that is, whether the Soviets today still harbor Vladimir Beloussov's original plan to occupy the continent's most prestigious location. While friendly, the Soviets do not speak of their intentions in Antarctica and their behavior is often mysterious. For example, in February 1981, as rumors flew

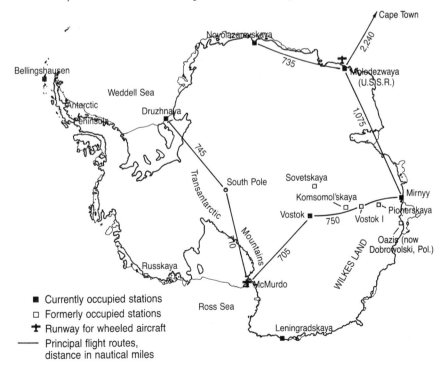

The Soviet's IGY effort was confined to a wedge in the Australian sector. Today, it has bases in every sector of the Antarctic coast.

round the stations that the United States, to save costs, would cut back drastically on its program, an Ilyushin-14 suddenly landed at the South Pole, just as the last U.S. C-130s were about to leave the Americans there for the winter. The Soviets stepped out of their plane and paid a brief visit, after which they returned to their base.[69]

There is no reason to suppose that the Soviet expansion in Antarctica will slow in the future. As we learned in chapter 3, the Soviet effort in Antarctica is largely an offshoot of its larger program for its own Arctic territory. There, the state attaches a high priority to keeping ice-free the northern sea routes from Murmansk to Vladivostok. Thus, the resources available to the Arctic and Antarctic Institute in Leningrad for deployment in Antarctica are enormous, especially in the austral summer, which corresponds to the Arctic winter. No other country active in Antarctica has as vast a fund of resources, manpower, and ice-worthy ships on which to draw. At the same time, much of the Soviet equipment used today in Antarctica is old and well may be replaced in coming years by more modern gear and aircraft.

Having downplayed the Soviet role in Antarctica for many years, other nations may notice a change in its presence as 1991 draws closer, especially if new Soviet stations appear along the coast, if Soviet aircraft and equipment are modernized, and flights to the continent are increased.

To conclude, U.S. logistics well-suited to the IGY twenty-five years ago and to the 1960s may be inadequate to deal with current and future developments in the region. Further, as we saw earlier, existing institutions have been incapable of serious review of U.S. logistical priorities. Most of the new activity in Antarctica—be it seismic exploration, increased Soviet influence, or the tentative ventures of new national players—is likely to occur where the U.S. "presence" is weakest, around Antarctica's coast and offshore. Several possibilities should be considered:

- At least one polar research vessel should be sent to the Antarctic for the austral season to conduct marine ecosystem research and to perform offshore geophysics studies. The NSF's present system of using a ship every other year for a few weeks is inadequate to the task. Not only do U.S. ship resources pale by comparison of those of other nations such as Japan, West Germany, and the Soviet Union, they cannot serve long-term U.S. scientific needs.

- The role of the U.S. station at the South Pole and its continental air presence needs to be reevaluated. The United States should consider sharing the costs of its South Pole station and the support it requires from McMurdo with other nations.

Indeed, U.S. allies, who also are Antarctic Treaty powers, utilize the South Pole station for their own research. If, for example, Great Britain, West Germany, and Australia bore some of these costs, U.S. resources could be freed up for new logistics efforts. Cost-sharing at the South Pole could become a precedent for sharing any nation's other Antarctic facilities and logistics.

- The possibility of important resource or environmental activities along Antarctica's coast suggests that the United States should find a way of locating temporary camps there as part of its ongoing research program. A program of baseline environmental or geological measurement could give the United States a reason to locate in different parts of Antarctica for discrete periods. If well-organized, such a program of research would entail a routine U.S. presence near areas of resource interest. Coastal stations of this kind could also complement and extend the research undertaken by a U.S. polar ship offshore. The present USARP offers many precedents for this kind of mobility—serving both science and geopolitics.

U.S. Interests in Antarctica

By the early 1980s, several factors on the international scene challenged the traditional post-IGY U.S. posture in Antarctica. First, resource and environmental impact issues presented the treaty powers with a new research agenda, in addition to science, which the United States was finding difficult to carry out. Second, Soviet expansion in Antarctica seemed likely to continue through 1991, giving the Soviets more ships, airplanes, and year-round stations than any other nation active in the region. Moreover, it seemed possible that by 1991 Soviet diplomatic influence might increase within the treaty group, since Poland had become a full consultative party in 1977 and East Germany seemed likely to seek consultative status in the near future. Romania, Bulgaria, and Czechoslovakia are all acceding parties as well. The Soviet presence posed a delicate challenge, as both common sense and treaty protocol precluded other treaty powers from audibly questioning Soviet motives. Third and last, several developing nations—including China, India, and Brazil—were becoming seriously interested in Antarctica for the first time. The accession to the treaty of China, Brazil, and other newcomers suggested that the balance of power in the treaty group would change. The developing nations' interest also suggested that the relationship between the treaty powers and the rest of the world could change as well.

These challenges suggest that the United States, as the historic leader of the group, needs a vision of its own role in Antarctica and what it wants the region to become.

The principal U.S. interest in Antarctica is to assure that the United States benefits from its enormous past investment in the region. Primarily, it needs to assure the region's military neutrality and to protect itself and its allies in Latin America and the South Pacific. After all, Sputnik and the threat of long-range Soviet missiles being placed in Antarctica was one reason that the rival Western nations concluded the treaty in 1959. Today the possibilities for conflict in Antarctica seem remote, but the 1982 Falkland Islands war served as a reminder that some nations, under certain circumstances, may be willing to fight there.[70]

Antarctica seems to have no more military usefulness now than it did twenty or thirty years ago. The continent's ice-clogged coasts and ports such as McMurdo cannot be used by an ordinary fleet, and its coastal waters are treacherous for submarines, as was shown by the rout of the submarine *Sennet* during Operation Highjump.[71] Although the Antarctic Peninsula at first glance seems more hospitable (the harbor at Deception Island had been used during World War II), there are very few good ports or runway sites. Indeed, the most likely military use for the region would be the commandeering of subantarctic islands as temporary ports in a large-scale conventional war. It is also possible that a country seeking to develop an atomic device secretly might test it over the vast ocean spaces somewhere between the southern continents and Antarctica. Indeed, it has been speculated that in 1979 South Africa tested a nuclear device north of the Antarctic Treaty area in the southwestern Indian Ocean.[72]

To assure the demilitarization and denuclearization of the region, the United States needs the Antarctic Treaty, or, at a minimum, a strong international agreement with the core provisions of the treaty. The treaty's importance to U.S. interests may be seen by glancing at what would happen should the treaty fail. The United States would revert to its position of the 1940s and 1950s, having both the basis for an enormous territorial claim and equally large diplomatic problems in asserting it. Any move to assert a claim would have to be weighed against the cost of defending it, and against the damage to diplomatic relations with the array of countries likely to oppose it. It is difficult to imagine circumstances under which a U.S. president would be willing to antagonize at a blow Argentina, Australia, Belgium, Chile, France, Great Britain, Japan, New Zealand, Norway, Poland, South Africa, the Soviet Union, and West Germany—to say nothing of China and India. In short, the U.S. interest in demilitarization and in keeping the territorial dispute "frozen" are of a piece. Both contribute to the overall U.S. goal of keeping the peace there and preventing it from becoming, in the treaty's fine phrase, the "scene or object of international discord."

Thus, a major U.S. concern is to ensure that the treaty, or at least its core provisions, will continue past 1991, indeed, indefinitely. Under Article XII of the treaty, only one consultative party must request the depository government—that is, the United States—to convene a conference to "review the operation" of the treaty. Only a majority vote is required to propose a change, but the change must be ratified by all the consultative parties in order for it to enter into force. Countries failing to ratify the change after two years may withdraw from the treaty. Their withdrawal takes effect after another two years, or by 1995 at the earliest. So, if a strong movement exists among the treaty powers for changing the treaty in 1991, or if there is discord in the group at that time, there could be substantial change in the status quo. Therefore, if the first U.S. interest is to assure the treaty's continuation, the United States should do everything possible to make it as broadly acceptable as possible between now and 1991.

Of great importance also is the preservation and enhancement of Antarctica's value as a laboratory for science. As we have seen, the United States has done much to further world interest in Antarctica as a scientific preserve and is preeminent in carrying out research there. It was also key in advocating that science be a centerpiece of the Antarctic Treaty. We also have seen that many new demands are being made of Antarctic science today, so the United States has an interest in seeing that these demands are met.

These primary interests have a direct bearing on the type of minerals regime the United States should be interested in negotiating. For the United States has a secondary interest in obtaining the benefits of its exploration, mapping, and study of Antarctica through nondiscriminatory access to all parts of the continent and the continental shelves and offshore waters. Indeed, the State Department lists "nondiscriminatory access for United States nationals and firms" as a condition of any minerals agreement.[73]

But the United States does not want to send its companies or its "nationals" to Antarctica to obtain minerals if the price of doing so is international discord, opposition, or conflict. Thus, the U.S. interest in minerals is tempered by the requirement that there be an orderly regime for minerals development that, in the course of negotiation and implementation, does not destroy the more fundamental arrangements that preserve the region's peace.

Thus, some action must be taken regarding Antarctic minerals in order to prevent the issue from disrupting the underlying political balance. Furthermore, this rules out exploration for minerals on a unilateral basis outside of the treaty. Were the United States to ignore the treaty system and work unilaterally, it would provoke the very "international discord" that it has promised to prevent. Gulf Oil's John Garrett, an industry representative on the State Department's advisory committee for the Antarctic minerals negotiations, insists that U.S. companies will not consider working in Antarctica without a secure legal arrangement under which they can operate.[74] Thus, if the United States seeks to permit or facilitate Antarctic minerals ventures by its own firms, it must assure order.

Another U.S. interest in the Antarctic potentially conflicts with its interest in assuring access to its minerals, wherever they are. This is its long-standing, often-affirmed promise to protect the Antarctic environment from damage caused by humans. Unlike other frontiers, such as the American West or Alaska, where development came first and environmental rules came later, Antarctic environmental rules are well established, although little basic surveying or exploration has taken place. Indeed, as chapter 4 explained, environmental rules make up the majority of Antarctic law as it has evolved from the treaty. They are one of the buttresses holding up the treaty system. Moreover, the treaty's mention of "conservation" as a fitting topic of consultations is the only explicit basis for the treaty powers to legislate Antarctic resources. Therefore, the United States cannot disregard its previous commitment to protect the Antarctic environment without undermining its main goal of upholding the Antarctic Treaty.

It is too early to guess how the tradeoff between the U.S. interest in assuring access to Antarctic minerals and its interest in environmental protection will come out. The environmental impacts of resource activities are little understood, and the location and type of economic minerals deposits are yet unknown. U.S. companies will not undertake minerals development unless it is economically feasible for them to do so. U.S. environmentalists are sufficiently alarmed about minerals development to want to thwart any scheme that does not preserve Antarctica's unique and fragile environment. They have not forgotten the controversy over the Alaskan pipeline and raise many of the same questions regarding Antarctica: How badly do we need these minerals? What substitutes are available? Where else could we get them? What are the alternative technologies for extraction and transportation? The only certain thing about this fight is that its outcome is unknown at present.

Moreover, as the 1977 Ohio State study of possible environmental impacts pointed out, decisions will be made at two levels regarding the value of environmental protection in Antarctica. At one level, a discussion will take place regarding local impacts on coastal areas or on a specific continental site. At another level will be Antarctica's aesthetic value as an intact, unspoiled wilderness. Does this value outweigh the value of any development, no matter how localized? Thus, the virgin character of Antarctica, the only continent untouched by humans in terms of economic development, may prove a key issue in future.

The United States also has a general interest in the Antarctic marine ecosystem and the fishery. When the prospect of huge Antarctic krill stocks was first raised, NOAA, in an attempt to determine the U.S. fishing industry's interest in krill, held discussions with industry. There was mild interest in using krill as salmon food, and manufacturers of shrimp-peeling devices retooled some machines for krill. Otherwise, U.S. industry showed little interest. On the other hand, fishermen often show no interest in alternative stocks until they are certain that the stocks they currently fish are collapsing. Thus, the NOAA officials who conducted the early meetings were keeping a weather eye on the krill fishery, in case U.S. shrimp fishermen—already in trouble—would need an alternative stock in future.

The United States has a specific stake in protection of Antarctic whale stocks. This derives from the public mandate which created the Marine Mammal Commission, an independent government agency reporting to Congress, to police U.S. activities to ensure that they do not further endanger whale populations. The commission has been concerned about the "krill problem" from the beginning, and has advocated a comprehensive, long-term research program. The U.S. interest in whales implies an interest in preventing overfishing for krill and maintaining the integrity of the Antarctic marine ecosystem. This would also maintain the long-term availability of the krill for eventual use by U.S. fishermen.

So the basic questions facing the United States are: What kind of diplomacy, logistics, and ongoing program will move Antarctica and the treaty through this critical period? Where, in Antarctica, should the United States have stations, ships, and aircraft to best ensure that the peace does not break down over some resource or political issue? What knowledge does the United States require to ensure that regulatory decisions within the krill convention, and the minerals–environment tradeoffs, are made smoothly and peaceably, so the treaty is seen as a useful system of administration? U.S. diplomacy has adjusted to these challenges, but it remains to be seen whether a comparable adjustment can be made in U.S. logistics and the U.S. program. The United States must begin soon to make adjustments by 1991.

The Sliding-Scale Strategy

To these basic interests others may be added, depending on one's political preferences or interpretation of the ambiguous U.S. legacy in Antarctica.

The United States could take a pro-conservation and antidevelopment stand on Antarctic minerals, if it concludes that Antarctic mineral deposits are marginal and substitutes easily found elsewhere. Once it acquired environmental expertise, the United States could educate other nations to the negative impacts of proposed development. Nothing in present U.S. policy—including the free access principle—precludes the United States from being the principal environmentalist within the treaty group. First, the United States has more stringent environmental impact rules than many other countries, which makes it uniquely qualified to develop similar rules for Antarctica. Second, the abundance of minerals and hydrocarbon resources in the United States, and of technologies for substitutes, makes the United States less likely than other nations to need Antarctic minerals. For example, the United States could plausibly counter a bid by Japan to exploit x barrels of oil from the Ross Sea shelf with an offer to help Japan to reduce its domestic demand for oil by conservation programs.

Or, the United States could become the major voice in whatever consortium of companies proposed developing Antarctic minerals under a regime. This would resemble the U.S. stance in the Law of the Sea Conference, in which the U.S. representatives on the seabeds committee echoed the views of U.S. mining companies seeking maximum freedom to mine the seabeds. Along with Norway and Great Britain, the United States has the most advanced offshore oil technology in the world. Thus, as in Law of the Sea, U.S. policy would have to emphasize the stake these companies have in their Antarctic investment.

Still another route for U.S. policy would be to interpret President Eisenhower's 1956 decision to downplay U.S. territorial claims and emphasize international cooperation in Antarctica as a first step toward true internationalization. This would make the Antarctic continent a *terra communis* and probably the "common heritage of mankind." Such a policy would mean turning away from the U.S. historic effort to maintain the validity of its territorial claim under the "constructive occupation" doctrine. If the United States were to simply abandon its claim and declare itself in favor of a *terra communis* (or a *terra nullius*) in Antarctica, it might face problems maintaining its "presence" if the treaty arrangement and the underlying power balance were to break down. Nonetheless, many kinds of internationalization exist, some of which could provide a new political balance.

Or, the United States could advocate broadening the treaty group by urging admission of more developing countries or changing the treaty's test for consultative status. Or it could advocate new agreements that broadened international participation. For example, it could urge the treaty group to establish a scientific convention, parallel to the krill convention, which developing nations could join if they wished to undertake scientific research in Antarctica. By easing membership requirements and by subsidizing research efforts the way it now contributes, say, to the World Bank, the United States would pave the way for a large number of developing countries to participate. Such an arrangement would allow greatly expanded scientific cooperation, and it would permit the United States to continue its historic emphasis on and interest in science in Antarctica.

The great, desolate Antarctic interior is, as it happens, one of the driest regions in the world—a snow-covered desert. Scientists from northern Africa or China might benefit from studying the meterology of the Antarctic interior, and dessication processes there. They might have some use for study of air-sea interfaces in the Southern Ocean and their impact on regional weather patterns. The U.S. government would have to regard part of its Antarctic program as a form of scientific foreign aid, but the diplomatic payoff could be handsome. The USARP budget should be expanded to allow these possibilities and many more. Innovative programs such as these could ease international tensions on resource and Third World issues in the years ahead. Moreover, an expanded USARP program to add resource and environmental studies seems, on its merits, overdue.

However, to implement these or any other changes during the forthcoming fluid time in Antarctic diplomacy will require authority and flexibility on the part of those in charge of U.S. Antarctic policy, two qualities that now are lacking. Part of the problem is cost, part is the twenty-five-year-old logistical plan, and another part is the institutional environment of the NSF. Another major inhibition is the prevalent assumption that present funding levels, logistical arrangements, and agency relationships cannot be changed, when obviously, if circumstances warrant, perhaps they could.

If present arrangements continue, look at what will happen if a major minerals cache were discovered, even accidentally, by another nation. If U.S. personnel were not included in the expedition that discovered the minerals, Washington might learn of the discovery through the press. The staff of the National Security Council would read through the guidance it has issued for Antarctica over the years and ask what is known about Antarctic minerals. It might be surprised to learn that no resource assessments as such exist because present arrangements preclude a thorough assessment.

Should the NSC look into how long it would take and what it would cost to carry out a resource assessment of the region in question, they would find it difficult to pull the needed resources, ships, and planes out of that season's scientific program for such a survey. Impatient at what is perceived to be yet another bureaucratic delay, the NSC staff or the president's national security adviser might try to strike a deal with the country making the discovery for its data. Our present inability to cope in a meaningful way with Antarctic resources under conditions of normalcy indicates how inadequate our response would be in such an emergency. Moreover, it would not be the first time that high officials, impatient with entrenched "Antarcticans" who fail to see the larger picture, made snap decisions.

Thus, the United States would be well served by adopting a "sliding scale" strategy organized around serious, but low-key, preparations for various contingencies in Antarctica. To continue its traditional role of broker, advisor, and peacemaker in the region, the United States needs to become more knowledgeable regarding Antarctic geology and to begin environmental baseline surveys. Finally, if it is not already too late, the United States should develop expertise in the applied aspects of the Antarctic marine ecosystem, in order to match and complement other nations' knowledge.

One way to do this is to broaden the existing USARP within the NSF. The addition of such studies would require many changes in the way the USARP does business, so that long-range more applied work could be planned and funded over periods of years. Broadening the present program would also require leadership by the USARP director, and agreement on USARP's new role among him, his naval counterpart, and his superiors at NSF. It would also require USARP officials and scientists to work more closely with SCAR and other international bodies than in the past—such as the new krill convention's scientific commission. The USARP has been unable to broaden its role in the past: whether it can do so in the future remains to be seen.

Alternatively, the United States could signal its interest in a broader Antarctic program by establishing a new, possibly more flexible vehicle such as a polar research institute, having an annual budget roughly double the present USARP budget. It would be made up of the existing NSF program and use experts from the USGS, NOAA, other agencies, and the universities. The proposed institute, besides providing Antarctic research and policy the coherence it now lacks, also would unify the country's diffuse programs of research in the Arctic, a topic outside the scope of this book.

The concept of a polar research institute stems from the fact that Antarctica places a peculiar set of interlocking demands on the U.S. government, demands which no existing federal agency—including the NSF—can meet. Each move in the Antarctic has some foreign policy significance and significant funding is required for such arcane purposes as an ice-strengthened research ship, or a temporary coastal base in an area of potential resource interest. And, as we have seen, Antarctic diplomacy differs from that exhibited elsewhere in the world and is loaded with subtleties that only diplomats with experience in the area can exploit to national advantage. It is no accident, for example, that the consistency and skill the United States showed in the resource diplomacy of the late 1970s and early 1980s has been based largely on the efforts of one individual, R. Tucker Scully, who under State Department rules would have normally been rotated out of the job after a few years, and his staff. Allowed to stay on the job, he has increased his expertise in the fashion of the British Antarctic diplomats Brian Roberts and John Heap, whose secure tenure and expertise helped make the treaty work in the first place.

In conclusion, a polar research institute would bring consistency, advance planning, centralized control and, above all, flexibility to U.S. Antarctic affairs. It could include:

• A basic research division, comprised of the present USARP program, which would be revamped to emphasize work on the most interesting problems, rather than slavishly following the present disciplinary structure. By consolidating effort on some central questions, money could be freed for needed hardware development. This is one message of the Polar Research Board's report on long-range plans for Antarctic research.[75]

• An applied research division that would carry out resource-related studies, such as geophysical surveys of the continental shelf, airborne magnetic and gravity surveys, and stock assessments of the marine ecosystem. The institute would utilize staff scientists, drawing together Antarctic experts who are now widely scattered without an organizational base. Because the division would have its own budget and logistical arm, policymakers in Washington would be able to "tune" it to reflect U.S. policy. This division, like the basic research division, also would contract research out to universities and private industry.

- An environment division to take responsibility for developing an environmental data base as the basis for policymaking that is separate from applied research on resource issues. In addition, if minerals development were to occur and require the regulation of U.S. nationals or firms, this division could assume a regulatory role.

- An international division to enhance the U.S. diplomatic role. A striking weakness in present U.S. arrangements for Antarctica is the fact that those who are most knowledgeable about the research activities of the other treaty powers are the university scientists who sit on the Polar Research Board forming the U.S. link to SCAR. Neither USARP officials, most U.S. grant recipients, nor staffers from other federal agencies have an easy way of keeping in touch with resource-related developments known to SCAR. In addition, SCAR and the treaty system face a work overload that is unlikely to be relieved anytime soon. The proposed international division could help fill this vacuum by allowing its staff to help shoulder the work of SCAR. This division also would have the responsibility of informing the rest of the institute about other nations' programs.

- An operations division to consist of NSF's present-day logistics section, plus as much of the naval effort as could be transferred from the Department of Defense, as well as the Antarctic work of the coast guard and the private contractor. The navy has never liked having the Antarctic mission, and the current tripartite arrangement is cumbersome. At first, this division would coordinate various efforts, much as the USARP logistics section does now. Eventually the division could employ its own mechanics, technicians, engineers, ship captains, and station managers. The experience of the British Antarctic Survey shows that costs go down if logistics are centralized within one organization. Unity of command will also lead to a more flexible, efficient U.S. Antarctic posture. It also might save money.

- A public relations division to organize materials relating to the U.S. presence in Antarctica. At present, the USARP has responsibility for collecting and storing information on U.S. activities; older material relating to Antarctica is stored in the Polar Regions Section of the National Archives. Since the work of the TANT committee in the 1950s, there has been little organized effort to collect and store information or obtain private papers, such as those of Byrd, which remain unavailable. Whereas the U.S. space program's achievements are commemorated in the Smithsonian Institution's National

Air and Space Museum, little publicity has been given to present or past U.S. Antarctic activities. A public relations division would have a broader charter than that of the USARP information office; it would collect material, commission films, develop museum exhibits, and sponsor high-quality contemporary and historical writing.

• A director who would assure the independence of the proposed institute by reporting directly to Congress on various missions, as the Marine Mammal Commission does, to assure that it does not become lost among more powerful, federal agencies. He could even report to the president's science adviser as a way of emphasizing the primary U.S. interest in science and the role of science in the treaty system. The key is to give the institute and its director authority and independence. The director's role would resemble that of the director of the British Antarctic Survey; he would be in charge of all the logistics, the staff, and outside grants and contracts. He would serve as the principal U.S. representative to SCAR, thus ending the isolation of the USARP from SCAR's concerns. But he would also be free, in a way that the director of the USARP is not, to shape and direct Antarctic research activities through staff assignments and planning and awarding of long-term contracts.

• A budget that would support an evolving polar research program. The polar research institute could start small and be expanded as circumstances require. Thus, the Antarctic portion of its budget would roughly double that of the USARP budget, to allow for the new studies and to pay for a ship. Another key role for the institute would be to take over the Division of Polar Program's Arctic activities as well as Arctic research programs currently in other agencies. It is noteworthy that Congress finally passed legislation creating an Arctic Research Commission to oversee and guide the myriad efforts of federal agencies working in the Arctic. The problems of poor coordination in the Arctic have been noted by many, including the Polar Research Board.[76]

The proposed polar research institute could remain at this level so long as the Antarctic situation did not change markedly. It would have a truly "balanced" program—that is, one balanced among resource, environmental, diplomatic, and scientific needs. It could make a fresh start on all these issues without the interagency friction that appears today.

Moreover, it would underpin the U.S. diplomatic needs, equally heighten U.S. interest in Antarctica as 1991 approaches, and help ease the strains on the treaty group and SCAR—in a way that the

USARP cannot. Finally, the removal of the tripartite carrying out of U.S. activities might create a more coherent U.S. "presence" in Antarctica itself. Central management of Antarctic operations could also remedy some of the human problems in Antarctica—such as the fact that the scientists, the navy, and the private contractor have chosen their personnel separately, preventing naval doctors from choosing the group which will be locked up together for four months in isolation for the winter.[77]

The advantage of such an institute is that it could move in whatever direction the international situation requires with minimum friction. Today, for example, a major USGS survey of likely Antarctic minerals sites means that the director of USGS must be convinced that a site in Antarctica is promising enough to transfer money from exploration in the United States. The institute, on the other hand, would be making small-scale resource assessments as a matter of course, which could be scaled up or redirected as necessary. Should U.S. industry become interested, the institute would be a more appropriate point of contact than the NSF. On the other hand, if the minerals issue becomes moot, this aspect of the institute's activities could remain small. Finally, if the United States were to take a strong conservationist role, the environment division would expand.

Alternatives

The main argument against the proposed institute is that there is nothing urgent enough in Antarctic diplomacy or resource activities to justify such a shift. After all, the last major shift in institutional roles took place after the IGY and the negotiation of the treaty, both of which were seminal developments. The counter-argument is that many disparate forces are impacting on the treaty now—the developing nations, possible secretiveness and rivalry over minerals, and the role of the Soviet Union. A real contingency then, is that after 1991 the treaty could be dealt a mortal blow or even break down, or that a period of highly ambiguous and uncertain diplomacy could modify or replace it. Should the United States wait until after an emergency occurs to consider changing its Antarctic programs? Can the government only respond to crisis, and not incremental change?

The principal alternative to an independent institute is inter-agency coordination, something being tried by NOAA, the NSF, and the Marine Mammal Commission. Is it practical to let each federal agency work separately in Antarctica with the Antarctic Policy Group as the coordinating body, as at present? The Antarctic Policy Group has shown itself incapable of moving any of its member agencies to action; indeed, it even failed to meet during some of the most critical events of the early 1970s; it could not alter NSF's priorities when the NSF rejected broadening its research programs in the late 1970s. Recently, NOAA Associate Administrator James W. Winchester pointed out in testimony: "The criteria by which these priorities [for Antarctic research] are established reflect the policy guidance of the President, but are internal to NSF and do not involve other agencies."[78] As to the 1982 White House guidance that other agencies may undertake Antarctic short-term research if it is "recommended" by the Antarctic Policy Group, Winchester said: "The APG has no fixed set of criteria to evaluate research proposed by Federal agencies other than the pursuit of U.S. interests in Antarctica generally."[79]

The strongest argument against relegating U.S. Antarctic efforts to an interagency coordinating committee or the present NSF alone is the fate of Arctic research, where both approaches have been tried. Arctic research entails much stronger industry interest, established programs in the navy, the Minerals Management Service in the Interior Department, the Department of Energy, and NOAA, as well as the NSF. At one time, the Office of Polar Programs was in charge, but it had little influence over what other agencies did. Since 1971, coordination has been in the hands of an Interagency Arctic Policy Group (IAPG), chaired by the assistant secretary of state for oceans and international environmental and scientific affairs. This group produced one classified report after its formation and lay dormant for years until producing a second report in 1982.[80] It has yet to influence the behavior of member agencies, whose interests differ in the Arctic. As mentioned, Congress has concluded that the only way to impose organization on U.S. Arctic research is through a new, independent commission. This suggests that it is not necessary to repeat another experiment in interagency coordination for the opposite pole.

The Future of Science in Antarctica

The contrast between the interagency friction in Washington and the joys of a successful Antarctic scientific expedition are striking. For twenty-five years now, the U.S. scientific community has enjoyed the fruits of the region's peace and stability. They have a stake, therefore, in the future.

Katherine Bouton, during her 1979 visit to Antarctica, went on a successful field trip to the remote Ellsworth Mountains at the head of the Weddell Sea, a project that Gerald Webers, a paleontologist from Macalester College in Saint Paul, Minnesota, and

John F. Splettstoesser of the Minnesota Geological Survey at the University of Minnesota had planned for more than a decade.

The construction crew flew ahead to prepare the temporary base. When the team of twenty-six scientists and students flew there in early December for their month of work, they found a small town waiting for them; five Jamesway huts (including the mess hall with familiar golden arches, called "Bebe's Golden Arches Restaurant"), several tents, helicopters, ten snowmobiles, a Caterpillar tractor, and regular, planned mail service from McMurdo. The "town's" population would be forty-nine men and three women. Scientists, pilots, and the camp's crew worked together in good spirits despite delays and snafus.

To study the geology of the mountains, the scientists flew in helicopters to the most promising exposed rock faces. Bouton recounts this journey:

> We followed the path of a glacier through rounded peaks and low ridges, many of them bare, with multiple parallel layers of metamorphic rock running through them in graceful curves and inclines.
>
> Altogether, the stratigraphic sequence here was a remarkable thirteen thousand metres in depth. We had seen in slides—and could now clearly see out the window of the helicopter—that the layers were as distinct, and as telling, as the rings of a tree.[81]

While the paleontologists looked for fossil trilobites and archaeocyathids, George Denton (a University of Maine scientist who has described the possible melting of the West Antarctic ice sheet in another two hundred years) hunted for clues to its past extent while exploring by helicopter.

> The pilot set down gently and bounced the helicopter to test the surface, and the crewman, wearing the safety harness, leaped out to check for crevasses. He immediately sank through the snow and up to the hip in a crevasse, so he got back in and we took off. The second place we stopped was a snow-covered ridge between two nunataks, and this time the surface was solid.[82]

On the eve of her departure from this strange, happy place, Bouton went skiing:

> [T]he sun was low on our right, almost level with the tops of the mountains. Far away on the left, forty-five miles across the glacier, the Vinson Massif, silhouetted against blue sky, was topped by a spiral plume of cloud. Behind the nunatak was a steep downhill slope. Ojakangas, the better skier, led the way. It was easy to imagine that nobody had ever been here before. For an hour or so, we were out of sight of the camp, and in the shadow cast by the nunatak it was cold. At

last, after a long climb, we came up and over a ridge into the sunshine. There below was the camp. How quickly one becomes accustomed to even the most exotic and hostile places. This camp—this fragile, temporary assemblage—seemed like the most secure place in the world.[83]

U.S. scientists have enjoyed using Antarctica as a international laboratory for twenty-five years. In a sense as modern "Antarcticans" they have the largest stake in seeing that their work can continue past 1991 and that the new challenges do not cause the peace and stability of the region to break down. Doubtless the traditional forms of science will continue to be important for the years to come in Antarctica. But there may be other things U.S. scientists can do—more innvotative programs, a broader effort, stronger international ties—to maintain the region's stability in the unpredictable times ahead.

eight

The Future of the Antarctic Treaty

The years 1982 and 1983 were as important as any since the Antarctic Treaty was negotiated in 1958–59. In 1982 the Antarctic Treaty powers, who then numbered fourteen, began their long-delayed negotiations regarding the appropriation of development rights to Antarctica's presumed riches of offshore oil and onshore minerals. This topic was discussed at three special meetings—in Wellington, Bonn, and Washington, with the usual lack of fanfare—except at one meeting where a group of environmentalists, dressed in penguin costumes, waved signs to protest Antarctica's imminent ruin. By the end of 1983, the likelihood that the powers would reach an agreement on this momentous, divisive question was in doubt, and subsequent meetings were planned. Meanwhile, the talks spurred renewed interest among developing nations in the treaty, and in the great white continent and its potential resources.

The attack on the treaty by nonsignatory nations finally came, having built up momentum from the 1973 oil embargo and resulting interest by the developing world in controlling minerals commodities. The treaty powers were thrown "for a loop," said one observer,[1] when, in September 1982, Malaysian Prime Minister Datuk Seri Matahir bin Mohamad rose at the Thirty-Seventh UN General Assembly to make a speech on Antarctica. Speaking for developing nations like his own former British colony, Matahir bin Mohamad argued that the "uninhabited lands" of Antarctica did not belong to the colonial powers claiming them. Now that the Law of the Sea Conference had been successfully concluded, it was time to negotiate "a new international agreement" for Antarctica—a prospect that was anathema to the treaty powers.[2]

He followed up this salvo with a memorandum to the meeting of nonaligned countries, held in New Delhi in March 1983, arguing, "Like the seas and the sea-beds, these uninhabited lands [Antarctica] belong to the international community. The countries presently claiming them must give them up so that either the United Nations administer[s] these lands or the present occupants act as trustees for the nations of the world."[3] Consequently, the delegates recommended that the next UN General Assembly take up the question of Antarctica.

Thus the Thirty-Eighth UN General Assembly, meeting in September 1983, had Antarctica on its agenda. The question subsequently was referred for study to the First Committee, which deals with disarmament matters. After extensive discussion, the committee proposed a resolution calling on the UN Secretary-General to make a study of the matter—which was unanimously approved by the General Assembly on December 15. Matahir bin Mohamad, in placing Antarctica on the UN agenda and in offering nonsignatories to the treaty a forum, had taken the issue well beyond the point where Indian Prime Minister Jawahalal Nehru had left off in 1958, when he initially proposed, then dropped, his suggestion that the UN take up the Antarctic question.[4]

Malaysia, attempting to gather together a bloc of developing countries to speak for the "outsiders" to the Antarctic Treaty club, was unable to convince its fellow members of the Association of South East Asian Nations (ASEAN)—including Indonesia, the Philippines, Singapore, and Thailand—to support its Antarctic initiative. It picked up some support from sub-Saharan African nations who objected to South Africa's presence in the treaty club and saw a UN discussion as an opportunity to protest against what one Peruvian diplomat labeled the treaty's "apartheid" in discriminating against developing nations.[5] Malaysia did find a co-sponsor, the Caribbean island nation of Antigua and Barbuda, for its move in the General Assembly. But Antigua and Barbuda, unable to corral other Caribbean neighbors to favor the move, found that few developing nations regarded the Antarctic question as the next battleground for fighting the industrial world.

Hence the frontal assault on the treaty by a bloc of developing nations never materialized. If Mahatir bin Mohamad's UN speech had been a shock, the treaty powers were not to be surprised a second time. When it became clear to them that the item would be on the General Assembly's formal agenda, the powers launched an "educational" effort among nontreaty nations, such as Malaysia, defending the "system." Working together, the treaty powers passed the word that each would oppose any UN initiative that threatened the Antarctic Treaty. Finally, public interest groups that followed Antarctic issues—notably the International Institute for Environment and Development—canvassed developing nation representatives extensively. As a result, some of the treaty's would-be attackers, during the First Committee's deliberations, backed off from insisting the treaty be scrapped.[6] And the committee's final resolution on Antarctica is so neutrally worded that not even the treaty powers can disagree with it. It urged that the "Secretary-General prepare a comprehensive, factual and objective study on all aspects of Antarctica, taking fully into account the Antarctic Treaty system and other relevant factors."[7]

Opposition to the treaty at the United Nations may have been dampened by events at the twelfth consultative meeting in Canberra, Australia, in September. As the meeting opened, the fourteen consultative parties admitted two more—India and Brazil. These nations were newcomers in more than a formal sense: both are large, politically influential developing nations that did not share the original treaty nations' historic interests in Antarctica, although they had sent scientific expeditions there. Evidently, the recent expedition of Brazil and India's two to the Antarctic indicated enough scientific interest in the region to meet the requirements of Article IX, the treaty's test for full membership.

The treaty powers made another move to persuade the world they were not so "exclusive," after all. They invited the twelve acceding parties to the Antarctic Treaty to the twelfth consultative meeting as observers. (All but Czechoslovakia attended.) Previously this group had played no role, having been assigned none by the treaty.

Both gestures showed that the treaty powers were sensitive to the charge of being a closed club, but the door was not flung wide, either. They declined to invite the acceding parties to observe their most sensitive talks, the special consultative meetings on minerals; and they also disregarded the demands of some public interest advocates to release the details of their closed-door meetings. After each consultative meeting, only the list of agreed recommendations and a terse report were issued.[8] The treaty powers preferred a low profile, perhaps figuring that Malaysia already was giving them more publicity than they wanted.

By the end of 1983, the developing world seemed divided over Antarctica; some countries preferred to attack the treaty group while others chose to join it. It is hard to say what specific nations were motivated by ideological considerations or by perceptions of the best means to obtain a share of Antarctica's presumed mineral riches. And the institutional debate over the relationship between the Antarctic Treaty "system" and the United Nations—overdue since 1959—had just begun.

This chapter takes up the question of nonsignatories to the treaty, who in 1982–83 were more actively interested in Antarctica than at any time before. It discusses the needs of these nations—mostly developing nations who cannot meet the treaty's qualifications for consultative status—against the present treaty system.

How important is the treaty's lack of equity in not giving all nations an equal vote in Antarctica's governance—as would happen if the United Nations accorded "common heritage" status to Antarctica? How important will Antarctic krill and mineral resources be in the next ten, twenty, or fifty years? What role, if any, is there for Antarctic scientific research in the rapidly evolving relationships between developing nations and the original treaty group? If the treaty remains in force, even after the 1991 deadline when it may be reviewed by its parties—which, as of this writing, looks likely—what role, if any, will nonsignatory nations play? How should the treaty's promise that all "mankind" should benefit from it be fulfilled? What specific "benefits" can the treaty powers offer nonsignatory nations and acceding parties? What should be the *quid pro quo*?

This chapter summarizes the Antarctic Treaty "system" as an unfinished, but politically useful institution facing twin crises concerning resources and nonsignatory nations. It concludes that emerging from behind the closed doors of the treaty powers' meetings and in the corridors of the United Nations is some new kind of North–South relationship, very different from the confrontational relationship of the 1970s. The details of this new relationship are undefined and unknown, like a new, unexplored land beyond some distant horizon. And willy-nilly, the treaty powers have begun their journey, making the treaty the linchpin for further agreements that widen its original scope and participation. There are special arrangements for the environment (see chapter 4), and there are the conventions for sealing and for exploitation of marine living resources. In the future there may be another agreement governing minerals and, perhaps one day, an

Debators in international forums on Antartica and its resources have difficulty putting a value on Antarctica's scenic beauty, such as the sculpture of drifting ice.

agreement that allows poorer nations to participate in Antarctic science.

The chapter does not extol the treaty system, which turns out to be neither as systematic nor as legitimate as its champions contend. Nonetheless, it is probably the only politically viable government for Antarctica in the future. And, despite the lumbering political efforts made to date by treaty powers and nonsignatories such as Malaysia, the system offers intriguing possibilities for new international institutions. Whether they like it or not, both the treaty powers and the outsiders are on a new journey as tantalizing and perilous as that at the end of any Antarctic explorer's flight over uncharted glacier, plain, or snow-covered peak. It could bring disaster, but possibly rewards as well.

The Impact of the Law of the Sea Conference

When Malaysia's Matahir bin Mohamad spoke to the General Assembly, he was carrying out a long-standing promise of some developing-world spokesmen that they would discuss Antarctica once the Third United Nations Conference on the Law of the Sea, begun in 1974, had concluded. As chapter 4 explained, from time to time during the Law of the Sea negotiations, representatives of the Group of 77 (developing countries seeking control over the deep seabed minerals) suggested that Antarctica should be included within the conference's jurisdiction. Antarctica then could be considered the "seabed beyond national jurisdiction," which a 1970 General Assembly resolution had declared the "common heritage of mankind."[9] But the treaty powers—whose number happened to include Chile and Argentina, two full-fledged members of the Group of 77—forestalled these moves and inspired the pledge by some leaders to take up Antarctica later.[10]

In 1975 Hamilton Shirley Amerasinghe, Sri Lanka's ambassador to the United Nations for many years and the president of the Law of the Sea Conference, called Antarctica "this last, dark continent," a place where "there are vast possibilities for a new initiative that would rebound to the benefit of all mankind. Antarctica is an area where the now widely accepted ideas and concepts relating to international economic cooperation with their special stress on the principle of equitable sharing of the world's resources, can find ample scope for application...."[11]

But the prospects of a 1970s-style confrontation over Antarctica diminished in 1981, when the government of Indira Gandhi decided to send an expedition to Antarctica, hinting that India also might join the treaty. Although at the time it was not even an acceding party, India leased a Norwegian icebreaker, the *Polar*

Circle, and assembled a team of scientists led by the Department of Ocean Development. They trained in the Himalayas, India's high-altitude, bad-weather analog to Antarctica. The *Polar Circle,* with a predominantly Indian crew, sailed from Goa on December 6, 1981. Stopping at the island of Mauritius, it continued southward, reaching the Antarctic coast on January 9, 1982. There the team constructed a weather station, carried out radiation and magnetic surveys, and planted the Indian flag. It called the base Dakshin Gangotri. The following season a second Indian expedition returned to Antarctica, where it "collected enough data to enable the setting up of a permanent manned station," noted one Indian publication.[12]

If the Gandhi government had clear motives, they were not apparent to the outside world. India's long-standing rivalry with China may have been one; indeed, India had picked up rumors in 1977 that China was considering the sending of a scientific team to Antarctica, "but it did not materialize," gloated one official account. Indian national pride was apparent in the statement that "the Indian success at the very first attempt is therefore recognized the world over as a remarkable achievement."[13]

India's 1981 Antarctic expedition was the first sent by any nation entirely outside of the treaty framework. And as leader of the nonaligned movement, India, which may have planned to lead the attack on the treaty on behalf of the developing world, may have wanted to show it could defy the former imperial powers on their own Antarctic turf. Another possible motive was to show the scientifically advanced nations that India, too, was capable of advanced, expensive, scientific research.

The Gandhi government was deliberately ambiguous toward both the treaty group and in the developing world. It scored points with the treaty powers by stating that the expedition was "purely scientific," that India had no "territorial ambitions" in Antarctica, and that India subscribed "to the principles of the treaty."[14] But before the expedition left, the Indian government notified other developing nations of its plan—in effect, testing the waters of developing world opinion.

A second event which diminished the chances that "mankind" would rise up in opposition to the treaty came when the People's Republic of China joined the Antarctic Treaty. This occurred without fanfare in May 1983, when China's ambassador to the United States presented notice of China's accession at the U.S. Department of State, the office of the depository government. China had, in fact, sent scientists to several Antarctic stations in the 1981–82 season. News accounts had said that China was considering joint research with Chile on fishing, oceanography, and geology in the Antarctic. And the Chinese Bureau of Oceanography had invited two prominent New Zealand scientists, George A. Knox and Robert B. Thompson, to lecture. In addition, Chinese scientists had attended several SCAR meetings, and one Peking news account indicated that China would like to set up a research station in Antarctica. China's accession to the treaty brought the number of acceding powers to thirteen. With its 1 billion people, it brought one-fourth of the world's population into the treaty club. Once India—with its 700 million people— joined as well, the treaty group could assert that most of "mankind" was inside, not outside, its tent.[15]

Brazil, which had once considered a claim of its own in Antarctica, showed new interest as well. Brazil had acceded to the treaty in 1975, and in 1982 it bought an older Danish research vessel, the *Thala Dan*, and sent it to the Antarctic for the 1982–83 season, with thirty-eight aboard. The expedition earned Brazil full consultative status the following September.[16]

The Impact of the New International Economic Order

It is by no means assured that a peaceful compromise between nontreaty nations and the Antarctic Treaty powers will take place. The merits and many demerits of a Law of the Sea-style regime for Antarctica were debated, for years, in many quarters. Indeed, the memo that Malaysia submitted to the nonaligned nations in March 1983 said that "the international community must therefore take the initiative now to ensure that the continent would become *the common heritage* of all the nations in this planet... instead of being the object of international discord" [italics added].[17]

The Law of the Sea Convention interprets common heritage to mean *terra communis,* meaning that no one could exploit the deep seabed without obtaining the permission of its communal owners, "all the nations in this planet." The United States takes the opposite view, which is that the deep seabeds are a *terra nullius* that anyone can exploit unilaterally. The United States likewise denies that Antarctica is a *terra communis.* It insists on the right of free access for anyone wishing to go there, including developers, so long as they do not violate the treaty. The other treaty powers appear to agree. (The United States also declines to term Antarctica a "global commons," for this, like *terra communis,* would undercut the basis of U.S. territorial rights there. When a recent government executive order accidentally referred to Antarctica as a global commons, the State Department quickly corrected the record, saying the order was not "a statement of U.S. Antarctic policy."[18])

The common heritage concept became refined during the debate on the New International Economic Order, proclaimed in a 1974 resolution of the UN General Assembly.[19] This doctrine argues that the free market works systematically against developing nations, whose main income comes from exporting minerals and other raw materials to the rich, industrial nations. But industrial nations vary in their purchases, while the developing nations have fixed needs for export revenue to buy finished goods. Thus, the free market in international trade is but a cover for exploitation of the many, poor nations by the few, rich ones. Argued in the early 1970s, against the background of OPEC's success with its oil cartel and the wild swings in world commodities prices, the platform of the New International Economic Order gained many developing-world adherents. It could arguably apply to Antarctica's minerals development too, on the assumption that only the rich developed nations, who are parties to the treaty, will be able to exploit them. From this perspective, a treaty-sponsored Antarctic minerals regime seemed unfair.

Such ideas inspired the International Sea-Bed Authority, created by the Law of the Sea Convention. In this new international organization, a General Assembly-type body, in which all nations have an equal say in whether the seabeds are mined, dominates key decisions. Although there are counterweights, in which mining states are supposed to have their say, there is still much controversy over whether this body will allow mining to take place. Indeed, the Reagan administration objected so strongly to the final terms for seabed mining in the convention that it declined to sign it, thus throwing down a gauntlet to those nations which did. By the end of 1983, there were nine ratifications of the document; sixty are required for its entry into force. The Reagan administration also challenged other mining states, such as France and Japan, to come over to its side.

The outline of this controversial new organization, including a communally run "Enterprise" to mine on behalf of the developing world with the aid of technology sharing with the mining nations,[20] illustrates what developing-world spokesmen—many of them veterans of the Law of the Sea negotiations—have in mind when they speak of applying common heritage principles to Antarctica.

Any New International Economic Order-type regime for Antarctica would aim at giving developing nations control over decisions on Antarctic minerals. It would be diametrically opposed to the treaty "system," whereby each Antarctic Treaty consultative party retains an absolute veto on key decisions by the mechanism of consensus voting. Both the new krill convention and the minerals regime proposed in the "Beeby draft" (see chapter 6) retain this veto power by consensus voting on key decisions. It is difficult to see how the treaty powers would ever agree to such diminution of their control, especially now that they have successfully extended it through the krill convention. So such a regime for Antarctica (sketched hypothetically here) only would come into being in defiance of the Antarctic Treaty or if the treaty broke down. In the International Sea-Bed Authority, policy is made by a supreme organ of all states that ratify the convention, and could include all 158 member states of the United Nations. This huge body, which is to meet once a year in Jamaica, will decide all matters of substance by a two-thirds vote. The assembly also elects an executive council to run the authority, consisting of thirty-six states and comprised of four mining states, four importers of seabed minerals, four exporters of seabed minerals, six developing countries, and eighteen more chosen for geographic distribution. The executive council will be advised by two technical commissions, which would do most of the real work of the authority. One would implement decisions and deal with market questions relating to the minerals to be exploited. The second would, among other things, review applications and assess environmental impacts.[21]

Were an organization like this to be substituted for the treaty or take on Antarctic minerals development, it might resemble our hypothetical sketch. Recent U.S. law literature contains several such proposals purporting to make Antarctic governance more equitable. A typical plan is that proposed by Edward E. Honnold in 1978 and his subsequent draft regime for Antarctic governance.[22] Under Honnold's scheme, Antarctica would be governed by a bicameral legislature of all nations, with an upper house balancing the demands of industrial and developing worlds. Since the treaty shows the "frailty" of multilateral arrangements with no executive, Honnold also advocates a strong executive organization staffed by international civil servants. The legislature would set guidelines balancing resource use against environmental protection, while consortia or investors from developing nations receive technical assistance. Revenues from development would be shared according to a "formula giving special benefit to developing countries," and the regime would be authorized to employ economic sanctions and peacekeeping forces. Amazingly, Honnold concludes, "The task of administering the Antarctic regime will not be impractical," adding that it would be less complex than the United Nations.

Like other authors of this school, Honnold expects that having a UN-style body in charge will improve the chances of Antarctica's remaining environmentally pure. For example, whether to make Antarctica a world park is high on the agenda of his bicameral legislature. This and other, similar, proposals seem to be motivated by a desire to prevent greedy treaty powers from grabbing the huge minerals wealth of Antarctica and despoiling its pristine environment. They reflect what transpired in the Law of the Sea meetings far more than the actual situation in Antarctica.[23]

Such proposals express a basic feeling about Antarctica, namely, that it is not fair for a small group of powers meeting in secret, with a history of rivalry over turf, to have exclusive control over the earth's seventh continent. However laudible or widely shared this sentiment, the mechanisms so far proposed in its name have serious practical and political problems. Most such proposals presume that man would remain in Antarctica, but offer neither the present treaty powers nor other nations any incentive for spending the huge sums required to maintain year-round stations, and to run weather, communications, and rescue services. In seeking to wipe out the basis for the treaty powers' exclusive control—namely, the dispute over territorial claims—such proposals also wipe out the treaty powers' incentive to mount Antarctic programs.

Therefore, a practical problem remains: If Antarctica were declared *terra communis* and turned over to the United Nations, thereby erasing any hope of political or economic advantage to the present treaty powers, who would go there? As was evident during the International Geophysical Year, while there is some reason to do science in Antarctica, politics—and not science—is the real reason behind the treaty powers' willingness to foot the bill for an Antarctic presence.

A variant of a "pure" New International Economic Order approach would be to grant the treaty powers some special benefits in return for their footing the bills. While denied ultimate control, they could be induced to remain in Antarctica to carry out the work of the international authority if they could be assured of some benefits. It is doubtful that the treaty powers would go along with such loss of control, however, and a primary goal of their present deliberations, and of their newfound "openness" toward developing countries and the United Nations, is to avoid such negotiation in the first place.

A final weakness of these new-style regimes for Antarctica is that most proposals breezily assume that the region can be policed by a UN force. Practically speaking, it is not clear that a UN-policed Antarctica would be as free from conflict as it has been under the demilitarization, unilateral inspection, and denuclearization provisions of the present treaty. When Argentina invaded South Georgia and then the Falkland Islands in 1982, it stopped short of the 60° south latitude boundary that divides the demilitarized Antarctic Treaty area from the rest of the globe. But would Argentina have stopped if only a UN police force had guarded the boundary? Which is a more powerful incentive to continued disarmament—a UN force or the treaty?

In sum, while trying to address issues like equitable resource development, New International Economic Order proposals for Antarctica tend to reopen old issues that have been solved by the present treaty, such as political stability, a continuing human presence, and demilitarization.[24]

At present it seems unlikely that such a new regime will be concluded for Antarctica—although this possibility cannot be ruled out. For one thing, many developing nations seem disillusioned with the New International Economic Order and its lack of practical success in other forums. Even Malaysia's prime minister, in his speech arguing that Antarctica should be a "common heritage," also warned the developing nations not to delude themselves that economic concessions would be forthcoming from the industrial world; instead, he advised they should use their own "resilience" in solving economic problems. In addition, the staggering international debts many of these countries now carry have changed the arguments of the mid-1970s. Most developing-world governments know full well that their fortunes depend more on how they cope with their debt problems, than on a politically inspired transformation of the world economy. Indeed, the lukewarm response of many governments to Malaysia's initiative on Antarctica may reflect their realistic view that having a $1/158$ share in an Antarctic gold mine hardly will be the answer to their economic woes.

Finally, there are key differences between the seabeds and the Antarctica issues. The treaty powers have never agreed that the General Assembly or the International Sea-Bed Authority has jurisdiction over Antarctica or its resources. Yet, with the seabeds issue, the United States and other would-be mining nations voted for the 1970 "common heritage" resolution that clearly put these resources under the jurisdiction of the Law of the Sea. The General Assembly can discuss anything it wishes, of course, but this does not give it jurisdiction over them—especially when the five permanent members of the Security Council are all Antarctic Treaty parties and opposed to such jurisdiction.

Indeed, the treaty powers specifically add that the UN Charter permits regional peacekeeping arrangements to solve its problems internally, *without* resorting to the United Nations. Arguably, the treaty is a regional peacekeeping arrangement like SEATO or NATO, as defined in Chapter VIII, Article 52, of the UN Charter. These are arrangements "dealing with such matters relating to the maintenance of international peace and security as are appropriate for regional action, provided that such arrangements or agencies and their activities are consistent with the Purposes and Principles of the United Nations."[25]

Article 52 further specifies that should a dispute arise concerning the region in question, the parties to the regional agreement must "make every effort" to achieve a settlement before referring it to the UN Security Council. But the Security Council can act only if the permanent members agree, and all five permanent members of the UN Security Council are parties to the Antarctic Treaty (see table 8-1). For even if one member, such as the Soviet Union, saw some advantage in having the dispute taken up there, the others—namely, the United States, Great Britain, France, and China—would probably all vote against. Thus, Antarctica is unlikely to be brought under the jurisdiction of the United Nations through the window of settling a dispute.

Robert D. Hayton, a prominent scholar of the Antarctic claims

Table 8–1. UN Security Council Membership and Antarctic Treaty Parties, as of May 1983

Security Council	Antarctic Treaty
Permanent members	Consultative (voting) parties
China	Argentina
France	Australia
Soviet Union	Belgium
United Kingdom	Chile
United States	France
	Federal Republic of Germany (1979, 1982)[a]
Nonpermanent members	Japan
Guyana	New Zealand
Jordan	Norway
Malta	Poland (1961, 1977)[a]
The Netherlands	Republic of South Africa
Nicaragua	Soviet Union
Pakistan	United Kingdom
Poland	United States
Togo	
Zaire	Acceding parties (year of accession)
Zimbabwe	Brazil (1975)
	Bulgaria (1978)
	China (1983)
	Czechoslovakia (1962)
	Denmark (1965)
	German Democratic Republic (1974)
	Italy (1981)
	The Netherlands (1967)
	Papua New Guinea (letter of succession)
	Peru (1981)
	Romania (1971)
	Spain (1982)
	Uraguay (1980)

Source: Treaty Office, U.S. Department of State, Washington, D.C.

[a] First year in parentheses indicates year of accession, second year indicates year moved to consultative status. All other Consultative Parties are in the group of twelve that negotiated the original treaty.

dispute, wrote, when the ink was barely dry on the Antarctic Treaty: "The Antarctic Treaty itself meets, technically, all criteria for a regional arrangement under Article 52, Chapter VIII, of the United Nations Charter. There are provisions for consultation and peaceful settlement of local disputes, a defined area, and dedication of that area to peaceful use. Action is required to be consistent with the Charter; the agreement appears not to violate any of the Charter's principles or purposes."[26]

And he added, "The North Atlantic Treaty in its original conception and wording was, after all, no more adequate."[27] Thus, even under UN rules the treaty may count as the legitimate forum for legislating on Antarctic matters.

Another key difference between the two issues is that two developing nations, Chile and Argentina, have always been in the exclusive treaty group. When united, Chile and Argentina can influence the behavior of the Latin American bloc at international meetings. And now that Uruguay and Peru have acceded to the treaty and Brazil is a voting party, assaults from Latin America are unlikely. With China acceding, a united Asian bloc also is unlikely. Without the Latin and Asian blocs, only the African nations would be in a position to mount a group attack on the Antarctic Treaty. Indeed, some have joined in Malaysia's and Antigua and Barbuda's actions at the Thirty-Eighth General Assembly in order to have a chance to reproach South Africa, an original treaty power. Alvaro de Soto, a Peruvian diplomat now working for the UN Secretary General, in 1979 called on the African nations to rise up in opposition to the treaty, as a form of "apartheid."[28] However, it is noteworthy that in his speech de Soto was careful to speak only for himself and not for the Peruvian government. Moreover, he specifically called on African, not Latin, nations to rise up in opposition. De Soto's caution may reflect the care other Latin governments show for the strong feelings of Chile and Argentina on the issue.

Finally, unlike the Law of the Sea negotiations—where would-be mining states were on one side and minerals-exporting states on the other—the Antarctic Treaty group includes many of the minerals-exporting states likely to be hurt if a huge supply of gold or diamonds were exploited in Antarctica. Such treaty nations as South Africa and the Soviet Union would have to defect from the treaty in order to unite with their fellow exporters of these minerals—an unlikely course.

Thus, a regime for Antarctica based on the International Sea-Bed Authority has appeal in terms of global equity, and as a massive remedy to the old "nonparties problem" of the Antarctic Treaty. But despite Mahatir bin Mohamad's rhetoric, a regime based on the New International Economic Order would have symbolic value only and, as proposed, could be so impractical as to be useless in Antarctica.

The Antarctic Treaty "System"

While denigrating the New International Economic Order approach and extolling the Antarctic Treaty system, the treaty powers face their own, internal deadline. For the treaty could be changed substantially in 1991, the date after which a conference to review its operation may be called by any of the consultative parties. When the treaty was negotiated, the twelve nations represented at the table had no idea how well it would work. The memory of the Argentines firing shots over the heads of the British landing party in 1948 was fresh in their minds; even more vivid was their awareness that the Soviet Union could assert a claim in the Australian sector after the IGY ended. Earlier, we also saw how the treaty came about as a solution to the perils of 1957–59.

As a hedge against the future, Article XII states that after the treaty has been in force for thirty years, any consultative party can request the holding of "a Conference . . . to review the operation of the Treaty," which all the parties—consultative and acceding—would attend. At this conference, a "modification or amendment" to the treaty may be approved if a majority of all the parties present, including a majority of the "original and active" parties, favors it.[29]

But to be adopted, any proposed change in the treaty must be ratified by all the consultative parties. Should they not have done so after two years, the change will not enter into force. At that time, any of the parties may give notice of intention to withdraw from the treaty. Such withdrawal takes effect after another two years. This procedure conceivably will make it easier to propose changes to the treaty and encourage those who object to a proposed change to withdraw. But in fact, it will be as procedurally difficult to modify the treaty in 1991—assuming a review conference is called—as it is at present. Now, under Article XII, all the consultative parties must ratify a change for it to take effect.[30]

Thus, even if the developing countries were not an issue, the treaty powers would face their own hour of reckoning. Thus, the issue for 1991 and beyond is how well the treaty system of law and administration will stack up against whatever alternative the developing nations propose. It is clear from the public record that the United States and the other signatory nations believe the treaty system is valid and should continue indefinitely.

Many international law papers poorly evaluate the treaty system because they consider only the text of the treaty itself. Given that document's brevity, its deliberate ambiguities and omissions, and the fact that the real "life" of the treaty lies in its implementation, it is hardly surprising that legal scholars often find it deficient. But those who examine the whole body of Antarctic administrative law—the treaty itself, the agreed recommendations of the consultative meetings, the Agreed Measures on the Conservation of Antarctic Fauna and Flora, the sealing and krill conventions, and even the weather and communications services—have tended to conclude it was both valid and viable.[31]

Two Latin American commentators fall into the latter category—Roberto E. Guyer, an Argentine who was under secretary general of the United Nations, and Fernando Zegers Santa Cruz, a senior Chilean diplomat. Both have published descriptions of the treaty system of law.

Writing in 1976, Guyer argued that there are fundamental differences between Antarctica and outer space—an analogy often made in the legal literature of the late 1950s when the treaty was drawn up.[32] The Antarctic continent is finite and controllable, he wrote; there was national activity there long before international law came along to try to set rules. On the other hand, outer space is infinite and uncontrollable, he argued, and law is being made for it in advance of national activities there. The treaty entered the Antarctic setting to minimize conflict and to preserve and heighten what Guyer called the "peculiarity" of the Antarctic, which is the prominence given to science.

Guyer quoted Jawaharlal Nehru, who said at the time India tried to have Antarctica taken up by the United Nations: "Broadly speaking, we are not challenging anybody's rights there. There are certain countries, which, according to them, have certain rights there. We are not challenging them. But it has become important more especially because of the possible experimentation of atomic weapons and the like, that the matter should be considered by the UN and not left in a chaotic state—various countries trying to grab the area."[33]

Echoing a familiar view of the treaty group, Guyer said it did not need the explicit consent of nontreaty nations to manage the region. He admitted that the treaty is nonbinding on nonsignatories, but, if one such nation began violating one of the provisions—by asserting a claim or taking military action—clearly the treaty parties would act together to prevent it. So, they have, in effect, "generated a certain legislative right in reference to the region and that therefore Article X [in which the parties agree to see that *no one* operates contrary to the Treaty in the Treaty area] is not only legal but legitimate."[34] Thus, however inchoate their rights may have been in 1959, the treaty powers have earned, by customary international law and responsible management, a right to manage the region on behalf of the others.

Zegers's argument was similar. He also mentioned the provision of weather reporting and communications services as further evidence of the Antarctic system. He listed

> A communications network which is integrated into the world-wide system and links Antarctica to the rest of the world, while at the same time easing communication across the continent. Also integrated into the world-wide system are the meteorological stations.
>
> Through the actions of the active parties to the Treaty, it has been possible to establish a terrestrial, naval, and aerial transportation network which, although limited, allows access to and permits human activity in the area. . . . All this has been accomplished with the help and support of the system established by the Treaty.[35]

Zegers also cites the existence of an environmental reserve in Antarctica managed by the group, with no parallel in the world. He concluded optimistically that the system "has shown that it has been able to face successfully, during nearly two decades, the problems arising in Antarctica." As for resources, "A convention originating in the Antarctic System, but open to all interested states . . . seems to be a realistic and just solution."[36]

These views echo those the treaty powers formalized at the ninth consultative meeting in 1977, at which they asserted their "special responsibilities" in the recommendations on living and mineral resources. The two papers describe the underlying legal doctrine behind the treaty powers' view that they have earned the right to manage the region south of 60° south latitude, and, in the case of the living resources of the surrounding ocean, to manage the entire Antarctic marine ecosystem out to the Antarctic Convergence.

An idea of the implications of the Antarctic Treaty system for 1991 and beyond was covered by the State Department official representing the United States at Antarctic diplomatic meetings, R. Tucker Scully, at a nongovernment meeting in October 1982, at Teniente Marsh, one of Chile's Antarctic stations on the Peninsula. The views in the paper are solely those of Scully, but they suggest the view of several treaty powers.

To the criticism that the treaty system lacks institutions to carry it out, Scully notes that Article IX, the treaty's brief reference to future meetings with its list of possible topics, has nonetheless brought about a "system" for identification and resolution of issues. The silence of the treaty on the outlines of the system is not a weakness, he argues; instead it reflects the assumption that there should be maximum flexibility to solve issues the treaty does not address, once they become pressing. The lack of explicit institutions, he adds, reflects the gradualist philosophy of the treaty in leaving things unsaid until they evolve.

Indeed, there has been institutional development, he argues. One example is SCAR, whose initial mandate was to plan and coordinate scientific research after the IGY. But SCAR has evolved into a general scientific advisory body to the treaty system and identifies issues the consultative parties should take up. SCAR's key role is the more remarkable because its connection to the treaty system is indirect, through agreed recommendations to individual governments, which in turn transmit their requests to SCAR representatives at their national academies of science. Another example of evolution is what Scully calls a "rotating secretariat" in the form of the work done by each government when its turn comes to host a consultative meeting. The practice of calling special consultative meetings on specific subjects—such as those where the krill and minerals regimes were negotiated—is another institutional innovation. Finally, the treaty system has established ties to various international organizations—the World Meteorological Organization, the International Telecommunications Union, the International Oceanographic Commission, and the International Civil Aviation Organization.

Scully concludes that the treaty has evolved so much of its own accord that there may be no need to make major amendments or modifications to it. He advises the treaty powers to follow the maxim, "If it ain't broke, don't fix it."[37]

Critique of the Antarctic Treaty "System"

But the champions of the system are premature in extolling it, for it is hardly an accepted fact. Many legal loopholes exist in such difficult matters as minerals resources and criminal jurisdiction. And Article IV, dealing with territorial claims, is profoundly unclear regarding the legal status of expeditions and stations built while the treaty is in force. It says that no actions taken by the parties while the treaty is in force shall constitute "a basis for asserting, supporting or denying a claim," a statement contradicted by the apparent belief of some treaty powers that, should the treaty fail, stations installed during the treaty period and other "acts or activities" could then be invoked to support a new claim or to extend a preexisting claim. Therefore, there is nothing to stop claimants and potential claimants from doing everything they can while the treaty is in force to support their preexisting legal position. Thus, if the treaty ever ceases to be in force, they could invoke these actions in a new round of rivalry over claims. Indeed, as 1991 approaches, many nations may begin building stations there to hedge against a treaty breakdown.

Practically speaking, the many ambiguities of the treaty—no matter how irritating to scholars—do not matter so long as the treaty enhances the underlying political stability of the region. Thus, as changes in the treaty are contemplated, the overriding consideration should not be literary or ideological neatness, but whether the proposed alternative would be as stable politically. It would be a sad commentary on our ability to govern ourselves if the effort to fix the treaty system in the name of international justice would lead to the very war in Antarctica that has been avoided for so long.

Further, the treaty system is nowhere explained in a fashion that outsiders can comprehend. It is understandably difficult for non-signatories to the treaty, its would-be critics, and developing-world spokespersons to penetrate its almost Talmudic mysteries. They may be forgiven if they find it hard to understand why a six-page document should govern one-fifteenth of the earth's surface.

If it were written down, the system's many loopholes would be exposed to public view. Worldwide publicity attended the crash of a New Zealand commercial airliner into Mount Erebus in 1979, in which all 257 persons on board—embarked on a routine tourist flight—were killed. But it is less well known that, at the time, no one had clear responsibility for air navigational safety south of 60° south latitude. The maps that display navigation and tracking points worldwide, issued by the International Civil Aeronautics Organization in Canada, stop at 50° south latitude. South of this boundary, responsibility is patchy at best. At one time Australia's Ministry of Transport announced it had responsibility for airspace south of 50° south latitude, including that over Australia's claim. But Australia has installed no equipment to effectively monitor air traffic there, although it has a plan for a wheeled aircraft runway at Davis station.[38]

Another serious problem is that, although the treaty requires open exchange of data and advance notice of expeditions, as F. M. Auburn has written, in practice these requirements are not always met.[39] Some governments try to be model adherents to these provisions in order to strengthen the treaty generally, but others—including the Soviet Union—are more lax. This discrepancy had few practical consequences until resource issues arose. How much krill did you catch last season? How many tons of fish? What are the results of your seismic surveys of the Antarctic continental shelf? Under the reporting requirements of the new krill convention, and under the policy of voluntary restraint from minerals exploration and exploitation adopted in 1977, this information should be available. For the most part, the results of these surveys have never been published. There are many potentially tough issues here. For example, does the data count as being publicly available if they are on reels of tape stored in an obscure library? What if they become publicly available—as the West German seismic data has become—only after some persistent country finally offers to swap data in return?[40]

Whether the free exchange of data, which has been so useful in other Antarctic activities, can continue during minerals development remains a serious issue. Lax compliance with the treaty's data exchange rules in the past, then, invites would-be minerals developers "inadvertently" not to publish, and so garner advantage. Indeed, perhaps the strongest argument for the United States, as the principal upholder of the treaty, to undertake serious offshore geological studies in Antarctica (a suggestion made in chapter 7) is this: such studies, openly published by the United States—as are the preliminary geologic surveys of any frontier such as Alaska or the Canadian north—would deter other countries from trying to tuck away their own geologic information. It would reassure everyone that no significant information is being hidden and would go far in preventing the growth of secretiveness and suspicion.

Finally, it is not clear, as of this writing, that a minerals regime *can* be successfully completed by the treaty powers. The issues are more difficult than those for krill, a common property resource inhabiting mostly international waters. One proposed regime, the Beeby draft (discussed in chapter 6),[41] would resolve some of the problems. An application would trigger the creation of a specific committee to regulate that situation. The committee would be made up of the superpowers, the claimants of the area, and the state sponsoring the developer. It would set rules and oversee development in a given region. But the Beeby draft—however ingenious—is silent on how "mankind"—that is, nonsignatory nations—would benefit. Philip W. Quigg writes: "Every conceivable procedure is being explored: leases, licenses, joint ventures, consortia, and public/private corporations. . . . One suggestion is that, at least initially, any revenues should be used to expand scientific research in Antarctica—the idea being, presumably, that this would take the sting out of demands by outsiders for sharing of benefits."[42]

While it sounds fine in theory, getting an agreement in practice may be very difficult. As Finn Sollie, director of the Fridtjof Nansen Foundation, wrote bluntly, "It is when these [economic exploitation] issues are met head on that the antarctic system must prove how much of a success it really is."[43]

Room for Compromise

Indeed, the success of the present Antarctic Treaty system may depend equally on how well it reaches out to meet the double challenge of the institution of the United Nations and the real interests—whatever they are—of developing nations in the Antarctic. Clearly, there is much room for compromise: both a minerals regime, and other parts of the treaty system can provide benefits beyond the club.

One obvious, specific benefit would be for a percentage of the revenues from an Antarctic offshore well or onshore mine to be paid automatically to "mankind." The Law of the Sea Treaty provides an analog in the provisions for oil and gas development beyond the coastal states' 320-kilometer (200 mile) exclusive economic zones. It recognizes that the coastal state has recognized jurisdiction over continental shelf minerals out to 320 kilometers from shore, and a presumed exclusive right to resources beyond this limit. On the other hand, both the 1970 General Assembly resolution and the text of the Law of the Sea Treaty state that resources "beyond the limits of national jurisdiction" are common heritage as well. In contrast to the stormy public battles between commercial seabed miners and the Group of 77 over deep seabed mining in the Law of the Sea negotiations, major oil companies quietly won agreement in the discussions on how continental shelf minerals—oil and gas—beyond the exclusive economic zone shall be handled. Control of the development rests with the coastal state—as it does for oil and gas development within the state's zone. But a percentage of the profits of offshore development beyond the zone is paid as a kind of royalty to the treasury of the International Sea-Bed Authority.[44]

Were the Antarctic Treaty powers to direct a portion of mining revenue to "mankind," they would have to decide whether recipients would be required to become parties to the minerals regime (or acceding parties to the Antarctic Treaty) to realize their share. Otherwise, the link could be institutional, with "mankind's" share going to an international organization, such as the International Sea-Bed Authority.

Unless international conditions change, the treaty group seems unlikely to cede real control over Antarctic decision making beyond limited royalty provisions. Quigg suggests a more drastic compromise.[45] In his plan, the criteria for consultative status would be made consistent by having the two consultative parties no longer active in Antarctica—Belgium and Norway—withdraw. Their seats would be given to two outside nations, who would represent the world community and be drawn from among the acceding parties at the start. They would sit for four-year terms (that is, two consultative meetings) until replaced by another pair. At first, selection of the outsiders would be made by the treaty powers but later could be turned over to some UN body. So long as consensus voting is retained in consultative meetings, the two could block certain actions, but only for the duration of their terms.

Quigg argues that this "compromise of the conflicting principles of universalism and exclusivity" would address many of the nonsignatory nations' objections to the Antarctic Treaty "system" and so forestall a crisis in 1991. He adds that even if, in 1991, the two outside representatives were determined to destroy the treaty system, the meeting would be "awkward" but manageable. Voting in the meeting will be by simple majority, so that the minority which loses—presumably including the would-be destroyers—would have to withdraw.

A more drastic compromise has been proposed by James N. Barnes, a public interest advocate who was an adviser to the U.S. State Department during the negotiation of the krill convention. Barnes defines the international community's interests in Antarctica as almost purely environmental and peaceful. They are "freedom of scientific research, continued demilitarization of the region, protection of the environment, preservation of endangered species, maintenance of the potential food production for future generations, and prevention of adverse climactic change due to human activities."[46]

The treaty powers' behavior in the Antarctic has been consistent with their role as international trustees, except that they have maintained their rival claims under Article IV, Barnes writes. If world opinion will not tolerate the maintenance of this exclusive and secretive system, he continues, they should agree to turn Antarctica over to a more representative international commission which would become the permanent trustee. As an interim step, the treaty powers should abolish Article IV and renounce all possible claims, leaving the rest of the treaty mechanism in place. As Barnes views it, "The present Antarctic Treaty Consultative Parties would be able to maintain their strong ties to the area and would in fact continue to have a major voice in all decisions. Moving to international control would defuse Third World concern over resource decisions in Antarctica, and would allow the Commission members to focus their attention on the same goals identified so wisely in the Antarctic Treaty."[47]

Barnes's model is the unimplemented draft of a treaty for the islands of Spitsbergen, whose sovereignty had been disputed and whose large iron and coal reserves became important in the late 1800s. Under the draft treaty—which was never implemented because of the outbreak of World War I—an international commission was slated to run the island and the special interests of certain nations were to be recognized. Although all sovereignty would be renounced, interested countries could exploit Spitsbergen's minerals under the commission's rules.[48]

Others—such as Gunnar Skagestad of the Fridtjof Nansen Foundation, F. M. Auburn, and even the British journal *Nature*—have looked to the final Spitsbergen Treaty of 1920 as the more relevant model. This treaty gave Norway sovereignty over the archipelago on the condition that it be kept demilitarized and permit equal access to the other parties for purposes of resource development. Applied to the Antarctic, this "Svalbard model" (Svalbard is the Norwegian name for the islands) would suggest an internal accommodation among the treaty powers that would give claimant states a role in managing their areas as recompense for "not flying their national flags too ostentatiously," as *Nature* put it.[49] Meanwhile other parties would have equal access for the purpose of resource development.

Evolution Through Accession

Two conclusions seem certain. One is that the Antarctic Treaty system will evolve, perhaps dramatically, in the next several years, regardless of whether exploitable Antarctic minerals are found, or even deemed important. This book has shown that the *anticipation* of Antarctic minerals wealth is important to the world political community and a forceful factor in the development of law for the region. Change has already occurred, as witnessed by the completion and early operation of the Convention for the Conservation on Antarctic Marine Living Resources (see chapter 6). The treaty powers now feel they must negotiate a minerals regime, too, lest the United Nations does it for them. Separate concerns, such as the claimants' maintenance of sovereignty and environmental protection, are being swept aside in the rush to flesh out the treaty system and prove its viability by 1991.

A second conclusion is that more and more countries will accede to the treaty, and perhaps gain consultative status by sending scientific expeditions down on the ice between now and 1991. As the treaty powers succeed in making their system look viable—and retain decision-making power over rights to presumed minerals riches—more countries will gravitate to the treaty to obtain their share. One key possibility— overlooked in the legal literature—is the likelihood of revolution by evolution: perhaps the system will change most dramatically from within. This is all the more likely now that Brazil and India have obtained consultative status because both may decide to speak at treaty meetings on behalf of those outside the group.

Consultative status in the treaty—and an accompanying scientific presence in the region—may offer great prestige value to developing nations. This may have been one of the Gandhi government's motives in its ambitious Antarctic plans. Indian scientists have taken pride in their activities with the United States and Great Britain in Antarctica. Chinese scientists, who now participate in SCAR and whose government has acceded to the treaty, may feel similarly. The real benefit of Antarctica to developing nations—long before mineral deposits are found or exploited—may be its value as a political symbol.

Few published schemes take into account the prestige value of an Antarctic scientific presence to such governments. They may well conclude that bloc action in the United Nations along the lines of the New International Economic Order is likely to disappoint. Some may conclude that unilateral action will bring the quickest, most direct rewards. In the end, these governments may prefer that their scientists fly a flag in Antarctica for the benefit of the media rather than engaging in a drawn-out struggle at meetings attended by more than one hundred nations for some hypothetical future economic gain. Such a presence, achievable for the price of a leased icebreaker and the salaries of crew and scientific staff, enables a government (particularly one wishing to project an image of progress) to appear advanced *and* have a voice in the evolution of the treaty system. Thus, participation in the Antarctic Treaty may come to have the same prestige value as a nuclear research reactor, a space program, or buying advanced fighters.

India's actions, therefore, may foreshadow the future in Antarctica more than all the law papers combined. India is the only major nation looking south across the ocean toward Antarctica that has no historic presence there. As chapter 4 shows, Nehru tried to introduce the subject of Antarctica in the United Nations at the time the treaty was being negotiated. His remarks at the time suggest that he was more concerned with keeping the peace than with elaborating Third World rights. India's two recent expeditions to Antarctica indicate the government is capable of major scientific undertakings. Its Antarctic program may be but the latest step in India's efforts to secure recognition as a serious technological power: It has launched its own satellite (taken to the launch-pad on a bullock-cart), detonated a nuclear device, and finally, it has expanded its oceanographic and offshore petroleum exploration.

Evolution through accession is already changing the web of national interests operating in the region. Since 1975 three Latin

American countries—Brazil, Uruguay, and Peru—have acceded to the Antarctic Treaty, and Brazil became a full party in 1983. All three have pretensions of becoming claimant states. In the past, Brazil has identified a "zone of influence" in Antarctica extending from 34° 40″ to 53° 20″ west longitude; and Peru at one time considered a sector claim extending from 75° 40″ to 81° 20″ west longitude. Uruguayan writers have said that their country has "rights and duties" in the part of Antarctica stretching from 53° 20″ to 56° 40″ west longitude.[50] The German Democratic Republic has acceded to the treaty and may try for full consultative status before 1991. Although hardly a developing country, it is not a rich one either.

Papua New Guinea, in effect, has acceded by giving what is formally known as a notice of succession. Australia may have had some influence in urging it to join. China's accession has already been discussed. The acceding parties are bound by the Antarctic Treaty but, at present, do not derive any benefit from having joined, nor have the treaty parties ever given them anything to do. Clearly, the consultative parties could induce more outsiders to join if they awarded acceding parties some benefits. The acceding parties were invited to attend the twelfth consultative meeting in September 1983; and this may have been a first step in making accession more attractive to the international community.

The treaty system could evolve dramatically if there were a large number of accessions from developing nations. Table 8-2 is a hypothetical outline of how it could look under these circumstances by 1991. States acceding are assumed to either fish in the Antarctic or to have a high per-capita GNP, or both, and as such they are capable of mounting Antarctic expeditions. Some northern African nations that were assumed capable of joining, did not, objecting to the presence of South Africa in the group. It was presumed that Middle Eastern nations were too preoccupied with local events to join. But northward-looking countries likely to develop Arctic oilfields or to have technology useful for Antarctic offshore exploration, joined. Finland, Switzerland, and Austria were excluded because of their claims of neutrality. The result is a hypothetical total of forty-seven acceding and consultative parties to the treaty or to its satellite agreements regarding krill and minerals, of which twenty could be termed developing nations (these include Chile and Argentina), nine Eastern bloc, and nineteen western industrial nations (including Japan). Obviously, this groups' balance of interests would be quite different from that of the group today. There would also be shifts in relations between the seven announced claimant states and the nonclaimants as the latter became a decisive majority.

Table 8–2. The Antarctic Treaty in 1991 Showing Possible Growth Through Association

Region	Original twelve consultative parties	New consultative and acceding parties	Possible new consultative or acceding parties
Europe	Belgium France Norway United Kingdom	Denmark Federal Republic of Germany The Netherlands Spain	Greenland Iceland Irish Republic Sweden
Eastern bloc	Soviet Union	Bulgaria Czechoslovakia German Democratic Republic Poland Romania	Hungary Yugoslavia
The Americas	Argentina Chile United States	Brazil Peru Uruguay	Canada Colombia Ecuador Mexico Paraguay
Southwest Pacific	Australia New Zealand	Papua New Guinea	Indonesia Philippines
Asia	Japan	China	India South Korea North Korea
Africa	South Africa		Nigeria Morocco Algeria
Totals	12	14	19

A Continent for Applied Science

What do the krill fishery, the growing number of offshore geophysical expeditions, and the interest of developing nations mean for the continent itself? Inevitably, the pattern of human activities will change down on the ice, perhaps bringing as profound a revolution as the IGY did in 1957–58.

Scientific research, the principal activity in the region, is the glue binding the Antarctic Treaty system together. If the Antarctic Treaty powers continue to encourage newcomers to join, there will be an increase in the number and variety of scientists going to the Antarctic each season. The new nations will probably use their scientists as "cover" for their political wish to have an Antarctic presence, as the treaty powers have done. The treaty group's effort to expand the treaty "system" with new members should entail long-range planning for scientific programs useful to the new members. What should Chinese, African, Indian, or, for that matter, Malaysian, scientists *do* in Antarctica? The same research the British or Japanese have done, such as ionosopheric studies, terrestrial biology, or meteorite searches? Surely the evolution and transformation of the treaty "system" implies an accompanying change—yet undefined—in Antarctic science.

What role will there be for science down on the ice if minerals exploration goes forward in the 1990s? Proponents of the view that Antarctica must be dedicated to pure, undirected, university-based research (a view held mainly by officials of the U.S. Antarctic Program) often remark that the resource issue will disappear, that the krill fishery will not expand much more, and that the age-old dream of Antarctic mineral riches will prove ephemeral. Therefore, they conclude, Antarctic scientific research can continue as it has in the past. The self-serving conclusion, ultimately, is that national scientific programs need change only as basic scientific discoveries dictate.

But recent history suggests that the other future for Antarctica is equally or more likely one in which its resources are assessed and are a continual subject of international political deliberations. In this scenario, fishing increases offshore, perhaps the Ross and Weddell seas have offshore oil exploration under discussion, and Antarctica becomes host to oil workers, managers, environmental inspection teams, fishermen, tankers, coastal ports, and the like. Under these circumstances, how well will the present logistics network of weather reporting and forecasting, navigation, communications, and rescue work? Will there still be scientists? What will they do?

Antarctica's history shows that different nations and interests can run smoothly in the region only when lubricated by common activities, rules, and adequate communication. This was the function that scientists served during the IGY; indeed, without them Chile's Escudero Declaration of 1948, which anticipated the Antarctic Treaty, would have died. This suggests that even the best-crafted treaties for Antarctica are unworkable unless there are some lubricators in the region itself. And the best candidates for the job are scientists, who already perform such functions under their common bond of science. If a resource age gets under way in Antarctica, will these scientists have to step aside? If that is the case, who will play this vital role?

A UN takeover could eliminate the motivation of the treaty powers to send scientists to Antarctica at vast expense. Although some Antarctic scientific research merits the enormous logistical trouble and cost required, most of it is motivated by a nation's need to symbolize and maintain its own interests. A successful move to make Antarctica "common heritage" under UN auspices could eliminate the incentives for the United States, the Soviet Union, the seven announced claimants, and others to maintain their national presences there. Thus, a UN-based regime might eliminate science as the principal activity in the region.

A UN-based international authority for Antarctica could sponsor research directly, or deputize agencies—such as the World Meteorological Organization or the Food and Agriculture Organization—to do so. But as we have seen, the developing nations' motive for UN control is international equity and a share in resources—not at present a wish to mount scientific expeditions there. Therefore, it seems unlikely that science would continue in an Antarctica administered by the United Nations.

At the same time, the new diplomatic agenda for the krill fishery and Antarctic minerals carries with it an enormous applied research agenda (see chapter 5). For both industrial and developing nations to benefit from the abundant, high-protein krill, a good deal of applied research must be done on the processing of krill for animal and human food. To date, because this effort has been scattered, its general lack of success is hardly a surprise. The genius of U.S. and European food companies, which can turn vegetable oil into catsup or once-spurned seafoods like tuna into household staples, has hardly been brought to bear on the subject. Krill processing, then, offers opportunities for partnerships between industry and government, and for joint scientific endeavors between industrialized nations and the developing world. And previous, disappointing UN efforts to produce wonder foods that could solve the world's food problem—such as "Pruteen," a commercially made, single-cell bacterial protein, and a fish protein concentrate, type B—should not discourage but inform this effort.[51]

Another problem requiring rigorous study is how Southern Ocean protein can be used to meet developing world human and animal food needs. The International Institute for Environment and Development has raised the possibility of locating krill-production facilities in the Southern Hemisphere countries, which are closer to the fishery, and the potential markets for krill products. These countries currently lack production facilities or an established infrastructure for distribution to those in food-poor countries.[52] This is not basic scientific research, but it is research nonetheless, and it could benefit mankind directly.

It is necessary to gather environmental data on a systematic basis in Antarctica itself and use it as a reference point for decisions on resource development. Already the BIOMASS program and various national marine research activities are gathering this kind of information as a step toward monitoring the fishery. Parallel efforts are required in environmentally sensitive coastal areas and potential port sites, so that if and when the treaty powers decide to allow an oil company or mining consortium into Antarctica, there will be a proper framework for regulation. Moreover, for countries seeking to deter minerals development in the Antarctic, organized, applied, environmental research could be very useful in proving how fragile the coastal ecosystem really is to human interference.

Beginning with the 1977 report of SCAR's Group of Specialists on the Environmental Impact Assessment of Mineral Resource Exploration and Exploitation in Antarctica (EAMREA),[53] many expert groups have outlined the research required. Indeed, EAMREA's list of research topics is so long and sweeping that it could justify almost any Antarctic scientific project. Nonetheless, environmental research is useful for management only if it is carried out in an integrated, coherent, and organized manner.

Geophysical surveys of the continental shelf must be completed; in fact, the area of Antarctica so far mapped is equivalent to mapping the state of Delaware to get an idea of the geology of the United States and Mexico. Customarily, governments sponsor such baseline surveys of unknown regions and also gather data about likely offshore oil deposits in order to increase their own expertise when it comes time to manage their leasing. This kind of research is applied, as well as being routine and logistics-intensive. The results can be made public without infringing on future proprietary rights of companies. Indeed, such surveys are a prerequisite for any commercial interest.

By its nature, much of this work would improve if it were a cooperative effort by several nations. If the environmental baseline studies reflect the confidence of more nations, they become a better management tool. Offshore geophysical surveys and the mapping of the continent are too expensive for any one nation to undertake alone, but unfortunately, the lack of organizational and international leadership has encouraged individual treaty nations to undertake offshore research on their own, with the result that their data have not been disseminated. An international program is needed to forestall a future knowledge race, in which each nation holds back its geophysical data in order to have something to barter with another, similarly secretive nation.

On the other hand, cooperative geological surveys must precede drilling for confirmation of an actual mineral or oil deposit, at least for the time being. The single event which could most upset the treaty system and the discussion at the United Nations would be the discovery of a rich minerals deposit or a giant oil basin offshore. Those nations engaging in *de facto* minerals exploration now are playing with fire. If they find something of interest, they cannot publish it for fear of encouraging international publicity that would undermine the negotiations for a minerals agreement. But should they fail to publish it, inevitably the news of the discovery would be rumored to the outside world, encouraging others to withhold their data.

An *international* geological survey program could do much to minimize these problems. It could work out an agreement on what voluntary restraint—the policy under which the treaty powers are supposedly holding off on minerals exploration—means in operational terms. It could enforce requirements to publish all findings through regular international conferences of participating geologists. And it could fairly distribute the costs of a geologic regional inventory. Finally, it would put teeth into the treaty powers' claim that what they do in Antarctica will benefit "mankind"— for national geological programs will acquire a more international, public cast. It could reach out to the other nations by accommodating individual teams of developing-country scientists—perhaps geologists from the acceding powers to the treaty. An international geological survey could lubricate the politically difficult, even explosive, situation that is developing as more nations seek a stake in Antarctic minerals.

Scientific surveys should be continued in Antarctica well into the 1990s. Not only do a certain number of basic science problems warrant attention—such as the discovery of past climates through ice cores, the mechanics and possible instability of the great ice sheets, obtaining new information from meteorites found there and the use of the South Pole for solar observation on a twenty-four-hour basis. A need exists for additional applied research on the marine ecosystem in light of the new krill convention; and coastline environmental research is needed as is an international geological survey. Together, these programs would resemble a latter-day IGY, and they could ease Antarctica's transition into a new political era and a resource age.

On this theme, many changes can be rung. Scientific research itself could be a vehicle for easing the irritations of those outside the treaty group and of the acceding parties. The treaty group could negotiate a separate "science convention," like the krill and minerals regimes, which would facilitate the use of Antarctica by scientists from developing nations. The requirement for membership could be far simpler than in the present treaty, so that governments not wanting to buy icebreakers or build Antarctic stations could join. Their scientists then could participate in the treaty nations' basic science programs. Those from Saharan Africa, for example, might wish to study the dessication of the Antarctic "desert" in the great interior. Those from southern Africa, Asia, or Latin America might study Antarctic oceanography or meteorology, since the waters off their coasts and the weather in their skies originate in the Antarctic region. Iceberg studies might be of interest to nations like Saudi Arabia who are in need of fresh

Worldwide economic difficulties, or a stalemate in international negotiations over Antarctica's future, could lead to the abandonment of the continent by man.

water. Their participation in Antarctic science could be paid for, perhaps, by the royalty from Antarctic minerals development— the treaty power's tithe to "mankind" mentioned earlier.

If Nothing Happens...

Conflict in the world, a global depression, or some stalemate in Antarctic diplomacy could shut the continent down for a time, as has happened before. The Norwegian Arctic explorer Fridtjof Nansen wrote that polar exploration occurs in the "slackness" of various ages, when countries have the leisure and wealth to sponsor faraway adventures and dream about polar mysteries. A prolonged world crisis could stall the current momentum in Antarctica's political evolution.

Then, all but the richest treaty powers would close down their stations. For years, their activities would wane in twilight like the long Antarctic evening when the autumn sun, casting long shadows on the violet ice, sinks slowly in a blue-gray sky, and birds fly to their winter nesting places. A few, vitally interested nations would maintain their stations, nestled on the fringes of the continent like the hardy emperor penguins who sit all winter, huddled in sleepy masses over their eggs.[54] As happened when the British occupied Deception Island during World War II, there would be volunteers for this lonely vigil, drawn by Antarctica's mystery. Their thoughts might echo the lines from Robert Browning's "Prospice," the epitaph for Shackleton's dead companions in McMurdo:

> Things done for gain are nought
> But great things done endure....
>
> Let me pay in a minute Life's glad
> arrears of pain, darkness & cold.[55]

Years, perhaps a decade, would pass, marked only by the regular return of the sun and the icebreakers to give staring winter eyes a chance to look on civilization. The years would tick by, resembling lines on a calendar marking the passing of the dead Antarctic winter. As Bond and Siegfried portrayed it: "Only dull rhythms of day and night mark the passing weeks, although the skies are sometimes bizarre with streaking auroras. When the wind is still the frozen sea lies flat and peaceful about the birds, gleaming under the moon, and even reflecting the stars which have colour in the hard winter sky."[56] Antarctica—never high on the priorities of nations—would sink into oblivion for a time.

But if the world stays somewhat prosperous, somewhat concerned about the disposition of the planet's resources, nations will go on toying with the dream of Antarctic resource riches. The international community could try—and succeed—in taking over the region's governance, or the treaty system could remain in place, even after 1991. Since real income from Antarctic resource exploitation, especially that from minerals, seems slight at present, the continent's value will be then—as it has been in history— mainly symbolic.

It was a symbol to the sponsors of Capt. James Cook, who sent him to find the gold and jewels of the legendary *terra australis incognita.* It was a symbol to Robert Falcon Scott and his party in their race for the South Pole, in which they were beaten by the Norwegians and paid "life's glad arrears of pain, darkness and cold." For years, it remained a symbol of national exploration and conquest. To the participants in the IGY and the treaty that grew out of it, Antarctica symbolized how well nations can get along with each other, given enough common understandings and dangers. In the future, Antarctica could symbolize something else— how wisely and peacefully humans can decide the fate of the Earth's seventh continent, which was the last to be discovered and will be the last to be exploited.

appendix A

The Antarctic Treaty

The Treaty

The Governments of Argentina, Australia, Belgium, Chile, the French Republic, Japan, New Zealand, Norway, the Union of South Africa, the Union of Soviet Socialist Republics, the United Kingdom of Great Britain and Northern Ireland, and the United States of America,

Recognizing that it is in the interest of all mankind that Antarctica shall continue forever to be used exclusively for peaceful purposes and shall not become the scene or object of international discord;

Acknowledging the substantial contributions to scientific knowledge resulting from international co-operation in scientific investigation in Antarctica;

Convinced that the establishment of a firm foundation for the continuation and development of such co-operation on the basis of freedom of scientific investigation in Antarctica as applied during the International Geophysical Year accords with the interests of science and the progress of all mankind;

Convinced also that a treaty ensuring the use of Antarctica for peaceful purposes only and the continuance of international harmony in Antarctica will further the purposes and principles embodied in the Charter of the United Nations;

Have agreed as follows:

Article I

1. Antarctica shall be used for peaceful purposes only. There shall be prohibited, *inter alia*, any measures of a military nature, such as the establishment of military bases and fortifications, the carrying out of military manœuvres, as well as the testing of any type of weapons.

2. The present Treaty shall not prevent the use of military personnel or equipment for scientific research or for any other peaceful purpose.

Article II

Freedom of scientific investigation in Antarctica and co-operation toward that end, as applied during the International Geophysical Year, shall continue, subject to the provisions of the present Treaty.

Article III

1. In order to promote international co-operation in scientific investigation in Antarctica, as provided for in Article II of the present Treaty, the Contracting Parties agree that, to the greatest extent feasible and practicable:

   (*a*) information regarding plans for scientific programs in Antarctica shall be exchanged to permit maximum economy and efficiency of operations;

   (*b*) scientific personnel shall be exchanged in Antarctica between expeditions and stations;

   (*c*) scientific observations and results from Antarctica shall be exchanged and made freely available.

2. In implementing this Article, every encouragement shall be given to the establishment of co-operative working relations with those Specialized Agencies of the United Nations and other international organizations having a scientific or technical interest in Antarctica.

Article IV

1. Nothing contained in the present Treaty shall be interpreted as:

   (*a*) a renunciation by any Contracting Party of previously asserted rights of or claims to territorial sovereignty in Antarctica;

   (*b*) a renunciation or diminution by any Contracting Party of any basis of claim to territorial sovereignty in Antarctica which it may have whether as a result of its activities or those of its nationals in Antarctica, or otherwise;

   (*c*) prejudicing the position of any Contracting Party as regards its recognition or non-recognition of any other State's right of or claim or basis of claim to territorial sovereignty in Antarctica.

2. No acts or activities taking place while the present Treaty is in force shall constitute a basis for asserting, supporting or denying a claim to territorial sovereignty in Antarctica or create any rights of sovereignty in Antarctica. No new claim, or enlargement of an existing claim, to territorial sovereignty in Antarctica shall be asserted while the present Treaty is in force.

Article V

1. Any nuclear explosions in Antarctica and the disposal there of radioactive waste material shall be prohibited.

2. In the event of the conclusion of international agreements concerning the use of nuclear energy, including nuclear explosions and the disposal of radioactive waste material, to which all of the Contracting Parties whose representatives are entitled to participate in the meetings provided for under Article IX are parties, the rules established under such agreements shall apply in Antarctica.

Article VI

The provisions of the present Treaty shall apply to the area south of 60° South Latitude, including all ice shelves, but nothing in the present Treaty shall prejudice or in any way affect the rights, or the exercise of the rights, of any State under international law with regard to the high seas within that area.

Article VII

1. In order to promote the objectives and ensure the observance of the provisions of the present Treaty, each Contracting Party whose representatives are entitled to participate in the meetings referred to in Article IX of the Treaty shall have the right to designate observers to carry out any inspection provided for by the present Article. Observers shall be nationals of the Contracting Parties which designate them. The names of observers shall be communicated to every other Contracting Party having the right to designate observers, and like notice shall be given of the termination of their appointment.

2. Each observer designated in accordance with the provisions of paragraph 1 of this Article shall have complete freedom of access at any time to any or all areas of Antarctica.

3. All areas of Antarctica, including all stations, installations and equipment within those areas, and all ships and aircraft at points of discharging or embarking cargoes or personnel in Antarctica, shall be open at all times to inspection by any observers designated in accordance with paragraph 1 of this Article.

4. Aerial observation may be carried out at any time over any or all areas of Antarctica by any of the Contracting Parties having the right to designate observers.

5. Each Contracting Party shall, at the time when the present Treaty enters into force for it, inform the other Contracting Parties, and thereafter shall give them notice in advance, of

   (a) all expeditions to and within Antarctica, on the part of its ships or nationals, and all expeditions to Antarctica organized in or proceeding from its territory;

   (b) all stations in Antarctica occupied by its nationals; and

   (c) any military personnel or equipment intended to be introduced by it into Antarctica subject to the conditions prescribed in paragraph 2 of Article I of the present Treaty.

Article VIII

1. In order to facilitate the exercise of their functions under the present Treaty, and without prejudice to the respective positions of the Contracting Parties relating to jurisdiction over all other persons in Antarctica, observers designated under paragraph 1 of Article VII and sci-

entific personnel exchanged under sub-paragraph 1 (*b*) of Article III of the Treaty, and members of the staffs accompanying any such persons, shall be subject only to the jurisdiction of the Contracting Party of which they are nationals in respect of all acts or omissions occurring while they are in Antarctica for the purpose of exercising their functions.

2. Without prejudice to the provisions of paragraph 1 of this Article, and pending the adoption of measures in pursuance of sub-paragraph 1 (*e*) of Article IX, the Contracting Parties concerned in any case of dispute with regard to the exercise of jurisdiction in Antarctica shall immediately consult together with a view to reaching a mutually acceptable solution.

Article IX

1. Representatives of the Contracting Parties named in the preamble to the present Treaty shall meet at the City of Canberra within two months after the date of entry into force of the Treaty, and thereafter at suitable intervals and places, for the purpose of exchanging information, consulting together on matters of common interest pertaining to Antarctica, and formulating and considering, and recommending to their Governments, measures in furtherance of the principles and objectives of the Treaty, including measures regarding:—

(*a*) use of Antarctica for peaceful purposes only;

(*b*) facilitation of scientific research in Antarctica;

(*c*) facilitation of international scientific co-operation in Antarctica;

(*d*) facilitation of the exercise of the rights of inspection provided for in Article VII of the Treaty;

(*e*) questions relating to the exercise of jurisdiction in Antarctica;

(*f*) preservation and conservation of living resources in Antarctica.

2. Each Contracting Party which has become a party to the present Treaty by accession under Article XIII shall be entitled to appoint representatives to participate in the meetings referred to in paragraph 1 of the present Article, during such time as that Contracting Party demonstrates its interest in Antarctica by conducting substantial scientific research activity there, such as the establishment of a scientific station or the despatch of a scientific expedition.

3. Reports from the observers referred to in Article VII of the present Treaty shall be transmitted to the representatives of the Contracting Parties participating in the meetings referred to in paragraph 1 of the present Article.

4. The measures referred to in paragraph 1 of this Article shall become effective when approved by all the Contracting Parties whose representatives were entitled to participate in the meetings held to consider those measures.

5. Any or all of the rights established in the present Treaty may be exercised as from the date of entry into force of the Treaty whether or not any measures facilitating the exercise of such rights have been proposed, considered or approved as provided in this Article.

Article X

Each of the Contracting Parties undertakes to exert appropriate efforts, consistent with the Charter of the United Nations, to the end that no one engages in any activity in Antarctica contrary to the principles or purposes of the present Treaty.

Article XI

1. If any dispute arises between two or more of the Contracting Parties concerning the interpretation or application of the present Treaty, those Contracting Parties shall consult among themselves with a view to having the dispute resolved by negotiation, inquiry, mediation, conciliation, arbitration, judicial settlement or other peaceful means of their own choice.

2. Any dispute of this character not so resolved shall, with the consent, in each case, of all parties to the dispute, be referred to the International Court of Justice for settlement; but failure to reach agreement on reference to the International Court shall not absolve parties to the dispute from the responsibility of continuing to seek to resolve it by any of the various peaceful means referred to in paragraph 1 of this Article.

Article XII

1.—(a) The present Treaty may be modified or amended at any time by unanimous agreement of the Contracting Parties whose representatives are entitled to participate in the meetings provided for under Article IX. Any such modification or amendment shall enter into force when the depositary Government has received notice from all such Contracting Parties that they have ratified it.

(b) Such modification or amendment shall thereafter enter into force as to any other Contracting Party when notice of ratification by it has been received by the depositary Government. Any such Contracting Party from which no notice of ratification is received within a period of two years from the date of entry into force of the modification or amendment in accordance with the provisions of sub-paragraph 1 (a) of this Article shall be deemed to have withdrawn from the present Treaty on the date of the expiration of such period.

2.—(a) If after the expiration of thirty years from the date of entry into force of the present Treaty, any of the Contracting Parties whose representatives are entitled to participate in the meetings provided for under Article IX so requests by a communication addressed to the depositary Government, a Conference of all the Contracting Parties shall be held as soon as practicable to review the operation of the Treaty.

(b) Any modification or amendment to the present Treaty which is approved at such a Conference by a majority of the Contracting Parties there represented, including a majority of those whose representatives are entitled to participate in the meetings provided for under Article IX, shall be communicated by the depositary Government to all the Contracting Parties immediately after the termination of the Conference and shall enter into force in accordance with the provisions of paragraph 1 of the present Article.

(c) If any such modification or amendment has not entered into force in accordance with the provisions of sub-paragraph 1 (a) of this Article within a period of two years after the date of its communication to all the Contracting Parties, any Contracting Party may at any time after the expiration of that period give notice to the depositary Government of its withdrawal from the present Treaty; and such withdrawal shall take effect two years after the receipt of the notice by the depositary Government.

Article XIII

1. The present Treaty shall be subject to ratification by the signatory States. It shall be open for accession by any State which is a Member of the United Nations, or by any other State which may be invited to accede to the Treaty with the consent of all the Contracting Parties whose representatives are entitled to participate in the meetings provided for under Article IX of the Treaty.

2. Ratification of or accession to the present Treaty shall be effected by each State in accordance with its constitutional processes.

3. Instruments of ratification and instruments of accession shall be deposited with the Government of the United States of America, hereby designated as the depositary Government.

4. The depositary Government shall inform all signatory and acceding States of the date of each deposit of an instrument of ratification or accession, and the date of entry into force of the Treaty and of any modification or amendment thereto.

5. Upon the deposit of instruments of ratification by all the signatory States, the present Treaty shall enter into force for those States and for States which have deposited instruments of accession. Thereafter the Treaty shall enter into force for any acceding State upon the deposit of its instruments of accession.

6. The present Treaty shall be registered by the depositary Government pursuant to Article 102 of the Charter of the United Nations.

Article XIV

The present Treaty, done in the English, French, Russian and Spanish languages, each version being equally authentic, shall be deposited in the archives of the Government of the United States of America, which shall transmit duly certified copies thereof to the Governments of the signatory and acceding States.

In witness whereof, the undersigned Plenipotentiaries, duly authorized, have signed the present Treaty.
Done at Washington this first day of December, one thousand nine hundred and fifty-nine.

For Argentina:
   Adolfo Scilingo
   F. Bello
For Australia:
   Howard Beale
For Belgium:
   Obert de Thieusies
For Chile:
   Marcial Mora M
   E. Gajardo V
   Julio Escudero
For the French Republic:
   Pierre Charpentier
For Japan:
   Koichiro Asakai
   T. Shimoda

For New Zealand:
   G. D. L. White
For Norway:
   Paul Koht
For the Union of South Africa:
   Wentzel C. du Plessis
For the Union of Soviet Socialist Republics:
   V. Kuznetsov (Romanization)
For the United Kingdom of Great
   Britain and Northern Ireland:
   Harold Caccia
For the United States of America:
   Herman Phleger
   Paul C. Daniels

Treaty Entered into Force, June 23, 1961.

Antarctic Treaty—Ratifications

Consultative Parties (Date)

Argentina	(June 23, 1961)
Australia	(June 23, 1961)
Belgium	(June 26, 1960)
Brazil	(Acceded May 16, 1975; consultative, Sept. 12, 1983)
Chile	(June 23, 1961)
Federal Republic of Germany	(Acceded Feb. 5, 1979; consultative, March 3, 1981)
France	(Sept. 16, 1960)
India	(Acceded Aug. 19, 1983; consultative, Sept. 12, 1983)
Japan	(Aug. 4, 1960)
New Zealand	(Nov. 1, 1960)
Norway	(Aug. 24, 1960)
Poland	(Acceded June 8, 1961; consultative, July 11, 1977)
Republic of South Africa	(June 21, 1960)
Soviet Union	(Nov. 2, 1960)
United Kingdom	(May 31, 1960)
United States	(Aug. 18, 1960)

Acceding Parties (Date)

Bulgaria	(Sept. 11, 1978)
Cuba	(Aug. 16, 1984)
Czechoslovakia	(June 14, 1962)

Denmark	(May 20, 1965)
Finland	(May 15, 1984)
Hungary	(Jan. 27, 1984)
Italy	(March 18, 1981)
German Democratic Republic	(Nov. 19, 1974)
Netherlands	(March 30, 1967)
Papua New Guinea	(March 16, 1981—letter of succession)
Peru	(April 10, 1981)
Romania	(Sept. 15, 1971)
Spain	(March 31, 1982)
Sweden	(April 24, 1984)
Uraguay	(Jan. 11, 1980)

Antarctic Treaty—Consultative Meetings

I. (1961)	Canberra, Australia
II. (1962)	Buenos Aires, Argentina
III. (1964)	Brussels, Belgium
IV. (1966)	Santiago, Chile
V. (1968)	Paris, France
VI. (1970)	Tokyo, Japan
VII. (1972)	Wellington, New Zealand
VIII. (1975)	Oslo, Norway
IX. (1977)	London, United Kingdom
X. (1979)	Washington, D.C., United States
XI. (1981)	Buenos Aires, Argentina
XII. (1983)	Canberra, Australia

appendix B

Convention on the Conservation of Antarctic Marine Living Resources

The Convention

The Contracting Parties,

Recognizing the importance of safeguarding the environment and protecting the integrity of the ecosystem of the seas surrounding Antarctica;

Noting the concentration of marine living resources found in Antarctic waters and the increased interest in the possibilities offered by the utilization of these resources as a source of protein;

Conscious of the urgency of ensuring the conservation of Antarctic marine living resources;

Considering that it is essential to increase knowledge of the Antarctic marine ecosystem and its components so as to be able to base decisions on harvesting on sound scientific information;

Believing that the conservation of Antarctic marine living resources calls for international co-operation with due regard for the provisions of the Antarctic Treaty and with the active involvement of all States engaged in research or harvesting activities in Antarctic waters;

Recognizing the prime responsibilities of the Antarctic Treaty Consultative Parties for the protection and preservation of the Antarctic environment and, in particular, their responsibilities under Article IX, paragraph 1(f) of the Antarctic Treaty in respect of the preservation and conservation of living resources in Antarctica;

Recalling the action already taken by the Antarctic Treaty Consultative Parties including in particular the Agreed Measures for the Conservation of Antarctic Fauna and Flora, as well as the provisions of the Convention for the Conservation of Antarctic Seals;

Bearing in mind the concern regarding the conservation of Antarctic marine living resources expressed by the Consultative Parties at the Ninth Consultative Meeting of the Antarctic Treaty and the importance of the provisions of Recommendation IX-2 which led to the establishment of the present Convention;

Believing that it is in the interest of all mankind to preserve the waters surrounding the Antarctic continent for peaceful purposes only and to prevent their becoming the scene or object of international discord;

Recognizing, in the light of the foregoing, that it is desirable to establish suitable machinery for recommending, promoting, deciding upon and co-ordinating the measures and scientific studies needed to ensure the conservation of Antarctic marine living organisms;

Have Agreed as follows:

Article I

1. This Convention applies to the Antarctic marine living resources of the area south of 60° South latitude and to the Antarctic marine living resources of the area between that latitude and the Antarctic Convergence which form part of the Antarctic marine ecosystem.

2. Antarctic marine living resources means the populations of fin fish, molluscs, crustaceans and all other species of living organisms, including birds, found south of the Antarctic Convergence.

3. The Antarctic marine ecosystem means the complex of relationships of Antarctic marine living resources with each other and with their physical environment.

4. The Antarctic Convergence shall be deemed to be a line joining the following points along parallels of latitude and meridians of longitude: 50° S, 0°; 50° S, 30° E; 45° S, 30° E; 45° S, 80° E; 55° S, 80° E; 55° S, 150° E; 60° S, 150° E; 60° S, 50° W; 50° S, 50° W; 50° S, 0°.

Article II

1. The objective of this Convention is the conservation of Antarctic marine living resources.

2. For the purposes of this Convention, the term "conservation" includes rational use.

3. Any harvesting and associated activities in the area to which this Convention applies shall be conducted in accordance with the provisions of this Convention and with the following principles of conservation:

(a) prevention of decrease in the size of any harvested population to levels below those which ensure its stable recruitment. For this purpose its size should not be allowed to fall below a level close to that which ensures the greatest net annual increment;

(b) maintenance of the ecological relationships between harvested, dependent and related populations of Antarctic marine living resources and the restoration of depleted populations to the levels defined in sub-paragraph (a) above; and

(c) prevention of changes or minimization of the risk of changes in the marine ecosystem which are not potentially reversible over two or three decades, taking into account the state of available knowledge of the direct and indirect impact of harvesting, the effect of the introduction of alien species, the effects of associated activities on the marine ecosystem and of the effects of environmental changes, with the aim of making possible the sustained conservation of Antarctic marine living resources.

Article III

The Contracting Parties, whether or not they are Parties to the Antarctic Treaty, agree that they will not engage in any activities in the Antarctic Treaty area contrary to the principles and purposes of that Treaty and that, in their relations with each other, they are bound by the obligations contained in Articles I and V of the Antarctic Treaty.

Article IV

1. With respect to the Antarctic Treaty area, all Contracting Parties, whether or not they are Parties to the Antarctic Treaty, are bound by Articles IV and VI of the Antarctic Treaty in their relations with each other.

2. Nothing in this Convention and no acts or activities taking place while the present Convention is in force shall:

(a) constitute a basis for asserting, supporting or denying a claim to territorial sovereignty in the Antarctic Treaty area or create any rights of sovereignty in the Antarctic Treaty area;

(b) be interpreted as a renunciation or diminution by any Contracting Party of, or as prejudicing, any right or claim or basis of claim to exercise coastal state jurisdiction under international law within the area to which this Convention applies;

(c) be interpreted as prejudicing the position of any Contracting Party as regards its recognition or nonrecognition of any such right, claim or basis of claim;

(d) affect the provision of Article IV, paragraph 2, of the Antarctic Treaty that no new claim, or enlargement of an existing claim, to territorial sovereignty in Antarctica shall be asserted while the Antarctic Treaty is in force.

Article V

1. The Contracting Parties which are not Parties to the Antarctic Treaty acknowledge the special obligations and responsibilities of the Antarctic Treaty Consultative Parties for the protection and preservation of the environment of the Antarctic Treaty area.

2. The Contracting Parties which are not Parties to the Antarctic Treaty agree that, in their activities in the Antarctic Treaty area, they will observe as and when appropriate the Agreed Measures for the Conservation of Antarctic Fauna and Flora and such other measures as have been recommended by the Antarctic Treaty Consultative Parties in fulfillment of their responsibility for the protection of the Antarctic environment from all forms of harmful human interference.

3. For the purposes of this Convention, "Antarctic Treaty Consultative Parties" means the Contracting Parties to the Antarctic Treaty whose Representatives participate in meetings under Article IX of the Antarctic Treaty.

Article VI

Nothing in this Convention shall derogate from the rights and obligations of Contracting Parties under the International Convention for the Regulation of Whaling and the Convention for the Conservation of Antarctic Seals.

Article VII

1. The Contracting Parties hereby establish and agree to maintain the Commission for the Conservation of Antarctic Marine Living Resources (hereinafter referred to as "the Commission").

2. Membership in the Commission shall be as follows:

(a) each Contracting Party which participated in the meeting at which this Convention was adopted shall be a Member of the Commission;

(b) each State Party which has acceded to this Convention pursuant to Article XXIX shall be entitled to be a Member of the Commission during such time as that acceding Party is engaged in research or harvesting activities in relation to the marine living resources to which this Convention applies;

(c) each regional economic integration organization which has acceded to this Convention pursuant to Article XXIX shall be entitled to be a Member of the Commission during such time as its States members are so entitled;

(d) a Contracting Party seeking to participate in the work of the Commission pursuant to sub-paragraphs (b) and (c) above shall notify the Depositary of the basis upon which it seeks to become a Member of the Commission and of its willingness to accept conservation measures in force. The Depositary shall communicate to each Member of the Commission such notification and accompanying information. Within two months of receipt of such communication from the Depositary, any Member of the Commission may request that a special meeting of the Commission be held to consider the matter. Upon receipt of such request, the Depositary shall call such a meeting. If there is no request for a meeting, the Contracting Party submitting the notification shall be deemed to have satisfied the requirements for Commission Membership.

3. Each Member of the Commission shall be represented by one representative who may be accompanied by alternate representatives and advisers.

Article VIII

The Commission shall have legal personality and shall enjoy in the territory of each of the States Parties such legal capacity as may be necessary to perform its function and achieve the purposes of this Convention. The privileges and immunities to be enjoyed by the Commission and its staff in the territory of a State Party shall be determined by agreement between the Commission and the State Party concerned.

Article IX

1. The function of the Commission shall be to give effect to the objective and principles set out in Article II of this Convention. To this end, it shall:

(a) facilitate research into and comprehensive studies of Antarctic marine living resources and of the Antarctic marine ecosystem;

(b) compile data on the status of and changes in population of Antarctic marine living resources and on factors affecting the distribution, abundance and productivity of harvested species and dependent or related species or populations;

(c) ensure the acquisition of catch and effort statistics on harvested populations;

(d) analyse, disseminate and publish the information referred to in sub-paragraphs (b) and (c) above and the reports of the Scientific Committee;

(e) identify conservation needs and analyse the effectiveness of conservation measures;

(f) formulate, adopt and revise conservation measures on the basis of the best scientific evidence available, subject to the provisions of paragraph 5 of this Article;

(g) implement the system of observation and inspection established under Article XXIV of this Convention;

(h) carry out such other activities as are necessary to fulfill the objective of this Convention.

2. The conservation measures referred to in paragraph 1(f) above include the following:

(a) the designation of the quantity of any species which may be harvested in the area to which this Convention applies;

(b) the designation of regions and sub-regions based on the distribution of populations of Antarctic marine living resources;

(c) the designation of the quantity which may be harvested from the populations of regions and sub-regions;

(d) the designation of protected species;

(e) the designation of the size, age and, as appropriate, sex of species which may be harvested;

(f) the designation of open and closed seasons for harvesting;

(g) the designation of the opening and closing of areas, regions or sub-regions for purposes of scientific study or conservation, including special areas for protection and scientific study;

(h) regulation of the effort employed and methods of harvesting, including fishing gear, with a view, inter alia, to avoiding undue concentration of harvesting in any region or sub-region;

(i) the taking of such other conservation measures as the Commission considers necessary for the fulfillment of the objective of this Convention, including measures concerning the effects of harvesting and associated activities on components of the marine ecosystem other than the harvested populations.

3. The Commission shall publish and maintain a record of all conservation measures in force.

4. In exercising its functions under paragraph 1 above, the Commission shall take full account of the recommendations and advice of the Scientific Committee.

5. The Commission shall take full account of any relevant measures or regulations established or recommended by the Consultative Meetings pursuant to Article IX of the Antarctic Treaty or by existing fisheries commissions responsible for species which may enter the area to which this Convention applies, in order that there shall be no inconsistency between the rights and obligations of a Contracting Party under such regulations or measures and conservation measures which may be adopted by the Commission.

6. Conservation measures adopted by the Commission in accordance with this Convention shall be implemented by Members of the Commission in the following manner:

(a) the Commission shall notify conservation measures to all Members of the Commission;

(b) conservation measures shall become binding upon all Members of the Commission 180 days after such notification, except as provided in sub-paragraphs (c) and (d) below;

(c) if a Member of the Commission, within ninety days following the notification specified in sub-paragraph (a), notifies the Commission that it is unable to accept the conservation measure, in whole or in part, the measure shall not, to the extent stated, be binding upon that Member of the Commission;

(d) in the event that any Member of the Commission invokes the procedure set forth in sub-paragraph (c) above, the Commission shall meet at the request of any Member of the Commission to review the conservation measure. At the time of such meeting and within thirty days following the meeting, any Member of the Commission shall have the right to declare that it is no longer able to accept the conservation measure, in which case the Member shall no longer be bound by such measure.

Article X

1. The Commission shall draw the attention of any State which is not a Party to this Convention to any activity undertaken by its nationals or vessels which, in the opinion of the Commission, affects the implementation of the objective of this Convention.

2. The Commission shall draw the attention of all Contracting Parties to any activity which, in the opinion of the Commission, affects the implementation by a Contracting Party of the objective of this Convention or the compliance by that Contracting Party with its obligations under this Convention.

Article XI

The Commission shall seek to co-operate with Contracting Parties which may exercise jurisdiction in marine areas adjacent to the area to which this Convention applies in respect of the conservation of any stock or stocks of associated species which occur both within those areas and the area to which this Convention applies, with a view to harmonizing the conservation measures adopted in respect of such stocks.

Article XII

1. Decisions of the Commission on matters of substance shall be taken by consensus. The question of whether a matter is one of substance shall be treated as a matter of substance.

2. Decisions on matters other than those referred to in paragraph 1 above shall be taken by a simple majority of the Members of the Commission present and voting.

3. In Commission consideration of any item requiring a decision, it shall be made clear whether a regional economic integration organization will participate in the taking of the decision and, if so, whether any of its member States will also participate. The number of Contracting Parties so participating shall not exceed the number of member States of the regional economic integration organization which are Members of the Commission.

4. In the taking of decisions pursuant to this Article, a regional economic integration organization shall have only one vote.

Article XIII

1. The headquarters of the Commission shall be established at Hobart, Tasmania, Australia.

2. The Commission shall hold a regular annual meeting. Other meetings shall also be held at the request of one-third of its members and as otherwise provided in this Convention. The first meeting of the Commission shall be held within three months of the entry into force of this Convention, provided that among the Contracting Parties there are at least two States conducting harvesting activities within the area to which this Convention applies. The first meeting shall, in any event, be held within one year of the entry into force of this Convention. The Depositary shall consult with the signatory States regarding the first Commission meeting, taking into account that a broad representation of such States is necessary for the effective operation of the Commission.

3. The Depositary shall convene the first meeting of the Commission at the headquarters of the Commission. Thereafter, meetings of the Commission shall be held at its headquarters, unless it decides otherwise.

4. The Commission shall elect from among its members a Chairman and Vice-Chairman, each of whom shall serve for a term of two years and

shall be eligible for re-election for one additional term. The first Chairman shall, however, be elected for an initial term of three years. The Chairman and Vice-Chairman shall not be representatives of the same Contracting Party.

5. The Commission shall adopt and amend as necessary the rules of procedure for the conduct of its meetings, except with respect to the matters dealt with in Article XII of this Convention.

6. The Commission may establish such subsidiary bodies as are necessary for the performance of its functions.

Article XIV

1. The Contracting Parties hereby establish the Scientific Committee for the Conservation of Antarctic Marine Living Resources (hereinafter referred to as "the Scientific Committee") which shall be a consultative body to the Commission. The Scientific Committee shall normally meet at the headquarters of the Commission unless the Scientific Committee decides otherwise.

2. Each Member of the Commission shall be a member of the Scientific Committee and shall appoint a representative with suitable scientific qualifications who may be accompanied by other experts and advisers.

3. The Scientific Committee may seek the advice of other scientists and experts as may be required on an ad hoc basis.

Article XV

1. The Scientific Committee shall provide a forum for consultation and co-operation concerning the collection, study and exchange of information with respect to the marine living resources to which this Convention applies. It shall encourage and promote co-operation in the field of scientific research in order to extend knowledge of the marine living resources of the Antarctic marine ecosystem.

2. The Scientific Committee shall conduct such activities as the Commission may direct in pursuance of the objective of this Convention and shall:

(a) establish criteria and methods to be used for determinations concerning the conservation measures referred to in Article IX of this Convention;

(b) regularly assess the status and trends of the populations of Antarctic marine living resources;

(c) analyse data concerning the direct and indirect effects of harvesting on the populations of Antarctic marine living resources;

(d) assess the effects of proposed changes in the methods or levels of harvesting and proposed conservation measures;

(e) transmit assessments, analyses, reports and recommendations to the Commission as requested or on its own initiative regarding measures and research to implement the objective of this Convention;

(f) formulate proposals for the conduct of international and national programs of research into Antarctic marine living resources.

3. In carrying out its functions, the Scientific Committee shall have regard to the work of other relevant technical and scientific organizations and to the scientific activities conducted within the framework of the Antarctic Treaty.

Article XVI

1. The first meeting of the Scientific Committee shall be held within three months of the first meeting of the Commission. The Scientific Committee shall meet thereafter as often as may be necessary to fulfill its functions.

2. The Scientific Committee shall adopt and amend as necessary its rules of procedure. The rules and any amendments thereto shall be approved by the Commission. The rules shall include procedures for the presentation of minority reports.

3. The Scientific Committee may establish, with the approval of the Commission, such subsidiary bodies as are necessary for the performance of its functions.

Article XVII

1. The Commission shall appoint an Executive Secretary to serve the Commission and Scientific Committee according to such procedures and on such terms and conditions as the Commission may determine. His term of office shall be for four years and he shall be eligible for reappointment.

2. The Commission shall authorize such staff establishment for the Secretariat as may be necessary and the Executive Secretary shall appoint, direct and supervise such staff according to such rules and procedures, and on such terms and conditions as the Commission may determine.

3. The Executive Secretary and Secretariat shall perform the functions entrusted to them by the Commission.

Article XVIII

The official languages of the Commission and of the Scientific Committee shall be English, French, Russian and Spanish.

Article XIX

1. At each annual meeting, the Commission shall adopt by consensus its budget and the budget of the Scientific Committee.

2. A draft budget for the Commission and the Scientific Committee and any subsidiary bodies shall be prepared by the Executive Secretary

and submitted to the Members of the Commission at least sixty days before the annual meeting of the Commission.

3. Each member of the Commission shall contribute to the budget. Until the expiration of five years after the entry into force of this Convention, the contribution of each Member of the Commission shall be equal. Thereafter the contribution shall be determined in accordance with two criteria: the amount harvested and an equal sharing among all Members of the Commission. The Commission shall determine by consensus the proportion in which these two criteria shall apply.

4. The financial activities of the Commission and Scientific Committee shall be conducted in accordance with financial regulations adopted by the Commission and shall be subject to an annual audit by external auditors selected by the Commission.

5. Each Member of the Commission shall meet its own expenses arising from attendance at meetings of the Commission and of the Scientific Committee.

6. A Member of the Commission that fails to pay its contributions for two consecutive years shall not, during the period of its default, have the right to participate in the taking of decisions in the Commission.

Article XX

1. The Members of the Commission shall, to the greatest extent possible, provide annually to the Commission and to the Scientific Committee such statistical, biological and other data and information as the Commission and Scientific Committee may require in the exercise of their functions.

2. The Members of the Commission shall provide, in the manner and at such intervals as may be prescribed, information about their harvesting activities, including fishing areas and vessels, so as to enable reliable catch and effort statistics to be compiled.

3. The Members of the Commission shall provide to the Commission at such intervals as may be prescribed information on steps taken to implement the conservation measures adopted by the Commission.

4. The Members of the Commission agree that in any of their harvesting activities, advantage shall be taken of opportunities to collect data needed to assess the impact of harvesting.

Article XXI

1. Each Contracting Party shall take appropriate measures within its competence to ensure compliance with the provisions of this Convention and with conservation measures adopted by the Commission to which the Party is bound in accordance with Article IX of this Convention.

2. Each Contracting Party shall transmit to the Commission information on measures taken pursuant to paragraph 1 above, including the imposition of sanctions for any violation.

Article XXII

1. Each Contracting Party undertakes to exert appropriate efforts, consistent with the Charter of the United Nations, to the end that no one engages in any activity contrary to the objective of this Convention.

2. Each Contracting Party shall notify the Commission of any such activity which comes to its attention.

Article XXIII

1. The Commission and the Scientific Committee shall co-operate with the Antarctic Treaty Consultative Parties on matters falling within the competence of the latter.

2. The Commission and the Scientific Committee shall co-operate, as appropriate, with the Food and Agriculture Organisation of the United Nations and with other Specialised Agencies.

3. The Commission and the Scientific Committee shall seek to develop co-operative working relationships, as appropriate, with intergovernmental and non-governmental organizations which could contribute to their work, including the Scientific Committee on Antarctic Research, the Scientific Committee on Oceanic Research and the International Whaling Commission.

4. The Commission may enter into agreements with the organizations referred to in this Article and with other organizations as may be appropriate. The Commission and the Scientific Committee may invite such organizations to send observers to their meetings and to meetings of their subsidiary bodies.

Article XXIV

1. In order to promote the objective and ensure observance of the provisions of this Convention, the Contracting Parties agree that a system of observation and inspection shall be established.

2. The system of observation and inspection shall be elaborated by the Commission on the basis of the following principles:

(a) Contracting Parties shall co-operate with each other to ensure the effective implementation of the system of observation and inspecting, taking account of the existing international practice. This system shall include, inter alia, procedures for boarding and inspection by observers and inspectors designated by the Members of the Commission and procedures for flag state prosecution and sanctions on the basis of evidence resulting from such boarding and inspections. A report of such prosecutions and sanctions imposed shall be included in the information referred to in Article XXI of this Convention;

(b) in order to verify compliance with measures adopted under this Convention, observation and inspection shall be carried out on board vessels engaged in scientific research or harvesting of marine living resources in the area to which this Convention applies, through observers and inspectors designated by the Members of the Commission and operating under terms and conditions to be established by the Commission;

(c) designated observers and inspectors shall remain subject to the jurisdiction of the Contracting Party of which they are nationals. They shall report to the Member of the Commission by which they have been designated which in turn shall report to the Commission.

3. Pending the establishment of the system of observation and inspection, the Members of the Commmission shall seek to establish interim arrangements to designate observers and inspectors and such designated observers and inspectors shall be entitled to carry out inspections in accordance with the principles set out in paragraph 2 above.

Article XXV

1. If any dispute arises between two or more of the Contracting Parties concerning the interpretation or application of this Convention, those Contracting Parties shall consult among themselves with a view to having the dispute resolved by negotiation, inquiry, mediation, conciliation, arbitration, judicial settlement or other peaceful means of their own choice.

2. Any dispute of this character not so resolved shall, with the consent in each case of all Parties to the dispute, be referred for settlement to the International Court of Justice or to arbitration; but failure to reach agreement on reference to the International Court or to arbitration shall not absolve Parties to the dispute from the responsibility of continuing to seek to resolve it by any of the various peaceful means referred to in paragraph 1 above.

3. In cases where the dispute is referred to arbitration, the arbitral tribunal shall be constituted as provided in the Annex to this Convention.

Article XXVI

1. This Convention shall be open for signature at Canberra from 1 August to 31 December 1980 by the States participating in the Conference on the Conservation of Antarctic Marine Living Resources held at Canberra from 7 to 20 May 1980.

2. The States which so sign will be the original signatory States of the Convention.

Article XXVII

1. This Convention is subject to ratification, acceptance or approval by signatory States.

2. Instruments of ratification, acceptance or approval shall be deposited with the Government of Australia, hereby designated as the Depositary.

Article XXVIII

1. This Convention shall enter into force on the thirtieth day following the date of deposit of the eighth instrument of ratification, acceptance or approval by States referred to in paragraph 1 of Article XXVI of this Convention.

2. With respect to each State or regional economic integration organization which subsequent to the date of entry into force of this Convention deposits an instrument of ratification, acceptance, approval or accession, the Convention shall enter into force on the thirtieth day following such deposit.

Article XXIX

1. This Convention shall be open for accession by any State interested in research or harvesting activities in relation to the marine living resources to which this Convention applies.

2. This Convention shall be open for accession by regional economic integration organizations constituted by sovereign States which include among their members one or more States Members of the Commission and to which the States Members of the organization have transferred, in whole or in part, competences with regard to the matters covered by this Convention. The accession of such regional economic integration organizations shall be the subject of consultations among Members of the Commission.

Article XXX

1. This Convention may be amended at any time.

2. If one-third of the Members of the Commission request a meeting to discuss a proposed amendment the Depositary shall call such a meeting.

3. An amendment shall enter into force when the Depositary has received instruments of ratification, acceptance or approval thereof from all the Members of the Commission.

4. Such amendment shall thereafter enter into force as to any other Contracting Party when notice of ratification, acceptance or approval by it has been received by the Depositary. Any such Contracting Party from which no such notice has been received within a period of one year from the date of entry into force of the amendment in accordance with paragraph 3 above shall be deemed to have withdrawn from this Convention.

Article XXXI

1. Any Contracting Party may withdraw from this Convention on 30 June of any year, by giving written notice not later than 1 January of the same year to the Depositary, which, upon receipt of such a notice, shall communicate it forthwith to the other Contracting Parties.

2. Any other Contracting Party may, within sixty days of the receipt of a copy of such notice from the Depositary, give written notice of withdrawal to the Depositary in which case the Convention shall cease to be in force on 30 June of the same year with respect to the Contracting Party giving such notice.

3. Withdrawal from this Convention by any Member of the Commission shall not affect its financial obligations under this Convention.

Article XXXII

The Depositary shall notify all Contracting Parties of the following:

(a) signatures of this Convention and the deposit of instruments of ratification, acceptance, approval or accession;

(b) the date of entry into force of this Convention and of any amendment thereto.

Article XXXIII

1. This Convention, of which the English, French, Russian and Spanish texts are equally authentic, shall be deposited with the Government of Australia which shall transmit duly certified copies thereof to all signatory and acceding Parties.

2. This Convention shall be registered by the Depositary pursuant to Article 102 of the Charter of the United Nations.

Drawn up at Canberra this twentieth day of May 1980.

In Witness Whereof the undersigned, being duly authorized, have signed this Convention.

[Here follow the signatures on behalf of the parties to the Agreement, including Austalia]

Annex for an Arbitral Tribunal

1. The arbitral tribunal referred to in paragraph 3 of Article XXV shall be composed of three arbitrators who shall be appointed as follows:

(a) The Party commencing proceedings shall communicate the name of an arbitrator to the other Party which, in turn, within a period of forty days following such notification, shall communicate the name of the second arbitrator. The Parties shall, within a period of sixty days following the appointment of the second arbitrator, appoint the third arbitrator, who shall not be a national of either Party and shall not be of the same nationality as either of the first two arbitrators. The third arbitrator shall preside over the tribunal.

(b) If the second arbitrator has not been appointed within the prescribed period, or if the Parties have not reached agreement within the prescribed period on the appointment of the third arbitrator, that arbitrator shall be appointed, at the request of either Party, by the Secretary-General of the Permanent Court of Arbitration, from among persons of internatioanl standing not having the nationality of a State which is a Party to this Convention.

2. The arbitral tribunal shall decide where its headquarters will be located and shall adopt its own rules of procedure.

3. The award of the arbitral tribunal shall be made by a majority of its members, who may not abstain from voting.

4. Any Contracting Party which is not a Party to the dispute may intervene in the proceedings with the consent of the arbitral tribunal.

5. The award of the arbitral tribunal shall be final and binding on all Parties to the dispute and on any Party which intervenes in the proceedings and shall be complied with without delay. The arbitral tribunal shall interpret the award at the request of one of the Parties to the dispute or of any intervening Party.

6. Unless the arbitral tribunal determines otherwise because of the particular circumstances of the case, the expenses of the tribunal, including the remuneration of its members, shall be borne by the Parties to the dispute in equal shares.

Convention on the Conservation of Antarctic Marine Living Resources

Ratifying Nation	Date
Argentina	(May 28, 1982)
Australia	(May 6, 1981)
Chile	(July 22, 1981)
France	(Sept. 16, 1982)
European Economic Community	(Acceded April 21, 1982)
Federal Republic of Germany	(April 23, 1982)
German Democratic Republic	(March 30, 1982)
Japan	(May 26, 1981)
New Zealand	(March 8, 1982)
Norway	(Dec. 6, 1983)
Poland	(March 28,1984)
South Africa	(July 23, 1981)
Soviet Union	(May 26, 1981)
United Kingdom	(Aug. 31, 1981)
United States	(Feb. 18, 1982)

Convention Entered into Force, April 21, 1982.

Acceding Parties	Date
India	(June 17, 1985)
Uraguay	(March 22, 1985)

Notes

Chapter 1

1. *The Voyages of Captain James Cook Round the World* (London, Richard Phillips, 1809) vol. IV, p. 209, see also pp. 208–210, 216, and 217 and C. Hartley Grattan, *The Southwest Pacific to 1900, A Modern History* (Ann Arbor, The University of Michigan Press, 1963) pp. 14–30 (hereafter cited as Grattan, *To 1900*).

2. Robert F. Scott, *Scott's Last Expedition* (New York, Dodd, Mead, 1913), diary entry for Jan. 17, 1912, p. 374 (hereafter cited as Scott, *Scott's Last Expedition*).

3. Testimony of Laurence M. Gould, *The Antarctic Treaty*, Hearings before the Committee on Foreign Relations, U.S. Senate, 86 Cong. 2 sess. (Washington, D.C., Government Printing Office, 1960) p. 75.

4. Central Intelligence Agency, *Polar Regions Atlas* (Washington, D.C., U.S. Government Printing Office, 1978) p. 35. Memorandum, "Antarctic Statistics," Division of Polar Programs, National Science Foundation, Washington, D.C., Sept. 27, 1983.

5. Scott, *Scott's Last Expedition*, diary entry for Dec. 2, 1910, pp. 7–10.

6. Grattan, *To 1900*, p. 521; see also Margery and James Fisher, *Shackleton* (London, James Barrie Books Ltd., 1957) pp. 376–377.

7. See, generally, Sir James Clark Ross, *A Voyage of Discovery and Research in the Southern and Antarctic Regions, During the Years 1839–43* (London, John Murray, 1847) 2 vol. See also Grattan, *To 1900*, pp. 239 and 240.

8. Grattan, *To 1900*, pp. 231 and 232; and C. Hartley Grattan, *The Southwest Pacific Since 1900, A Modern History* (Ann Arbor, The University of Michigan Press, 1963) pp. 584 and 585.

9. Creina Bond and Roy Siegfried, *Antarctica, No Single Country, No Single Sea* (New York, Mayflower Books, 1979) p. 169 (hereafter cited as Bond and Siegfried, *Antarctica*).

10. CIA, *Polar Regions Atlas*, pp. 35.

11. See, generally, G. H. Denton and T. J. Hughes, eds., *The Last Great Ice Sheets* (New York, John Wiley & Sons, 1981). Also see "Tales the Ice Can Tell," *Mosaic* vol. 9, no. 5 (Sept.–Oct.) 1978 (Washington, D.C., National Science Foundation). Also see, Brian Roberts, "The Place Names 'Greater Antarctica' and 'Lesser Antarctica' versus 'East Antarctica' and 'West Antarctica,'" *British Antarctic Survey Bulletin* 53, 1981.

12. I. O. Norton and J. G. Sclater, "A Model for the Evolution of the Indian Ocean and the Breakup of Gondwanaland," *Journal of Geophysical Research* vol. 84, no. B12, pp. 6803–6830; Bond and Siegfried, *Antarctica;* pp. 23 and 24; and Walter Sullivan, *Quest for a Continent* (New York, McGraw-Hill, 1957) p. 15 (hereafter cited as Sullivan, *Quest*).

13. Michael Kuhn, "Optical Phenomena in the Antarctic Atmosphere," paper no. 9, in Meteorological Studies at Plateau Station, Antarctica, *Antarctic Research Series*, vol. 25 (Washington, D.C., American Geophysical Union, 1978).

14. Edward A. Wilson, *Diary of the Terra Nova Expedition to the Antarctic, 1910–12* (New York, Humanities Press, 1972) diary entry for Dec. 11, 1910, p. 74.

15. S.I. Akasofu, "Auroras" in *Encyclopedia Britannica* (15 ed., Chicago, Ill., William Benton, 1981) vol. 2, pp. 373–377; and Robert H. Eather, *Majestic Lights: The Aurora in Science, History, and the Arts* (Washington, D.C., American Geophysical Union, 1980).

16. Scott, *Scott's Last Expedition,* entry for June 22, 1911, p. 227.

17. Apsley Cherry-Garrard, *The Worst Journey in the World, Antarctic, 1910–1913* (London, Constable and Co. Ltd., 1922) pp. 232 and 237–238.

18. The most fantastic alleged link is the Piri Reis maps. Piri Reis was a Turkish admiral of the sixteenth century sailing in the service of the Ottoman Empire. About 1500, he allegedly drew two maps of the world based on his own voyages and ancient records. The maps were unearthed at the Topkapi Museum at Istanbul in 1929. They have been of interest to scholars of pre-Columbian America because of their outlines of "American" continents.

They also picture Antarctica and Greenland with coastlines they might have had before they were glaciated over. An American, Arlington H. Mallery, put forward the theory that since the glaciation was believed (at that time) to have occurred 10,000 years ago, during the era of Cro-Magnon man, there must have existed on earth at that time an advanced civilization, having aviation, hydrography, and other advances required to explore these regions. Mallery's theory was taken up by the French polar explorer, Paul-Emile Victor, who gave the impression he agreed with it, including the chance that the mysterious civilization had been "completely wiped out from one day to the next by a catastrophe." Victor's original article, "L'Enigme Piri Reis," appeared in *Planete* no. 29 (July) 1966. An English version, "The Story of the 'Impossible Maps'," appears in Louis Pauwels and Jacques Bergier, *Eternal Man* (London, Souvenir Press, 1972) pp. 40–51.

19. David Sugden, *Arctic and Antarctic* (Totowa, N.J., Barnes & Noble, 1982) p. 207.

20. See, for example, Bond and Siegfried, *Antarctica,* pp. 79 and 80. See also Alan Moorehead, *The Fatal Impact: An Account of the Invasion of the South Pacific; 1767–1840* (New York, Harper & Row, 1966).

21. Kenneth J. Bertrand, *Americans in Antarctica 1775–1948* (New York, American Geographical Society, 1971) Special Publication No. 39., pp. 66–73 and 77. See also chap. II.

22. William H. Hobbs, "The Discoveries of Antarctica Within the American Sector, as Revealed by Maps and Documents," *Transactions of the American Philosophical Society* vol. 31, pt. 1 (January) 1939 (Philadelphia, The American Philosophical Society, 1939) pp. 12, 13, and 16.

23. Sullivan, *Quest,* pp. 28–34. Also, Walter Sullivan, *Assault on the Unknown* (New York, McGraw-Hill, 1961) p. 6.

24. Sullivan, *Quest,* pp. 28–34.

25. Sir Douglas Mawson, *The Home of the Blizzard: Being the Story of the Australasian Antarctic Expedition, 1911–1914,* vol. 1 (Philadelphia, Pa., J.B. Lippincott, 1914) p. 297.

26. Grattan, *Since 1900,* p. 570; CIA, *Polar Regions Atlas,* pp. 52–53; Bond and Siegfried, *Antarctica,* pp. 109, 112, 114, 115, 118, and 119.

27. Scott, *Scott's Last Expedition;* Roald Amundsen, *The South Pole,* translated from Norwegian by A. G. Chater (London, John Murray, 1913) 2 vol.; Cherry-Garrard, *The Worst Journey in the World;* Wilson, *Diary of the Terra Nova Expedition;* and Ernest H. Shackleton, *The Heart of the Antarctic; Being the Story of the British Antarctic Expedition 1907–1909* (Philadelphia, J. B. Lippincott, Co., 1909) 2 vol.; *South! The Story of Shackleton's Last Expedition 1914–1917* (New York, Macmillan, 1920); Sir Douglas Mawson, *The Home of the Blizzard: Being the Story of the Australasian Antarctic Expedition, 1911–1914* (London, W. Heinemann, 1914) 2 vol.; Lennard Bickel, *Mawson's Will: The Greatest Survival Story Ever Written* (New York, Stein and Day, 1977).

28. See, for example, Edward Shackleton, *Nansen, The Explorer* (London, H. F. & G. Witherby, Ltd., 1959). See also *Encyclopedia Britannica,* 1967, p. 1168.

29. Grattan, *To 1900,* p. 520.

30. James Morris, *Farewell the Trumpets: An Imperial Retreat* (New York, Harcourt Brace Jovanovich, 1980) pp. 423–425.

31. Fridtjof Nansen, *In Northern Mists,* vol. 1 (New York, Frederick A. Stokes Co., 1911) pp. 1–4.

32. Roland Huntford, *Scott and Amundsen* (London, Hodder and Stoughton, 1979).

33. Fisher, *Shackleton,* pp. 351–397, quote p. 387.

34. H. R. Mill, *Geographical Journal* vol. 33, p. 570 (1909); quoted in ibid., p. 105.

35. Charles Neider, "Argentina Covets the Antarctic, Too," *New York Times,* April 9, 1982.

36. Deborah Shapley, "Antarctica: Up For Grabs," *Science 82* vol. 3, no. 9 (November) 1982, p. 75.

Chapter 2

1. Paul A. Siple, *90° South* (New York, G. P. Putnam's, 1959) p. 89.

2. U.S. Congress. Senate Committee on Foreign Relations, *Hearings on the Antarctic Treaty.* 86 Cong., 2 sess. (Washington, D.C., Government Printing Office, 1960) p. 15.

3. A list of claims markers (mostly sheets of paper, signed by U.S. personnel, that were encased in brass canisters and dropped from aircraft or buried in cairns) put down in Antarctica by U.S. expeditions through October 23, 1957, was compiled by the U.S. Antarctic Projects Office about 1957, probably by Henry Dater, the office's historian from 1956–65. It is found in Records of the Office of the Secretary of Defense, Records of the United States Antarctic Projects Officer, Deputy's subject files, "Claims, Antarctica," RG 330, NA. The list is the basis for the "x" marks on the figures in Chapter 2.

4. C. Hartley Grattan, *The Southwest Pacific to 1900, A Modern History* (Ann Arbor, University of Michigan Press, 1963) pp. 226 and 231.

5. Author interview with Edouard A. Stackpole, June 17, 1981. See also Edouard A. Stackpole, *The Sea-Hunters: The New England Whalemen During Two Centuries 1635–1835* (Philadelphia, J. B. Lippincott, 1953) pp. 357–365.

6. William H. Hobbs, "The Discoveries of Antarctica Within the American Sector, as Revealed by Maps and Documents," *Transactions of the American Philosophical Society* vol. 31, pt. 1 (January) 1939 (Philadelphia, Pa., The American Philosophical Society, 1939) pp. 12, 13, and 16 (hereafter cited as Hobbs, "Discoveries").

7. Kenneth J. Bertrand, *Americans in Antarctica 1775–1948,* Special Pub. No. 39 (New York, American Geographical Society, 1971) pp. 66–73 (hereafter cited as Bertrand, *Americans*).

8. Hobbs, "Discoveries," pp. 16, 18, and 23–26; and Edmund Fanning, *Voyages Round the World* (New York, Collins and Hannay, 1833) pp. 435–438.

9. Edouard A. Stackpole, *The Voyage of the Huron and the Huntress* (Hartford, Conn., Connecticut Printers, Inc., 1955) no. 29, pp. 10 and 28–32; and Hobbs, "Discoveries," pp. 24–26 and 13–16.

10. Bertrand, *Americans,* p. 84, note 1. Also see *Foreign Relations of the United States, 1950,* vol. 1, p. 912; and Hobbs, "Discoveries," p. 21.

11. Hobbs, "Discoveries," pp. 18 and 19.

12. Ibid., pp. 19 and 20.

13. Hugh R. Mill, *The Siege of the South Pole* (London, Alston Rivers Ltd., 1905) pp. 101 and 102; Bertrand, *Americans,* pp. 78–80; and Hobbs, "Discoveries," p. 21.

14. Author interview with Edouard A. Stackpole. Also see Stackpole, *Voyage of The Huron,* pp. 51, 52; and Bertrand, *Americans,* pp. 89–101, including p. 100, note 2.

15. Bertrand, *Americans,* pp. 144 and 145; and William Stanton, *The Great United States Exploring Expedition of 1838–1842* (Berkeley, University of California Press, 1975) pp. 8–15 (hereafter cited as Stanton, *The Great United States*).

16. Bertrand, *Americans,* p. 145.

17. Ibid., pp. 145–147.

18. Ibid., pp. 148–151. Joel W. Hedgpeth, "James Eights of the Antarctic (1798–1882)," and essays by Eights relating to the voyage are found in Louis O. Quam, ed., *Research in the Antarctic,* Pub. no. 93 (Washington, D.C., American Association for the Advancement of Science, 1971) pp. 3–40 (hereafter cited as Quam, *Research*).

19. Quam, *Research.*

20. Bertrand, *Americans,* pp. 160–162.

21. Ibid., pp. 160–167.

22. Generally, see Stanton, *The Great United States.*

23. David Jaffe, *The Stormy Petrel and the Whale: Some Origins of Moby Dick* (Baltimore, Md., Port City Press, 1976). Jaffe also suggests that Melville might have hit upon Moby Dick's name from an article by Jeremiah N. Reynolds, the promotor of the Wilkes expedition, titled, "Mocha-Dick; or the White Whale of the Pacific," which appeared in *Knickerbocker Magazine* (May) 1839.

24. The accuracy of Wilkes's land sightings in Antarctica was also an issue at the court-martial. See Bertrand, *Americans,* pp. 166, and 173–184.

25. Ibid., pp. 219–234.

26. Ibid., pp. 198–205.

27. Ibid., pp. 267 and 268. See Robert Cushman Murphy, "South Georgia, an Outpost of the Antarctic," *National Geographic* vol. 41, no. 4 (April) 1922.

28. Roald Amundsen, *My Life as an Explorer* (New York, Doubleday, 1927) pp. 27 and 28; and Walter Sullivan, *Quest for a Continent* (New York, McGraw-Hill, 1957) pp. 36 and 37 (hereafter cited as Sullivan, *Quest*).

29. J. P. Ault, "Sailing the Seven Seas in the Interest of Science," *National Geographic* vol. 42, no. 6 (December) 1922, pp. 646, 649, and 651.

30. Bertrand, *Americans,* p. 284.

31. See note 53 below.

32. Bertrand, *Americans,* pp. 276, 284–288. Sir Hubert Wilkins, "The Wilkins-Hearst Antarctic Expedition, 1928–29" *Geographical Review* vol. 19, no. 3 (July) 1929, pp. 353–376; and Charles J. V. Murphy, *Struggle, The Life of Commander Byrd* (New York, Frederick A. Stokes, 1928) p. 1 (hereafter cited as Murphy, *Struggle*).

33. Murphy, *Struggle,* p. 243. Also see *The Antarctican Society Newsletter* vol. 81–82, no. 6 (April) 1982, p. 2; and Bertrand, *Americans,* p. 291.

34. Richard E. Byrd, *Little America* (New York, G. P. Putnam's, 1930) p. 24 (hereafter cited as Byrd, *Little America*).

35. Bertrand, *Americans,* pp. 279, 292, and 302–304; and ibid., pp. 329, 332, 335, and 336.

36. Byrd, *Little America,* pp. 341 and 342.

37. Ibid., pp. 348–357; and Bertrand, *Americans,* pp. 290–312.

38. Laurence M. Gould, *Cold* (New York, Brewer, Warren & Putnam, 1931) p. 21.

39. Ibid., pp. 25–28.

40. Bertrand, *Americans,* p. 277.

41. Author interview with Peter Anderson.

42. Bertrand, *Americans,* pp. 295, 304–307.

43. See Paul A. Siple, *A Boy Scout With Byrd* (New York, G. P. Putnam's, 1931).

44. Richard E. Byrd, *Alone* (New York, G. P. Putnam's, 1938) pp. 34 and 35 (hereafter cited as Byrd, *Alone*).

45. Bertrand, *Americans,* pp. 292, 313, and 314.

46. *Congressional Record,* 71 Cong., 2 sess., 1930, pp. 12179 and 12180.

47. Richard E. Byrd, *Discovery* (New York, G. P. Putnam's, 1935) p. 1 (hereafter cited as Byrd, *Discovery*).

48. Bertrand, *Americans,* maps on pp. 314, 317, and 321.

49. Byrd, *Alone,* pp. 31 and 32; and ibid., pp. 316, 317, and 329.

50. Margery Fisher and James Fisher, *Shackleton and the Antarctic* (Boston, Mass., Houghton Mifflin, 1958) pp. 217 and 218.

51. Byrd, *Alone,* pp. 88 and 89.

52. See Siple's description of Byrd on Operation Deepfreeze and the indignities they both suffered at the hands of Adm. George Dufek's staff (Siple, *90° South,* pp. 119–123). Other incidents are recounted in Lisle A. Rose, *Assault on Eternity* (Annapolis, Md., Naval Institute Press, 1980) (hereafter cited as Rose, *Assault*).

53. Balchen based his accusation on the work of G. H. Liljequist of the University of Uppsala in Sweden. See Richard Montague, *Oceans, Poles and Airmen* (New York, Random House, 1971) app. A. An account of the Byrd family's attempt at censorship is found in app. B of this same volume. Bernt Balchen's sanitized account is published in *Come North With Me* (New York, E. P. Dutton, 1958).

54. Lincoln Ellsworth, *Beyond Horizons* (New York, The Book League of America, 1938) p. 255 (hereafter cited as Ellsworth, *Beyond*).

55. Bertrand, *Americans,* pp. 386 (map), 362, and 389. See also ibid., p. 12.

56. Bertrand, *Americans,* pp. 395–398, 403.

57. National Archives. Record Group 59. Files of State Department. "Secretary of State Charles E. Hughes to Norwegian Minister H. Bryn, April 2, 1924." SDDF 857.014/6. Also see *Foreign Relations of the United States, 1924,* vol. 2, p. 519.

58. National Archives. Record Group 59. Files of State Department. "Secretary of State to A. W. Prescott, May 13, 1924." SDDF 811.014/101.

59. *Newsweek,* March 25, 1957, p. 70.

60. National Archives. Record Group 59. Files of State Department. "Correspondence of Ronald Lindsay, British Ambassador, with Secretary of State Cordell Hull, November 14, 1934"; includes Postmaster General James A. Farley's authorization of Postmaster Charles F. Anderson on Byrd expedition. SDDF 031 Byrd South Polar Expedition, document 161; see also documents 158, 159, 160, 162, and 166; "Memorandum of Conversation with British Ambassador" by Under Secretary of State William Phillips, January 3, 1935, SDDF 031 Byrd South Polar Expedition, document 168; and "Prohibition Relating to Sealing at Little America," by S. W. Boggs, July 30, 1935, SDDF 031 Byrd South Polar Expedition document 202.

61. Senator Harry Byrd pestered the State Department to intercede with Norway regarding Commander Hjalmer Gjertsen. National Archives, Record Group 59. Files of State Department. "Correspondence involving Secretary of State Cordell Hull, Senator Byrd, William Phillips, Hoffman Philip (American Minister in Oslo), April 3, 1934 Through July 19, 1934." SDDF 031 Byrd South Polar Expedition, documents 145–153; also, "Letter with attachment, S. E. Duncan, U.S. Senate staff, to Hugh S. Cumming, Jr., January 25, 1935." SDDF 031 Byrd South Polar Expedition, document 173; "Letter, William Phillips from Bernon S. Prentice, September 27, 1934"; and "Memorandum, William Phillips, October 2, 1934;" SDDF 031.11 Ellsworth Antarctic Expedition, documents 5 and 6.

62. National Archives. Record Group 59. File of the State Department. "Memorandum of Conversation between Joseph Ulmer and S. W. Boggs, June 16, 1938." File 031.11

Ellsworth Antarctic Expedition, document H/AC. Also see "S. W. Boggs (Study, 9/21/33)." File 800.014, Arctic, document 31.

63. National Archives. Record Group 59. File of the State Department. "Letter, Cordell Hull to James Orr Denby, American Consul, Capetown, Union of South Africa, August 30, 1938," SDDF 031.11 Ellsworth Antarctic Expedition, document 89; and "Telegram, Cordell Hull to American Consul, Capetown, Union of South Africa, October 22, 1938," SDDS 031.11 Ellsworth Antarctic Expedition, document 93; "James Orr Denby to Cordell Hull, October 1, 1938." SDDF, 031.11 Ellsworth Antarctic Expedition, document 94; and "Arthur L. Richards, American Vice Consul, Capetown, South Africa to Secretary of State, November 7, 1938." SDDF 031.11 Ellsworth Antarctic Expedition, document 95.

64. Ibid.

65. National Archives. Record Group 59. Files of State Department. "Memorandum of S. W. Boggs, April 18, 1939." SDDF 031.11 Ellsworth Antarctic Expedition.

66. National Archives. Record Group 59. Files of State Department. "Memorandum, H. S. Cumming, Jr., July 28, 1938." SDDF 800.014 Antarctic, document 126, US/LW.

67. National Archives. Record Group 59. Files of State Department. "Sumner Welles to President Roosevelt, January 6, 1939," SDDF 800.014 Antarctic, document 129A US/LW; "Memorandum, Franklin D. Roosevelt to Acting Secretary of State Sumner Welles, January 7, 1939," SDDF 800.014 Antarctic, document 135 US/KFC; and "Letter, Franklin D. Roosevelt to Secretary of State, July 7, 1939," SDDF, 800.014 Antarctic, document 190 US/KFC.

68. Rose, *Assault,* p. 28.

69. Ibid.; National Archives, Record Group 59, Files of State Department. "Memorandum of Conversation, Richard E. Byrd, Samuel W. Boggs, and Hugh S. Cumming, Jr., January 28, 1939," SDDF 800.014 Antarctic, document 154 US/LW; and "Letter, Cordell Hull to the President, February 13, 1939," SDDF 800.014 Antarctic, document 155A.

70. Siple, *90° South,* p. 64.

71. Order, Franklin D. Roosevelt to Rear Adm. Richard E. Byrd, November 25, 1939; reprinted in full in Bertrand, *Americans,* pp. 472–474.

72. See Sullivan, *Quest,* p. 266. Also, National Archives. Record Group 59. Files of State Department. "Memorandum, Hugh S. Cumming to Pierrepont Moffat, March 4, 1939," SDDF 800.014 Antarctic, documents 156 1/2 US/KFC; and "Sumner Welles, Acting Secretary of State to American Diplomatic Officers in the American Republics, August 8, 1939," SDDF 800.014 Antarctic, document 217 US/LW.

See National Archives. Record Group 330. Files of the U.S. Antarctic Projects Officer, Records of the Office of the Secretary of Defense. "History and Current Status of Claims in Antarctica, M-1." (Central Intelligence Agency, March 1948) p. 24 (hereafter cited as CIA, "M-1," 1948).

See also National Archives. Record Group 330. Files of the U.S. Antarctic Projects Officer, Records of the Office of the Secretary of Defense "National Intelligence Survey-Antarctica (NIS 69)," (Central Intelligence Agency, January 1956) p. V-72 (hereafter cited as NIS-69).

73. Bertrand, *Americans,* pp. 414 and 422. Thus, East Base was in West Antarctica and West Base was closer to East Antarctica. To go east in Antarctica, one travels right from zero-degree longitude, or clockwise around a map centered on the South Pole. To go west, one travels left from zero degree longitude, or counterclockwise.

74. Sullivan, *Quest,* pp. 139–141 and 149.

75. Finn Ronne, *Antarctica My Destiny* (New York, Hastings House, 1979) pp. 112–118; and Bertrand, *Americans,* pp. 411, 421, 444, and 460–468.

76. See, for example, National Archives, Record Group 126, File of the Office of Territories, U.S. Antarctic Service, 9-13-2 Administrative Minutes of the Executive Committee. "Minutes" of the 11th meeting of the Executive Committee, U.S. Antarctic Service."

77. *New York Times,* February 21, 1947, p. 3; and of March 3, 1955, p. 12. See also "Atoms

Could Make Antarctic Inhabitable," *Science News Letter*, Sept. 15, 1956, p. 168; and Raymond T. Ellickson, "Nuclear Energy and the Polar Icecap," *Science* vol. 103, no. 2671, 1946, p. 316.

78. Rose, *Assault*, pp. 41–51; and Siple, *90° South*, p. 77.

79. Bertrand, *Americans*, pp. 483–485; Rose, *Assault*, pp. 101–112; and Sullivan, *Quest*, pp. 173–248. Submarines, however, have gone under the Arctic pack ice since Sir Hubert Wilkins's first plunge in 1931. See Lowell Thomas, *Sir Hubert Wilkins, His World of Adventure* (New York, McGraw-Hill, 1961) pp. 270–271.

80. National Archives, Record Group 59, Files of the State Department. "Letter, Acting Secretary of State Dean Acheson to Secretary of the Navy, December 14, 1946," SDDF 800.014, Antarctic, document 12-1446. See also *Report of Operation Highjump*, vol. 1, p. 2.

81. See CIA, "M-1," 1948, p. 28.

82. Rose, *Assault*, p. 250; and Bertrand, *Americans*, pp. 533 and 534.

83. Bertrand, *Americans*, pp. 540–542; and Sullivan, *Quest*, pp. 249–261.

84. Ronne, *Antarctica*, pp. 9 and 135–178, the incident is described on pp. 137–139. See also Bertrand, *Americans*, pp. 521–523.

85. Ronne, *Antarctica*, p. 178; and Bertrand, *Americans*, pp. 520 (map) and 525.

86. *New York Times Index 1947, 1948* (New York, The New York Times Co., 1948, 1949) entries for these countries.

87. NIS-69, p. V-65; and *Foreign Relations of the United States, 1947* vol. 1, p. 1050.

88. Author interview with Peter J. Anderson, assistant director, Institute of Polar Studies, The Ohio State University, Columbus.

89. NIS-69, p. V-67; the study is summarized in "Study in Modern Military section, Records of the Joint Chiefs of Staff," File 1830/1.

90. NIS-69, pp. V-67, V-68, V-69; *Foreign Relations of the United States, 1948*, vol. 1, pp. 997–1001; and CIA, "M-1," 1948. See *New York Times Index 1948*, for Chile's response (August 5), Great Britain's response (August 28), Australia's response (August 31), and Norway's response (November 23).

91. NIS-69, pp. V-68 and V-69; "Operation Highjump II, An Exercise in Planning," Monograph No. 2 in "History and Research Division, U.S. Naval Support Force, Antarctica" (Washington, D.C., March 1970) pp. 12 and 13.

92. Siple, *90° South*, pp. 81 and 82; Rose, *Assault*, pp. 251 and 252.

93. Memorandum, The Embassy of the Soviet Union to the Department of State, *Foreign Relations of the United States, 1950*, vol. 1, pp. 911–913.

94. NIS-69, pp. V-69 and V-70.

95. Ibid., p. V-67.

96. Siple, *90° South*, p. 88.

97. Ibid., p. 89. Also, National Archives. Record Group 401/4. Gift Collection of Materials Relating to Polar Regions. "Meetings" and "Memoranda" files in Paul A. Siple, "Committees, TANT Meetings, 1954–55," in Papers of Paul A. Siple, family collection.

98. Author interview with Paul R. Schratz.

99. NIS-69, p. V-70; and *Foreign Relations of the United States, 1948*, vol . 1, pt. 2, p.1001.

100. Walter Sullivan, *Assault on the Unknown* (New York, McGraw-Hill, 1961) pp. 20–23 and 26 (hereafter cited as Sullivan, *Assault*).

101. Ibid., p. 21; Siple, *90° South*, App. 2. Berkner is largely responsible for "Antarctic Research, Elements of a Coordinated Program" (Washington, D.C., National Academy of Sciences, May 2, 1949, reprinted for the U.S. National Committee for the IGY, May 1954).

102. Siple, *90° South*.

103. Author interviews with Alan Shapley, Neil Carothers, Hugh Odishaw, Laurence M. Gould. See also Siple, *90° South,* p. 95.

104. National Archives. Record Group 313. Files of Naval Operating Forces, U.S. Naval Support Forces, Antarctica, Group A. NSC 5424/1 "Statement of Policy by the National Security Council on Antarctica," July 16, 1954.

105. National Archives. Record Group 401/4. Gift Collection of Materials Relating to Polar Regions. "Meetings" and "Memoranda" files in Paul A. Siple, "Committees, TANT Meetings, 1954–55," in Papers of Paul A. Siple.

106. Siple, *90° South,* pp. 126 and 127.

107. National Archives. Record Group 313. Files of Naval Operating Forces, U.S. Naval Support Forces, Antarctica. "Revisions to NSC 5424/1 pursuant to NSC 5528, Action memo 1500, Approved by the President January 16, 1956, Group A"; and author's notes.

108. *Atka* cruise orders are "Guidance for the Commanding Officer of the Antarctic Expedition," November 2, 1954, in Operation Order, COMSEVRON FOUR, No. 1, 1955 pp. C-VIII, 1-3; and Deepfreeze II orders issued August 1, 1956 are Operation Plan CTF-43 No. 1-56, pp. A-1 through A-5. All are in National Archives, Record Group 313, Records of Naval Operating Forces, U.S. Naval Support Forces, Antarctica.
Change in orders is mentioned in Walter Sullivan, "The International Geophysical Year," *International Conciliation* no. 521 (January) 1959 (New York Carnegie Endowment for International Peace) pp. 321 and 322.

109. List of claims attributed to Dater (see note 3) has as its last entries two claims dropped from plane by Finn Ronne. One is dated March 16, 1957 at 80° 30′ south latitude and 25° 00′ west longitude. The second is dated October 23, 1957 (after the IGY had begun), at 85° 10′ south latitude and 49° 00′ west longitude.

110. Siple, *90° South,* pp. 152–189. See *Time Magazine,* vol. LXVIII, (Dec. 31, 1956).

Chapter 3

1. Walter Sullivan, *Quest for a Continent* (New York, McGraw-Hill, 1957) p. 269.

2. E. W. Kevin Walton, *Two Years in the Antarctic* (New York, Philosophical Library, 1955) pp. 107 and 108.

3. See "Falkland Islands," *New York Times Index, 1982,* pp. 340–349. (New York, The New York Times Co., 1983).

4. A survey of this literature to 1959, with abstracts of foreign language articles translated into English, is found in Robert D. Hayton, *National Interests in Antarctica* (Washington, D.C., United States Antarctic Projects Officer, 1959) (hereafter cited as Hayton, *National Interests*).

5. Charles Evans Hughes, Secretary of State, to H. Bryn, Norwegian minister, April 2, 1924, *Foreign Relations of the United States, 1924,* vol. 2, p. 519.

6. "Judicial Decisions Involving Questions of International Law—Arbitral Award—The Island of Palmas (or Miangas)" (United States-Netherlands) April 4, 1928, *American Journal of International Law* vol. 22, no. 4 (October) 1928, pp. 867–912.

7. "Arbitral Award on the Subject of the Difference Relative to the Sovereignty over Clipperton Island" (Mexico–France) 1931, *American Journal of International Law* vol. 26, no. 2, (April) 1932, pp. 390–394. Philip C. Jessup and Howard J. Taubenfeld, *Controls for Outer Space and the Antarctic Analogy* (New York, Columbia University Press, 1959) pp. 141 and 142 (hereafter cited as Jessup and Taubenfeld, *Controls*).

8. "Legal Status of Eastern Greenland" (Norway–Denmark) April 5, 1933, *World Court Reports,* vol. III, no. 43, 1932–35.

9. See chap. 2, this volume, note 65.

10. Robert D. Hayton, "The 'American' Antarctic," *American Journal of International Law* vol. 50, no. 3, (July) 1956, p. 603, note 88; and Hayton, *National Interests*, item 200, p. 20.

11. Hayton, *National Interests*, item 567, p. 42.

12. Hayton, "The 'American' Antarctic," pp. 583 passim and 603–607.

13. C. Hartley Grattan, *The Southwest Pacific to 1900, A Modern History* (Ann Arbor, University of Michigan Press, 1963) pp. 15–17 (hereafter cited as Grattan, *To 1900*).

14. C. Hartley Grattan, *The Southwest Pacific Since 1900, A Modern History* (Ann Arbor, University of Michigan Press, 1963) pp. 570 and 571 (hereafter cited as Grattan, *Since 1900*).

15. Grattan, *Since 1900*, pp. 571 and 572. Hayton, *National Interests*, item 679, p. 48.

16. Grattan, *Since 1900*, pp. 621 and 622.

17. Ibid., pp. 622–625.

18. Hayton, "The 'American' Antarctic," pp. 591, 593 and 607; and Grattan, *Since 1900*, pp. 662, 667, and 668.

19. Grattan, *Since 1900*, pp. 582–583. *British Antarctic Survey, 1977* (Cambridge, England) pp. 1 and 2. Author interview with Richard M. Laws, director, British Antarctic Survey.

20. Grattan, *Since 1900*, pp. 583, 625, 664, 665, 707, 708. See, for example, Records of Debate on Antarctic Marine Living Resources in the House of Lords, May 21, 1980, pp. 955–1004, Lord Shackleton is quoted on p. 962 passim.

21. Grattan, *Since 1900*, pp. 610–612. See also, "National Intelligence Survey— Antarctica (NIS-69)," (McLean, Va., Central Intelligence Agency, January 1956) p. V-36 (hereafter cited as NIS-69).

22. Barbara Mitchell and Richard Sandbrook, *The Management of the Southern Ocean* (London, International Institute for Environment and Development, 1980) p. 25; and Barbara Mitchell and Jon Tinker, *Antarctica and Its Resources* (London, Earthscan, House of Print, 1980) p. 58.

23. Grattan, *Since 1900*, pp. 640–642.

24. NIS-69, pp. V-50–V-51. Central Intelligence Agency, *Polar Regions Atlas* (Washington, D.C., Government Printing Office, 1978) p. 43.

25. NIS-69, p. V-52. Kenneth J. Bertrand, *Americans in Antarctica 1775–1948* (New York, American Geographical Society, 1971) Special Publication No. 39, p. 9.

26. NIS-69, pp. V-50-V-52.

27. See generally, Lennard Bickel, *Mawson's Will: The Greatest Survival Story Ever Written* (New York, Stein and Day, 1977); and Grattan, *Since 1900*, pp. 640–643, 615, 616.

28. Grattan, *Since 1900*, pp. 617.

29. Ibid., pp. 618–620.

30. NIS-69, p. V-55.

31. Grattan, *Since 1900*, pp. 565, 566, 584, 585 and 599; Grattan, *To 1900*, pp. 514 and 516; and NIS-69, p. V-89.

32. Grattan, *Since 1900*, pp. 652 and 653; NIS-69, p. V-88. A report on Ritscher's expedition by U.S. Office of Naval Intelligence, "The German Antarctic Expedition of 1938–39," dated September 1946, is found in the papers of Rear Admiral George Dufek, George Arents Research Library for Special Collections, Syracuse University, Syracuse, N.Y.

33. Grattan, *Since 1900*, p. 653; and NIS-69, pp. V-88 and V-89.

34. See West German reports to the Scientific Committee on Antarctic Research, Cambridge, England, for 1978 onward. Author correspondence with Gotthilf Hempel.

35. F. M. Auburn, *Antarctic Law and Politics* (London, C. Hurst and Co., 1982) p. 49; and Grattan, *Since 1900*, pp. 568, 621–625, 667, 668, and 669.

36. Grattan, *Since 1900,* pp. 665–668. NIS-69, p. V-15. Argentine claim is discussed in Hayton, "The 'American' Antarctic," pp. 583–610.

37. Creina Bond and Roy Siegfried, *Antarctica, No Single Country, No Single Sea* (New York, Mayflower Books, 1979) p. 175; and Jessup and Taubenfeld, *Controls,* p. 147.

38. Grattan, *Since 1900,* p. 623; Hayton, *National Interests,* item 213, p. 21; and Hayton, "The 'American' Antarctic," note 40 on p. 592.

39. The Chilean claim is discussed in Hayton, "The 'American' Antarctic." The Beagle Channel dispute is discussed in *Chilean–Argentine Relations: The Beagle Channel Controversy, Some Background Papers* (2 ed., Geneva, Switzerland, 1978).

40. See, for example, numerous Chilean papers listed in Hayton, *National Interests,* items 224–314 on pp. 20–27.

41. Grattan, *Since 1900,* pp. 682 and 683.

42. Ibid., pp. 682, 683, 700, and 704.

43. Hayton, *National Interests,* item 195, p. 20; NIS-69, pp. V-68, V-83, and V-84.

44. Grattan, *Since 1900,* pp. 583 and 584; Richard E. Byrd, "The Flight to Marie Byrd Land," *Geographical Review,* American Geographical Society, vol. XXIII, no. 2 (April) 1933, p. 180.

45. Hayton, *National Interests,* items 488 and 489 on p. 37.

46. Ibid., item 1156, p. 81. Author interview with Adam Urbanek, Polish Academy of Sciences.

47. Hayton, *National Interests,* item 570, p. 42 and items 573–587, pp. 42–43.

48. Ibid., item 567, p. 42. Peter A. Toma, "Soviet Attitude towards the Acquisition of Territorial Sovereignty in the Antarctic," *American Journal of International Law* vol. 50, no. 3 (July) 1956, pp. 611–626.

49. NIS-69, pp. V-73–V-83. Soviet base patterns shown in Chap. VI, fig. VI-5. Soviet IGY program described in Walter Sullivan, *Assault on the Unknown: The International Geophysical Year* (New York, McGraw-Hill, 1961) pp. 1, 28, 29, and 325–335 (hereafter cited as Sullivan, *Assault*).

50. Ibid., pp. 400–408, 411, 412, and 415.

51. Robert D. Hayton, "The Nations and Antarctica," *Osterreichische Zeitschrift fur Offentliches Recht* (Vienna, Springer-Verlag, 1960) pp. 385 and 386. NIS-69, pp. V-39, V-60, and V-61.

52. J. Tuzo Wilson, *IGY, The Year of the New Moons* (New York, Alfred A. Knopf, 1961) pp. 288–309 (hereafter cited as Wilson, *IGY*).

53. Sullivan, *Assault,* pp. 1–19.

54. Ibid., p. 20; and Grattan, *Since 1900,* pp. 696 and 697.

55. Sullivan, *Assault,* pp. 26 and 30. Robert H. Eather, *Majestic Lights* (Washington, D.C., American Geophysical Union, 1980).

56. Laurence M. Gould, *The Polar Regions in their Relation to Human Affairs,* Bowman Memorial Lectures (New York, American Geographical Society, 1958) p. 33.

57. Sullivan, *Assault,* pp. 292 and 293; NIS-69, p. V-81; Siple, *90°-South,* (New York, Putnam's, 1959) p. 99.

58. Walter Sullivan, "The International Geophysical Year," *International Conciliation* (Carnegie Endowment for International Peace, New York) no. 521 (January) 1959, pp. 318–323.

59. NIS-69, p. V-74.

60. Sullivan, *Assault,* pp. 293 and 294. Author interview with Walter Sullivan.

61. Sullivan, *Assault,* pp. 412 and 413. Laurence M. Gould, "The History of the Scientific Committee on Antarctic Research (SCAR)" in Louis O. Quam, ed., *Research in the Antarctic* (Washington, American Association for the Advancement of Science, 1971) pp. 47–55*ff.*

"SCAR Manual" (2 ed., Cambridge, England, Scientific Committee on Antarctic Research, 1972) pp. 13–19.

62. Author interview with Alan Shapley, former vice chairman of the U.S. National Committee for the International Geophysical Year. Also, see Wilson, *IGY*.

63. Sullivan, *Assault,* pp. 294 and 413.

Chapter 4

1. Walter Sullivan, *Assault on the Unknown: The International Geophysical Year* (New York: McGraw-Hill, 1961) p. 414 (hereafter cited as Sullivan, *Assault*); "National Intelligence Survey—Antarctica, NIS-69, ch. V 'Political'" (McLean, Va., Central Intelligence Agency, January 1956) p. V-47 (hereafter cited as NIS-69).

2. The text of the Escudero Declaration, as revised by United States and Great Britain and circulated to other governments, is found in *Foreign Relations of the United States, 1950,* vol. I, pp. 905–906; *Foreign Relations of the United States, 1948,* vol. I, pt. 2, p. 962–1016. See also NIS-69, p. V-69.

3. Dufek-Herter Briefing, National Archives. See also Papers of Rear Adm. George Dufek, George Arents Research Library for Special Collections at Syracuse University, Syracuse, N.Y. (hereafter cited as Dufek files, Syracuse). Paul Siple, *90° South* (New York, G. P. Putnam's, 1959) p. 83.

4. Sullivan, *Assault,* p. 414. Statement by the President, May 3, 1958, in Dufek files, Syracuse.

5. Author interview with Paul C. Daniels.

6. NIS-69, p. V-47. Sullivan, *Assault,* pp. 1 and 2, 414. See also ch. VIII, note 31.

7. Dufek files, Syracuse. Author interview with Paul C. Daniels.

8. Dufek files, Syracuse, Memorandum of meeting held June 24, 1958.

9. Ibid., Memorandum of meeting held July 31, 1958. Author interview with Paul C. Daniels.

10. Absence of title in Antarctica is argued by M. C. W. Pinto, "The International Community and Antarctica," *University of Miami Law Review* vol. 33, no. 2 (December) 1978, pp. 475ff. Possibility of title, including U.S. title, to land in Antarctica is described in Robert D. Hayton, "The Nations and Antarctica," in *Osterreichische Zeitschrift Fur Offentliches Recht* (Vienna, Springer-Verlag, 1960) pp. 368ff. The *terra nullius* and *terra communis* arguments are compared in Steven J. Burton, "New Stresses on the Antarctic Treaty: Toward International Legal Institutions Governing Antarctic Resources," *Virginia Law Review* vol. 65, no. 3 (April) 1979, pp. 462 and 463.

11. Creina Bond and Roy Siegfried, *Antarctica, No Single Country, No Single Sea* (New York, Mayflower Books, 1979) p. 169. Legal status of the icecap as high seas is argued in J. Peter A. Bernhardt, "Sovereignty in Antarctica," *California Western International Law Journal* vol. 5, no. 2 (Spring) 1975, pp. 297ff. (hereafter referred to as Bernhardt, "Sovereignty"). See also Frank C. Alexander, Jr., "A Recommended Approach to the Antarctic Resource Problem," *University of Miami Law Review* vol. 33, no. 2 (December) 1978, pp. 371ff.; see especially pp. 382–387.

12. Condominium is discussed in F. M. Auburn, *The Ross Dependency* (The Hague, The Netherlands, Martinus Nijhoff, 1972). Common heritage is argued in J. Kish, *The Law of International Spaces* (Leiden, The Netherlands, A. W. Sitjhoff, 1973) p. 1. Gradual *de facto* internationalization under the treaty is argued by Gunner Skagestad and Kim Traavik, "New Problems—Old Solutions," *Cooperation and Conflict,* 1974, pp. 39ff.; see especially p. 46.

13. Bernhardt, in "Sovereignty," pp. 310–317, discusses the resulting legal situation if the treaty breaks down. The legal weight of acts or activities taking place while the treaty is in force is discussed in Hayton, "The Nations and Antarctica," p. 404; and Robert D.

Hayton, "The Antarctic Settlement of 1959," *American Journal of International Law* vol. 54, no. 2 (April) 1960, p. 359.

14. How living and mineral resource jurisdiction relates to the treaty is outlined in Frank Pallone, "Resource Exploitation: The Threat to the Legal Regime of Antarctica," *International Lawyer* vol. 12, no. 2 (Summer) 1978, pp. 547ff. Ways to dissociate sovereignty from rights to minerals in Antarctica are discussed in Alexander, "A Recommended Approach," pp. 417–421.

15. Author interview with Paul C. Daniels.

16. Author interview with Paul C. Daniels. Memorandum of meeting of November 18, 1958, Dufek files, Syracuse.

17. William P. Manchester, *The Glory and the Dream* (New York, Bantam, 1975) p. 750.

18. Siple, *90° South*, pp. 120–122.

19. See, generally, *Extraterritorial Criminal Jurisdiction*, Hearings before the Subcommittee on Immigration, Citizenship, and International Law of the Committee on the Judiciary, U.S. House of Representatives, 95 Cong., 1 sess., July 21 (Washington, D.C., Government Printing Office, 1977). The hearing record includes a discussion of *U.S.* v. *Escamilla*, concerning a murder in the Arctic relevant to unsolved Antarctic jurisdictional questions. Fortunately, there have been no murders in the Antarctic to date, let alone one involving persons of different nationalities. Richard S. Lewis and Philip M. Smith, eds., *Frozen Future* (New York, Quadrangle Books, 1973) p. 44 (hereafter cited as Lewis and Smith, *Frozen Future*).

20. Lewis and Smith, *Frozen Future*, p. 61.

21. Author interview with Paul C. Daniels.

22. Author interview with R. Tucker Scully, director of the Office of Oceans and Polar Affairs, U.S. Department of State.

23. Lewis and Smith, *Frozen Future*, p. 62, 63.

24. Dufek files, Syracuse. Memoranda of meetings of June 24, July 1, August 12, 1958.

25. Dufek files, Syracuse. Memorandum of meeting of September 3, 1958, with attached draft articles. Draft article VIII deals with nonsignatories.

26. Author interview with R. Tucker Scully, director of the Office of Oceans and Polar Affairs, U.S. Department of State.

27. "United States Policy and International Cooperation in Antarctica, Message from the President," (Washington, D.C., Government Printing Office, 1964) p. 10 (hereafter cited as "Message of the President").

28. *The Antarctic Treaty*, hearings before the Committee on Foreign Relations, U.S. Senate, 86 Cong., 2 sess. (Washington, D.C., Government Printing Office, 1960) p. 78. (hereafter cited as *Hearings, 1960*).

29. Ibid.

30. There are many accounts of this growth of influence. See, for example, Daniel S. Greenberg, *The Politics of Pure Science* (New York, New American Library, 1967).

31. Lewis and Smith, *Frozen Future*, pp. 62–63.

32. *Hearings*, 1960, p. 75.

33. Siple, *90° South*, p. 93.

34. Walter Sullivan, "The International Geophysical Year," *International Conciliation* (New York, Carnegie Endowment for International Peace, January 1959) no. 521, p. 324.

35. Ibid., p. 328; and Lewis and Smith, *Frozen Future*, p. 252.

36. U.S. Arms Control and Disarmament Agency, *Arms Control and Disarmament Agreements, Texts and Histories of Negotiations* (1980 ed.) p. 21. Inspections were also conducted by New Zealand in 1963, by Australia and Great Britain in 1963, and by Argentina in 1966 and 1977. See Central Intelligence Agency, *Polar Regions Atlas* (May 1978) p. 44; and data on recent inspections from the U.S. Arms Control and Disarmament Agency.

37. Quartermaster Intelligence Agency of Quartermaster Corps, U.S. Army, "Foreign Clothing and Equipment IGY Operations, Antarctica, 1956–57," Antarctic Projects Office Records, Box 341, RG 313 (RG 330) USNS FA (USAPO), p. 101. The submarine story is found in Lisle Rose, *Assault on Eternity* (Annapolis, Md., Naval Institute Press, 1980) pp. 99–124.

38. See, generally, "Message of the President."

39. A. P. Crary, "The Antarctic," *Scientific American* vol. 207, no. 3 (September 1962) p. 73.

40. U.S. Department of State, *Antarctic Treaty, Handbook of Measures in Furtherance of the Principles and Objectives of the Antarctic Treaty* (2 ed., Washington, D.C., U.S. Department of State, September 1979) p. 2301 (hereafter cited as *Handbook*).

41. Ibid., pp. 2101 and 2301.

42. Ibid., pp. 2101–2121, 3101 and 3102.

43. Ibid., pp. 2402 and 2403; and Bond and Siegfried, *Antarctica,* p. 127.

44. *Handbook,* p. 2401.

45. Ibid., pp. 9301–9310.

46. Richard B. Bilder, "The Present Legal and Political Situation in Antarctica," in Jonathan I. Charney, ed., *The New Nationalism and the Use of Common Spaces* (Totowa, N.J., Allanheld, Osmun, 1982) pp. 193, 194.

47. *Handbook,* pp. 1102–1104.

48. Ibid., p. 1104.

49. Ibid., p. 1106.

50. Louis J. Halle, *The Sea and the Ice, a Naturalist in Antarctica* (Boston, Mass., The Audubon Library, 1973) p. 135 (hereafter cited as Halle, *The Sea*).

51. George E. Watson, *Birds of the Arctic and Sub-Antarctic* (Washington, D.C., American Geophysical Union, 1975) pp. 64–66; and Halle, *The Sea,* pp. 135–137.

52. Charles Neider, *Edge of the World: Ross Island, Antarctica* (Garden City, N.Y., Doubleday, 1974) pp. 325 and 326.

53. Lewis and Smith, *Frozen Future,* pp. 40–45; and author interview with Paul C. Daniels.

Chapter 5

1. Martin W. Holdgate, ed., *Antarctic Ecology,* vol. 1 (London, The Academic Press for the Scientific Committee on Antarctic Research, 1970) pp. xi, xii (hereafter cited as Holdgate, *Antarctic Ecology*).

2. Ibid., p. xi.

3. Creina Bond and Roy Siegfried, *Antarctica, No Single Country, No Single Sea* (New York, Mayflower Books, 1979) pp. 131–135 (hereafter cited as Bond and Siegfried, *Antarctica*).

4. P. A. Moiseev, "Some Aspects of the Commercial Use of the Krill Resources of the Antarctic Seas," in Holdgate, *Antarctic Ecology,* pp. 213–216.

5. Ibid.; and John L. Bengtson, "Review of Information Regarding the Conservation of Living Resources of the Antarctic Marine Ecosystem," Report to the U.S. Marine Mammal Commission, July 1978 (hereafter cited as Bengston, "Review").

6. Moiseev in Holdgate, *Antarctic Ecology,* p. 216.

7. See George A. Knox, "Antarctic Marine Ecosystems," *Ecology,* pp. 76 and 78; Sayed Z. El-Sayed, "On the Productivity of the Southern Ocean," pp. 119–135; and J. A. Gulland, "The Development of the Resources of the Antarctic Seas," in Holdgate, *Antarctic Ecology,* pp. 217–223.

8. See N. A. Mackintosh, "Whales and Krill in the Twentieth Century," in Holdgate, *Antarctic Ecology*, pp. 195–212; Richard M. Laws, "The Significance of Vertebrates in the Antarctic Marine Ecosystem," 1974 SCAR Symposium III; Adaptations in Vertebrates (Physiology), p. 414; and Bond and Siegfried, *Antarctica*, pp. 114 and 115.

9. Holdgate, *Antarctic Ecology*, p. 193.

10. Author interview with Richard M. Laws.

11. Bond and Siegfried, *Antarctica,* p. 127.

12. R. M. Laws, "Seals and Whales of the Southern Ocean," *Philosophical Transactions of the Royal Society* (London) vol. 279, 1977, p. 94.

13. Inigo Everson, *The Living Resources of the Southern Ocean* (Rome, UN Food and Agriculture Organization, 1977) GLO/SO/77/1, p. 118; and John R. Beddington and Robert M. May, "The Harvesting of Interacting Species in a Natural Ecosystem," *Scientific American* vol. 247, no. 5 (November) 1982, p. 65.

14. Beddington and May, *Scientific American*, p. 65. R. M. Laws, "Seals and Whales," p. 89. J. R. Gilbert and A. W. Erickson, "Distribution and Abundance of Seals in the Pack Ice of the Pacific Sector of the Southern Ocean," in G. A. Llano, ed., *Adaptations Within Antarctic Ecosystems, Proceedings of the Third SCAR Symposium on Antarctic Biology* (Washington, D.C., Smithsonian Institution, 1977) pp. 703–740.

15. Beddington and May, *Scientific American*, pp. 62–69.

16. Ibid., p. 66.

17. Laws, "Significance," pp. 415 and 419–421.

18. Robert J. Hofman, "Conservation of Living Resources in Antarctica," *Transactions of the 44th North American Wildlife and Natural Resources Conference, 1979* (Washington, D.C., Wildlife Management Institute) p. 474. Also, Bengston, "Review," p. 114.

19. Bond and Siegfried, *Antarctica*, pp. 131 and 134; J. W. S. Marr, *The Natural History and Geography of Antarctic Krill (Euphausia superba Dana),* 1962 *Discovery* Reports (32); and C. Hartley Grattan, *The Southwest Pacific Since 1900, A Modern History,* (Ann Arbor, University of Michigan Press, 1963) p. 625.

20. Bond and Siegfried, *Antarctica,* p. 135.

21. M. A. McWhinnie, "Antarctic Marine Living Resources With Special Reference to Krill, *Euphausia superba:* Assessment of Adequacy of Present Knowledge," Report to National Science Foundation, December 1978, p. 44 (hereafter cited as McWhinnie, "Antarctic Marine Living Resources").

22. Personal Communication: Mary Alice McWhinnie to Tucker Scully, January 20, 1979. (Author's files); ibid., p. 44.

23. See, for example, William M. Hamner, Peggy P. Hamner, Steven W. Strand, and Ronald W. Gilmer, "Behavior of Antarctic Krill, *Euphausia superba:* Chemoreception, Feeding, Schooling, and Molting," *Science* vol. 220, no. 4595 (April 22) 1983; and William M. Hamner, "Krill—Untapped Bounty From the Sea," *National Geographic* vol. 165, no. 5 (May) 1984, pp. 626–643. Author interview with Peggy Hamner, University of California, Los Angeles; and with Langdon B. Quetin, University of California, Santa Barbara.

24. SCAR/SCOR Group of Specialists on Living Resources of the Southern Ocean, Biological Investigations of Marine Antarctic Systems and Stocks (BIOMASS), Volume 1: Research Proposals (Cambridge, England, University Library, 1977) pp. xi and 1–4.

25. Ibid., p. 5.

26. Scientific Committee for the Conservation of Antarctic Marine Living Resources, Report of the First Meeting of the Scientific Committee, held in Hobart, Australia, June 7–11, 1982, SC-CAMLR, Appendix 4. National Science Foundation news release, "Largest School of Sea Animals Spotted by Scientists off Antarctica," March 24, 1981.

27. *Antarctic Treaty: Handbook of Measures in Furtherance of the Principles and Objectives of the Antarctic Treaty* (2 ed., Washington, D.C., Department of State, September 1979) pp. 2501–2504 (hereafter cited as *Handbook*).

28. D. E. Hayes, "Leg 28 Deep-Sea Drilling in the Southern Ocean," *Geotimes,* vol. 18, no. 6 (June) 1973, pp. 19–24.

29. Ibid.

30. N. A. Wright and P. L. Williams, "Mineral Resources of Antarctica," Geological Survey Circular 705 (Reston, Va., USGS, 1974) (hereafter cited as Wright and Williams, "Mineral Resources"). John C. Behrendt, ed., "Are There Petroleum Resources in Antarctica?" in *Petroleum and Mineral Resources of Antarctica,* Geological Survey Circular 909 (Alexandria, Va., USGS, 1983) (hereafter cited as Behrendt, ed., *Petroleum*).

31. Deborah Shapley, "Antarctica: World Hunger for Oil Spurs Security Council Review," *Science* vol. 184 (May 17) 1974, pp. 776–780; and Charles D. Masters, "Estimating the Antarctic Oil Resources," *The Washington Post,* March 12, 1975.

32. Behrendt, ed., *Petroleum,* p. 1502.

33. Wright and Williams, "Mineral Resources," pp. 1–3.

34. David H. Elliot, *A Framework for Assessing Environmental Impacts of Possible Antarctic Mineral Development,* Report prepared at Institute of Polar Studies, The Ohio State University for the U.S. Department of State, January 1977, pt. I, p. VIII-1 (hereafter referred to as Elliot, *A Framework*).

35. Wright and Williams, "Mineral Resources," p. 22.

36. Elliot, *Framework,* pt. II, pp. A5 and A6.

37. Walter Sullivan, *Continents in Motion* (New York, McGraw-Hill, 1974) pp. 1–15 and 176 (hereafter cited as Sullivan, *Continents in Motion*).

38. George A. Doumani and William E. Long, "The Ancient Life of the Antarctic," *Scientific American* vol. 207, no. 3 (September) 1962, p. 184 (hereafter cited as Doumani and Long, "The Ancient Life").

39. Ibid., pp. 171, 173, 175 and 176; and Sullivan, *Continents in Motion,* p. 181.

40. Doumani and Long, "The Ancient Life," p. 169.

41. John C. Behrendt, "Speculations on the Petroleum Resources of Antarctica," draft manuscript, in J. Splettstoesser, *Mineral Resource Potential of Antarctica* (Austin, University of Texas Press, In Press; hereafter cited as Behrendt, "Speculations"). G. L. Johnson, J. R. Vanney, and D. Hayes, "The Antarctic Continental Shelf," in Campbell Craddock, *Antarctic Geoscience* (Madison, Wisc., University of Wisconsin Press, 1982) pp. 995–1002 (hereafter cited as Craddock et al., *Antarctic Geoscience*). Behrendt, ed., *Petroleum,* p. 19. See also L. F. Ivanhoe, "Antarctica—Operating Conditions and Petroleum Prospects," *Oil and Gas Journal,* vol. 78, no. 52 (December 29) 1980, pp. 212–220.

42. Elliot, *A Framework,* pt. II, pp. B1, B2, B37, and B43.

43. Author interview with John Behrendt. See D. E. Hayes, "Leg 28 Deep Sea Drilling in the Southern Ocean," *Geotimes* vol. 18, no. 6 (June) 1973, pp. 19–24. D. E. Hayes and F. J. Davey, "A Geophysical Study of the Ross Sea, Antarctica," in D. E. Hayes and L. A. Frakes, *Initial Reports of the Deep Sea Drilling Project,* vol. 28 (Washington, D.C., Government Printing Office, 1975); Wright and Williams, "Mineral Resources," pp. 16 and 17; and Behrendt, *Petroleum,* pp. 16 and 17.

44. R. A. Cooper, C. A. Landis, W. E. LeMasurier, and I. G. Speden, "Geologic History and Regional Patterns in New Zealand and West Antarctica—Their Paleotectonic and Paleogeographic Significance," in Craddock, et al. *Antarctic Geoscience,* p. 46. Also see, *International Petroleum Encyclopedia 1980* (Tulsa, Penn Well Publishing Co., 1980) p. 143; and W. E. Scott, "Energy Search Turns to Antarctic," *Energy International* vol. 15, no. 6 (June) 1978, pp. 17–20. Author interview with Simon Murdock, New Zealand Embassy.

45. Behrendt, ed., *Petroleum,* pp. 9, 12 and 13; John L. LaBrecque and Peter Barker, "The Age of the Weddell Basin," *Nature* vol. 290, no. 5806 (April 9) 1981, pp. 489–492; and Wright and Williams, "Mineral Resources," p. 17.

46. Behrendt, "Speculations," pp. 12 and 13; and Behrendt, ed., *Petroleum,* pp. 3, 17, 22 and 23.

47. Wright and Williams, "Mineral Resources," p. 19; and Brian J. Skinner, *Earth Resources* (2 ed., Englewood Cliffs, N.J., Prentice-Hall, 1976) pp. 86, 97, 98, 60, and 61 (hereafter referred to as Skinner, *Earth Resources*).

48. L. V. Fedorov and M. G. Ravich, "Geologic Comparison of Southeastern Peninsular India and Sri Lanka with a Part of East Antarctica (Enderby Land, Mac. Robertson Land, and Princess Elizabeth Land)," in Craddock, *Antarctic Geoscience*, pp. 73 and 74. Author interview with Campbell Craddock, University of Wisconsin.

49. M. G. Ravich, L. V. Fedorov, O. A. Tarutin, "Precambrian Iron Deposits of the Prince Charles Mountains," in Craddock, *Antarctic Geoscience*, p. 853.

50. Behrendt, ed., *Petroleum*, p. 42; U.S. Department of Interior, *Mineral Facts and Problems*, 1975 ed., Bureau of Mines Bulletin 667, pp. 327, 328; Skinner, *Earth Resources*, pp. 115–117; and Wright and Williams, "Mineral Resources," p. 19.

51. Behrendt, ed., *Petroleum*, p. 40; Wright and Williams, "Mineral Resources," p. 15; and Elliot, *A Framework*, pt. II, p. A-8.

52. Wright and Williams, "Mineral Resources," pp. 21 and 22.

53. Campbell Craddock, "Antarctica and Gondwanaland (Review Paper)," in Craddock, *Antarctic Geoscience*, p. 8; Behrendt, ed., *Petroleum*, p. 29.

54. Campbell Craddock, "Antarctica and Gondwanaland (Review Paper)," in Craddock, *Antarctic Geoscience*, pp. 3, 8; and Ian W.D. Dalziel and David H. Elliot, "West Antarctica: Problem Child of Gondwanaland," *Tectonics* vol. 1, no. 1 (February) 1982, pp. 8, 9, and 11–13 (hereafter cited as Dalziel and Elliot, *Tectonics*).

55. Behrendt, ed., *Petroleum*, pp. 67 and 73; Wright and Williams, "Mineral Resources," p. 17; Skinner, *Earth Resources*, pp. 72, 95, and 100; and Elliot, *A Framework*, pt. II, p. A-17.

56. John C. Behrendt, David J. Drewry, Edward Jankowski, and Muriel S. Grim, "Aeromagnetic and Radio Echo Ice-Sounding Measurements Show Much Greater Area of the Dufek Intrusion, Antarctica," *Science* vol. 209, no. 4460 (August 29) 1980, pp. 1014–1017; and Arthur B. Ford, "The Dufek Intrusion of Antarctica and a Survey of Its Minor Metals and Possible Resources," in Behrendt, ed., *Petroleum*, pp. 51–75.

57. Author interview with Arthur Ford. Also, Behrendt, ed., *Petroleum*, p. 69; Wright and Williams, "Mineral Resources," pp. 17 and 18; Elliot, *A Framework*, pt. II, pp. A-17 and A-18.

58. Wright and Williams, "Mineral Resources," p. 18; and Edward Zeller "Radioactivity Survey in Antarctica, 1978–79," *Antarctic Journal of the United States* vol. XIV, no. 5 (October) 1979, p. 39.

59. For "copper" names in the peninsula region, see pp. 173–174 of the National Science Foundation, U.S. Board on Geographic Names, *Geographic Names of the Antarctic* (Washington, D.C., National Science Foundation, 1980), compiled by Fred G. Alberts. Author interview with Adam Urbanek, Polish Academy of Sciences. See also Peter D. Rowley, Paul L. Williams, and Douglas E. Pride, "Mineral Occurrences of Antarctica," in Behrendt, ed., *Petroleum*, pp. 33–39 and 43.

60. Ian W. D. Dalziel and David H. Elliot, "The Scotia Arc and Antarctic Margin," in Alan E. M. Nairn and Francis G. Stehli, *The Ocean Basins and Margins: Vol. 1, The South Atlantic* (New York, Plenum Press, 1973) pp. 171–246; and Ian W. D. Dalziel, "The Early (Pre-Middle Jurassic) History of the Scotia Arc Region: A Review and Progress Report (Review Paper)" in Craddock, *Antarctic Geoscience*, p. 124., More recent views are expressed in Dalziel and Elliot, "The Scotia Arc," pp. 3–19, and quote is found on p. 3.

61. Neal Potter, *Natural Resource Potentials of the Antarctic*, A Resources for the Future Study (The American Geographical Society, 1969) pp. 3 and 4.

62. See chapter 4 of this volume. See also Richard B. Bilder, "The Present Legal and Political Situation in Antarctica," in Jonathan I. Charney, ed., *The New Nationalism and the Use of Common Spaces* (Totowa, N.J., Allanheld, Osmun, 1982) pp. 196, 197 and 199 (hereafter cited as Belder, "The Present").

63. *Handbook*, p. 1504.

64. James H. Zumberge, ed., *Possible Environmental Effects of Mineral Exploration and Exploitation in Antarctica* (Cambridge, England, Scientific Committee on Antarctic Research, 1979).

65. Elliot, *A Framework*.

66. U.S. Department of State, "Final Environmental Impact Statement on the Negotiation of an International Regime for Antarctic Mineral Resources" (Washington, D.C., U.S. Department of State, Office of Environment and Health, 1982).

67. M. W. Holdgate and Jon Tinker, "Oil and Other Minerals in the Antarctic," for the Scientific Committee on Antarctic Research (London, House of Print, 1979); and Holdgate, *Antarctic Ecology*, p. xiii.

68. Central Intelligence Agency, *Polar Regions Atlas*, p. 35.

69. Deborah Shapley, "Canadians Pioneer Offshore Arctic Drilling Technology," *New York Times*, Jan. 15, 1980, p. C3.

70. Ibid.

71. Ibid.; and "Milestone Reached in Offshore Oil Production Technology," *Exogram and Oil & Gas* vol. 25, no. 11, 1979, p. 14; "Going Even Deeper in Offshore Exploration," *Exogram and Oil & Gas* vol. 25, no. 12, p. 9; Roger Vielvoye and Richard Wheatley, "Oil Prices Hold Key to Growth of Industry Action in Deep Water," *Oil & Gas Journal*, Jan. 11, 1982, pp. 25–30; and Robert J. Brown, "Here Are the Latest Advances in Pipeline Trenching by Plowing," *Oil & Gas Journal*, May 2, 1983, pp. 106–111.
Submarines for tanker transport of hydrocarbons directly from wellheads in the Arctic are proposed in William H. Kumm, "Non-nuclear Submarine Tankers," *Oil & Gas Journal*, March 5, 1984, pp. 76–79.

72. Elliot, *A Framework*, pt. I, pp. VII-2, VII-3; and pt. II, pp. H-9–H-18.

73. Ibib., pt. I, pp. VII-27–VII-32.

74. Bond and Siegfried, *Antarctica*, pp. 79, 84, 85, 159, and 173, and fig. 40 and 41.

75. Elliot, *A Framework*, pt. I, p. VII-1.

76. Holdgate and Tinker, "Oil and Other Minerals," pp. 19 and 20.

77. Elliot, *A Framework*, pt. I, pp. VI-2–VI-12.

78. Ibid., pt. I, pp. VII-8–VII-10.

79. Ibid., pt. I, pp. VII-16, VII-17.

80. Ibid., pt. I, p. VII-1 and VII-3.

81. Ibid., pt. I, p. VII-3.

82. James H. Zumberge, "Potential Mineral Resource Availability and Possible Environment Problems in Antarctica," in Charney, *New Nationalism*, p. 116.

Chapter 6

1. Brian B. Roberts, "International Cooperation for Antarctic Development: The Test for the Antarctic Treaty," *Polar Record* vol. 19, no. 119 (May) 1978, p. 112 (hereafter referred to as Roberts, "International Cooperation").

2. Ibid., p. 109.

3. Testimony of Leigh Ratiner in *Law of the Sea Conference*, Hearings before the Subcommittee on Domestic and International Scientific Planning, Analysis and Cooperation, Committee on Science and Technology, U.S. House of Representatives, 95 Cong., 1 sess. (April 26, 27, 28, 1977) pp. 81–85, 93 and 94.

4. Barbara Mitchell, *Frozen Stakes: The Future of Antarctic Minerals* (London, International Institute for Environment and Development, 1983) p. 42 (hereafter cited as Mitchell, *Frozen Stakes*).

5. Declaration of Principles Governing the Sea-Bed and Ocean Floor, and the Subsoil Thereof, Beyond the Limits of National Jurisdiction, G.A. Res. 2749, (XXV) 25 U.N. GAOR, Supp. (No. 28) 24 U.N. Doc. A/8028 (1970).

6. Richard B. Bilder, "The Present Legal and Political Situation in Antarctica," in Jonathan I. Charney, ed., *The New Nationalism and the Use of Common Spaces* (Totowa, N.J., Allanheld, Osmun, 1982) pp. 184–185 (hereafter referred to as Charney, *New Nationalism*). An excellent paper describing the geologic and legal status of the Antarctic continental shelf, slope, and rise is that of J. Michel Marcoux, "Natural Resource Jurisdiction on the Antarctic Continental Margin," *Virginia Journal of International Law* vol. 11, no. 3 (May) 1971, pp. 374–405.

7. F. M. Auburn, *Antarctic Law and Politics* (London, C. Hurst, 1982) p. 206. See also Arlen J. Large, "Rules on Oil Exploration in Antarctica To Be Set at Meeting in Bonn Next Month," *Wall Street Journal,* June 15, 1983; and Mitchell, *Frozen Stakes,* pp. 42 and 45.

8. FAO interest in Barbara Mitchell and Richard J. Sandbrook, *The Management of the Southern Ocean* (London, International Institute for Environment and Development, 1980) pp. 27–29 (hereafter referred to as Mitchell and Sandbrook, *Management*). The three reports are Inigo Everson, *The Living Resources of the Southern Ocean* (Rome, Food and Agriculture Organization of the UN, 1977, GLO/SO/77/1); G. C. Eddie, *Harvesting of Krill* (Rome, FAO, 1977, GLO/SO/77/2); and G. J. Grantham, *The Utilization of Krill* (Rome, FAO, 1977, GLO/SO/77/3).

9. Text of the Convention can be found in Appendix B.

10. U.S. Department of State, *Antarctic Treaty: Handbook of Measures in Furtherance of the Principles and Objectives of the Antarctic Treaty* (2 ed., Washington, D.C., U.S. Department of State, September 1979) p. 2504 (hereafter referred to as *Handbook,* 1979).

11. James N. Barnes, "The Emerging Convention on the Conservation of Antarctic Marine Living Resources: An Attempt to Meet the New Realities of Resource Exploitation in the Southern Ocean," in Charney, *New Nationalism,* pp. 250, 252, and 262 (hereafter referred to as Barnes, "Emerging Convention").

12. Ibid., pp. 261 and 262. See also Article I of the Convention found in Appendix B.

13. Barnes, "Emerging Convention," pp. 264 and 265; and Mitchell and Sandbrook, *Management,* pp. 127–139.

14. Barbara Mitchell and Jon Tinker, *Antarctica and Its Resources* (London, Earthscan, 1980) pp. 66–68 (hereafter referred to as Mitchell and Tinker, *Antarctica*). Also see Barnes, "Emerging Convention," pp. 265–266.

15. Barnes, "Emerging Convention," pp. 245 and 246; and *Handbook*, 1979, p. 9306. See also Article VII (2) of Convention found in Appendix B.

16. Barnes, "Emerging Convention," pp. 245, 255, 256, and 259.

17. See Article III, IV, V(1) of Convention found in Appendix B.

18. Barnes, "Emerging Convention," pp. 251 and 262.

19. Commission for the Conservation of Antarctic Marine Living Resources, "Report of the First Meeting of the Commission," Hobart, Australia, May 25–June 11, 1982, SC-CCAMLR. Also see Deborah Shapley, "Antarctica Up for Grabs," *Science 82* vol. 3, no. 9 (1982) p. 75.

20. Scientific Committee for the Conservation of Antarctic Marine Living Resources, "Report of the First Meeting of the Scientific Committee," Hobart, Australia, June 7–11, 1982, SC-CCAMLR.

21. Author interview with Robert J. Hofman, U.S. Marine Mammal Commission.

22. Robert J. Hofman, "Conservation of Living Resources in Antarctica," *Transactions of the 44th North American Wildlife and Natural Resources Conference* (Washington, D.C., Wildlife Management Institute, 1979) pp. 477 and 478.

23. John R. Beddington and Robert M. May, "The Harvesting of Interacting Species in a Natural Ecosystem," *Scientific American* vol. 247, no. 5 (November) 1982, p. 69.

24. Report of meeting held at the Fridtjof Nansen Foundation, May 30–June 10, 1973, Polhogda, Norway, and reprinted in *U.S. Antarctic Policy*, Hearings before the Subcommittee on Oceans and International Environment, Committee on Foreign Relations, U.S. Senate, 94 Cong., 1 sess. (May 5, 1975) pp. 68–84.

25. Ibid., pp. 80–84.

26. Ibid., p. 74.

27. Ibid., pp. 78 and 79.

28. Roberts, "International Cooperation," pp. 111 and 112.

29. Mitchell and Tinker, *Antarctica*, pp. 60 and 89. See Testimony of R. Tucker Scully in *U.S. Antarctic Program*, Hearings before the Subcommittee on Science, Research and Technology of the Committee on Science and Technology, U.S. House of Representatives, 96 Cong., 1 sess. (May 1 and 3 1979) pp. 2–5 and 33–41.

30. Pieces of this story have been recounted in Deborah Shapley, "Antarctica: World Hunger for Oil Spurs Security Council Review," *Science* vol. 184 (May 17) 1974, pp. 776–780. Also see Mitchell and Kimball, *Antarctica* and; "Telecon with Dr. Dixie Lee Ray," memo by Robert E. Hughes, assistant director, National and International Programs, National Science Foundation, February 24, 1975. See, generally, *U.S. Antarctic Policy*, Hearings, 1975. Later U.S. policy stance can be found in Scully testimony, *U.S. Antarctic Program*, Hearings, 1979.

31. *SCAR Bulletin* (Cambridge, England, Scott Polar Research Institute) no. 69 (September) 1981, pp. 98–100.

32. *Handbook*, 1979, pp. 1503 and 1504.

33. Ibid.

34. "Antarctic Treaty. Report of the Tenth Consultative Meeting," (Washington, D.C., U.S. Department of State, October 1979) p. 12.

35. Recommendation XI-1 is reprinted in Charney, *New Nationalism;* and Appendix G quotation is found on p. 331.

36. Keith G. Brennan, "Criteria for Access to the Resources of Antarctica: Alternatives, Procedure and Experience Applicable." Paper presented at a meeting sponsored by Universidad de Chile, Instituto de Estudios Internacionales, October 6–9, 1982, at Teniente Marsh Station, Antarctica, pp. 14 and 18.

37. For example, a discussion of law for North Sea oil and gas fields is found in William T. Onorato, "Apportionment of an International Common Petroleum Deposit," *International and Comparative Law Quarterly* vol. 26 (April) 1977, pp. 324–337; and in Paul Milner, "Petroleum Legislation and Licensing Procedures in the North Sea and Some Associated International Problems," *Natural Resources Journal* vol. 18 (July) 1978, pp. 545–568. Also see J. C. Woodliffe, "International Unitisation of an Offshore Gas Field," *International and Comparative Law Quarterly* vol. 26 (April) 1977, pp. 338–353.

38. The argument that the treaty group needs a minimum of permanent institutions is carried farthest by R. Tucker Scully in "Alternatives for Cooperation and Institutionalization in Antarctica: Outlook for the 1990's." Paper presented at a meeting sponsored by Universidad de Chile, Instituto de Estudios Internacionales, October 6–9, 1982, at Teniente Marsh Station, Antarctica. Scully argues that periodic consultative meetings, and the special meetings convened for the krill and minerals negotiations, constitute a floating secretariat for the treaty system. See also discussion in Chapter 8.

39. Philip W. Quigg, *A Pole Apart*, A Twentieth Century Fund Report (New York, McGraw-Hill 1982) pp. 199 and 200.

40. Jonathan I. Charney, "Future Strategies for an Antarctic Mineral Resource Regime—Can the Environment Be Protected?" in Charney, ed., *New Nationalism*, pp. 227–229.

41. "Antarctic Minerals Regime, Beeby's Slick Solution," *Eco* vol. 23, no. 1, 1983.

42. Ibid., pp. 3 and 15; and Articles X-XV and XXI.

43. Ibid., Articles XVI, XVII, XVIII, XX, XXIII–XXIX, and p. 15.

44. Ibid., Articles XXX–XXXII.

45. Ibid., Articles XXXIV and XXXV.

46. Ibid., pp. 1–4, 15; and Article XVI.

47. Ibid., p. 1.

48. Robert F. Scott, *Scott's Last Expedition* (New York, Dodd, Mead, 1913), entry for February 22, 1911, pp. 146 and 147.

49. Argentine Republic, National Report to SCAR for the seasons 1980–81, including plans for the forthcoming seasons; Federal Republic of Germany, National Reports to SCAR for the seasons 1979–80 to 1981–82; and Robert Reinhold, "Antarctic Explorers Shift Goal to Hidden Resources," *New York Times,* December 20, 1981.

50. Author interview with John Behrendt, U.S. Geologic Survey. Also, Mitchell and Tinker, *Antarctica,* p. 89.

51. D. E. Hayes and F. J. Davey, "A Geophysical Study of the Ross Sea, Antarctica," in *Initial Reports of the Deep Sea Drilling Project,* vol. 28, (Washington, D.C., Government Printing Office, 1975) pp. 887–907; and Japan Antarctic Research, Reports to SCAR for 1979, 1980, 1981, and 1982.

52. John C. Behrendt, "Speculations on the Petroleum Resources of Antarctica," in J. Splettstoesser, *Mineral Resource Potential of Antarctica* (Austin, University of Texas Press, in press) pp. 8ff.

53. Author interview with John C. Behrendt, USGS.

54. Deborah Shapley, "Arctic, Antarctic: A Polarization of Interests," in *The Almanac of Sea Power 1983* (Arlington, Va., Navy League of the United States, 1983) pp. 104 and 105.

55. Federal Republic of Germany, Reports to SCAR, on file at Polar Research Board, U.S. National Academy of Sciences.

56. Argentine Republic, National Reports to SCAR on Antarctic Scientific Activities (Buenos Aires, Argentina) for seasons 1980–81, 1981–82, and 1982–83; and Australian Antarctic and Sub-Antarctic Research Programme, Reports to SCAR for the seasons 1979–80, 1980–81, and 1981–82.

Chapter 7

1. Barbara Mitchell and Lee Kimball, "Conflict Over the Cold Continent," *Foreign Policy* no. 35 (Summer) 1979, pp. 124–141.

2. Barbara Mitchell, *Frozen Stakes: The Future of Antarctic Minerals* (London, International Institute for Environment and Development, 1983) p. 44.

3. Author conversation with R. Tucker Scully, director of the Office of Oceans and Polar Affairs, U.S. Department of State.

4. Data from the Office of Public Information, National Science Foundation, Washington, D.C.

5. An article on women's achievements in Antarctica is Mildred Rodgers Crary's, "It's About Time!" (Washington, D.C., The Antarctican Society, 1978) pp. 5 and 6.

6. Katherine Bouton, "A Reporter at Large: South of 60 Degrees South," *The New Yorker,* March 23, 1981, pp. 68 and 78 (hereafter cited as Bouton, "A Reporter").

7. *Antarctica Journal* vol. 16, no. 1 (March) 1981, p. 5; and ibid., p. 75.

8. Bouton, "A Reporter," p. 56. Also, author interview with Katherine Bouton.

9. *U.S. Antarctic Research Program Personnel Manual* (Washington, D.C., National Science Foundation, 1981) p. 34; and see also the 1979 ed., p. 27. Author interview with Guy Guthridge, manager of the Polar Information Program, Division of Polar Programs, National Science Foundation.

10. *Antarctic Journal* vol. 17, no. 5, 1982, pp. 264–270; *U.S. Activities in Antarctica*, Hearing Before the Committee on Energy and Natural Resources, U.S. Senate, 96 Cong., 1 sess., April 23, 1979, Pub. No. 96-21 (Washington, D.C., Government Printing Office) p. 3; *Polar Living Marine Resources Conservation Act of 1978,* Report of the Committee on Merchant Marine and Fisheries, U.S. House of Representatives, 95 Cong., 2 sess., May 15, 1978, p. 6.

11. "Memorandum for the President from the Director, Bureau of the Budget, Executive Office of the President," in Appendix V of Report of Adm. George Dufek to the Secretary of Defense, Aug. 1959. Dufek Papers in George Arents Research Library for Special Collections, University of Syracuse, Syracuse, N.Y. (hereafter cited as Dufek files, Syracuse).

12. "Comments on the Bureau of the Budget's Recommendations of the Organizational Arrangements for Planning and Conducting Antarctic Programs," in Appendix V of Report of Adm. George Dufek to the Secretary of Defense, August 31, 1959. Dufek files, Syracuse.

13. Box 3 of Dufek Papers. Dufek files, Syracuse.

14. *Message From the President: United States Policy and International Cooperation in Antarctica,* Special Report to the Committee on Foreign Affairs, U.S. House of Representatives, 88 Cong., 2 sess., 1964, p. 13 (hereafter cited as *Message of the President*).

15. Undated letter from the president to Alan T. Waterman, director, National Science Foundation, in Appendix V of Report of Adm. George Dufek to the Secretary of Defense, August 31, 1959. Dufek files, Syracuse.

16. "Planning and Conduct of the United States Program for Antarctica," Bureau of the Budget Circular No. A-51, from Maurice H. Stans, Director, to the Heads of Executive Departments and Establishments, August 3, 1960.

17. *Message of the President*, p. 13. *Antarctica Report-1965,* Hearings before the Subcommittee on Territorial and Insular Affairs of the Committee on Interior and Insular Affairs, U.S. House of Representatives, 89 Cong., 1 sess., April 12–13, May 6–7, and June 15, 1965, p. 138.

18. Office of the White House Press Secretary, Press Release, May 1, 1965.

19. James E. Heg, "U.S. Interest and Policy in Antarctica," Office of Polar Programs, National Science Foundation, April 1974, p. 56.

20. "United States Antarctic Policy and Programs," Memorandum from the President, February 5, 1982.

21. J. Merton England, *A Patron for Pure Science* (Washington, D.C., National Science Foundation, 1982) pp. 109, 131, 135, 143, 144, 154, 155, 215, 255–266, 292–297, and 347–351.

22. Undated letter from the president to Alan T. Waterman, Aug. 31, 1959.

23. Press Release, Office of the White House Press Secretary, May 1, 1965; and James E. Heg, "U.S. Interest and Policy in Antarctica," April, 1974, p. 57.

24. Author interview with George Llano, formerly with Office of Polar Programs, National Science Foundation.

25. Barbara Land, *The New Explorers, Women in Antarctica* (New York, Dodd, Mead, 1981) pp. 55–64 (hereafter cited as Land, *New Explorers*). *U.S. Activities in Antarctica*, Hearing before the Committee on Energy and Natural Resources, U.S. Senate, 96 Cong., 1 sess., April 23, 1979, pp. 16, 17, and 35 (hereafter cited as *U.S. Activities in Antarctica*).

26. "Position Paper on Antarctic Krill," in Memorandum from Robert H. Rutford, head of the Office of Polar Programs, National Science Foundation, Oct. 14, 1975, with distribution list.

27. Ibid.

28. "Summary of Antarctic Policy Group (APG) Meetings," from R. Tucker Scully, director of the Office of Oceans and Polar Affairs, U.S. Department of State to author. Author interview with Robert Rutford, former head of the Office of Polar Programs,

National Science Foundation; and with Edward P. Todd, director of the Division of Polar Programs.

29. Charles Neider, *Beyond Cape Horn* (San Franciso, Calif., Sierra Club Books, 1980) pp. 146, 147, and 156 (hereafter cited as Neider, *Beyond Cape Horn*).

30. Mary Alice McWhinnie to R. Tucker Scully, January 20, 1979. Mary Alice McWhinnie to "Everyone," January 30, 1979. Author's files.

31. Author interview with Edward P. Todd, director of the Division of Polar Programs. Author interview with Sayed Z. El-Sayed of Texas A&M University.

32. Author interview with Edward P. Todd.

33. Author interview with Sayed Z. El-Sayed.

34. Osmund Holm-Hansen and Theodore D. Foster, "A Multidisciplinary Study of the Eastern Scotia Sea," *Antarctic Journal* vol. 16, no. 5, 1981, pp. 159–160. Related work aboard the *Melville* in the Antarctic is described on pp. 160–172.

35. *U.S. Activities in Antarctica*, p. 8.

36. *Program Report* vol. 3, no. 6 (September) 1979, pp. 4, 5, 23, and 24.

37. Ibid.

38. *An Evaluation of Antarctic Marine Ecosystem Research*, Report of the Committee to Evaluate Antarctic Marine Ecosystem Research, A Joint Committee of the Polar Research Board and the Ocean Sciences Board, Assembly of Mathematical and Physical Sciences, National Research Council (Washington, D.C., National Academy Press, 1981) pp. vi, xiii, xiv, 15.

39. Ibid.

40. Ibid.

41. Ibid., pp. 4, 44, 45–47, 49, and App. A. See also pp. 65–84 (ice edge study) and pp. 85–94 (swarming study).

42. *Antarctic Journal* vol. 16, no. 3 (September) 1981, pp. 2–5; *Antarctic Journal* vol. 17, nos. 2 and 3, (September) 1982, pp. 1, 4, and 5; and *Antarctic Journal* vol. 17, no. 5, 1982 Review, pp. 166, 167. William M. Hamner, Peggy P. Hamner, Steven W. Strand, Ronald W. Gilmer, "Behavior of Antarctic Krill, *Euphausia superba:* Chemoreception, Feeding, Schooling and, Molting," *Science* vol. 220 (April 22) 1983, pp. 433–435. Author interviews with Peggy Hamner and Donald B. Siniff.

See also Edward P. Todd, Statement before the Subcommittee on Fisheries and Wildlife Conservation and the Environment, Committee on Merchant Marine and Fisheries, U.S. House of Representatives, June 30, 1983 (hereafter cited as Todd statement, June 30, 1983).

43. Testimony of James W. Winchester, Associate Administrator, National Oceanic and Atmospheric Administration before the House Subcommittee on Fisheries and Wildlife Conservation and the Environment, of the Committee on Merchant Marine and Fisheries, June 30, 1983. Author interview with Thomas Laughlin, Office of International Affairs, NOAA.

44. *Research Emphases for the U.S. Antarctic Program*, Report of the Polar Research Board, Commission on Physical Sciences, Mathematics, and Resources, National Research Council (Washington, D.C., National Academy Press, 1983) pp. 2, 7, 22–24, 26, 27, 43, 44, 51, and 52.

45. Deborah Shapley, "Antarctica: World Hunger for Oil Spurs Security Council Review," *Science* vol. 184, May 17, 1974, pp. 776–780; and Joseph O. Fletcher, "U.S. Antarctic Policy," speech dated October 27, 1972, pp. 8 and 9.

46. "Memorandum: Significant Energy R&D Efforts for Supplemental Funding," H. Guyford Stever, director of the NSF, to assistant directors, etc., NSF, August 3, 1973, with attachments.

"Memorandum: Significant Energy R&D Effort in Antarctica for Supplemental Funding" from Joseph O. Fletcher, head of the Office of Polar Programs, to Deputy Assistant

Director, National and International Affairs, National Science Foundation, August 9, 1973, with attachments. Author interview with Joseph O. Fletcher, former head of the Office of Polar Programs, NSF.

47. Author interview with James E. Heg, former acting head of the Office of Polar Programs, NSF.

48. W. O. Bazhaw, manager of the Aquatic Exploration Co., letter to Al Fowler, deputy head of the Office of Polar Programs, Aug. 19, 1975. *U.S. Antarctic Policy,* Hearings before the Subcommittee on Oceans and International Environment, Committee on Foreign Relations, U.S. Senate, 94th Cong., 1 sess., May 15, 1975, p. 18.

49. "International Academic-Industry-Government Geophysical Reconnaissance and Resource Assessment of Antarctic Continental Margins," Gulf Research and Development Company, reproduced and forwarded by A. N. Fowler, acting division director, Division of Polar Programs, January 12, 1979 (hereafter cited as International Academic-Industry-Government proposal).

50. Ibid.; and author interview with Edward P. Todd.

51. See comments on Gulf proposal attached to letter to Edward P. Todd from John C. Behrendt, chairman of the Geologic Division, Antarctic Research Committee, USGS, Jan. 31, 1979.

52. *Program Report,* 1979, p. 27. Dreschhoff's work is described in Land, *New Explorers,* pp. 161–175. Author interview with Gisela Dreschhoff, University of Kansas.

53. Neider, *Cape Horn,* pp. 147 and 160.

54. Author interview with Guy Guthridge. Also, *Antarctic Journal* vol. 15, no. 3 (September) 1980; *Antarctic Journal* vol. 16, no. 4 (December) 1981, p. 14.

55. Neider, *Cape Horn,* pp. 109–125. Author interview with Guy Guthridge.

56. *U.S. Activities in Antarctica,* Hearings, 1979, pp. 43, 44. Office of Technology Assessment, *Technology and Oceanography, An Assessment of Federal Technologies for Oceanographic Research and Monitoring* (Washington, D.C., Government Printing Office, 1981) pp. 47, 50, 53, and 54 (hereafter cited as OTA, *Technology and Oceanography*).

57. Information supplied by Division of Polar Programs, NSF.

58. Steele Report, pp. 36–40. Author interview with Chester C. Langway, Jr., State University of New York at Buffalo, and Richard E. Cameron, program officer in Glaciology, Division of Polar Programs, NSF.

59. Author interview with Terence N. Edgar of the USGS and former chief scientist of the U.S. Deep Sea Drilling Project.

60. Ibid.; also, OTA, *Technology and Oceanography,* pp. 56 and 116–124; Colin Norman, "Ocean Drilling Program Loses Oil Industry Funds," *Science* vol. 214, November 6, 1981, p. 637; and John C. Behrendt, *Petroleum and Mineral Resources of Antarctica,* U.S. Geological Survey Circular 909 (Alexandria, Va., USGS, 1983) p. 1.

61. Robert Reinhold, "Antarctic Explorers Shift Goal to Hidden Resources," *The New York Times,* Dec. 20, 1981; Robert Reinhold, "As Others Seek to Exploit Antarctic, U.S. Takes the Scientific Approach," *The New York Times,* Dec. 21, 1982.

62. Todd statement, June 30, 1983, p. 13.

63. *Program Report of the National Science Foundation* vol. 5, no. 7 (October) 1981, p. 33. Author interview with Guy Guthridge.

64. *Antarctic Journal* vol. 14, no. 5 (October) 1979, pp. 233, 234. Land, *New Explorers,* pp. 214–216.

65. *Antarctic Journal* vol. 14, no. 5 (October) 1979, pp. 98–101.

66. Author interview with Joseph Bennett, head of the Polar Coordination and Information Section, Division of Polar Programs, NSF. Author interview with Kendall Moulton, associate manager of the Polar Operations Section, Division of Polar Programs, NSF.

67. Report by Soviet representative to sixteenth meeting of SCAR, October 1980, Queenstown, New Zealand. On file at the Division of Polar Programs, NSF.

68. The Central Intelligence Agency keeps track of Soviet activities in Antarctica but does not make this information available to the public. Evolution of the Soviet logistical posture in Antarctica may be seen by comparing Soviet stations shown on maps in *Scientific American* vol. 207, no. 3, (September) 1962, pp. 62 and 63 with CIA, *Polar Regions Atlas*, 1978 and 1979 editions.

69. The Visit to Vostok is described in Charles Neider, *Edge of the World: Ross Island, Antarctica* (Garden City, N.Y., Doubleday, 1974) pp. 179–192; Frank Sechrist, "With the Soviets in Antarctica" *Antarctica Journal* vol. 12, nos. 1 (March) and 2 (June), 1977, pp. 4–11. See also Edward S. Grew, "With the Soviets in Antarctica, 1972–1974," *Antarctic Journal* vol. 10, no. 1 (January–February) 1975, pp. 1–8.

The incident of Soviet plane visiting the South Pole is recounted in Message 090853Z, February 1981, from COMNAVSUPFOR ANTARCTICA TO RHHMTRA/COMTHIRDFLEET, on file, Division of Polar Programs, NSF. For one interpretation of Soviet motives, see Jack Anderson, "Soviets Seeking to Displace U.S. at South Pole," *Washington Post*, February 5, 1982.

70. Charles Neider, "Argentina Covets the Antarctic, Too," *The New York Times*, April 9, 1982.

71. Lisle A. Rose, *Assault on Eternity* (Annapolis, Md., Naval Institute Press, 1980), pp. 101–109.

72. Bernard Gwertzman, "U.S. Monitors Signs of Atom Explosion Near South Africa," *The New York Times*, Oct. 26, 1979; Bernard Gwertzman, "Report of A-Blast off Africa Is Called Inconclusive by U.S.," *The New York Times*, Oct. 27, 1979.

73. U.S. Department of State, "Final Environmental Impact Statement on the Negotiation of an International Regime for Antarctic Mineral Resources," 1982, p. x, for example.

74. Author interview with John Garrett, manager of the Business Research Group, Planning Department, Gulf Oil Exploration and Production Co.

75. *Research Emphases for the U.S. Antarctic Program.*

76. "Arctic Research and Policy Act of 1984," in *Congressional Record;* U.S. Congress, April 24, 1984, pp. H2962–2964. The United States Commitment to Arctic Research, Committee on Arctic Research Policy of the Polar Research Board, Assembly of Mathematical and Physical Sciences, National Research Council (Washington, D.C., National Academy Press, 1982).

77. Author interview with Noel Howard, Bureau of Medicine and Surgery, U.S. Navy; and Bouton, "A Reporter at Large," p. 75.

78. Testimony of James W. Winchester, associate administrator, National Oceanic and Atmospheric Administration, before the Subcommittee on Fisheries and Wildlife Conservation and the Environment of the Committee on Merchant Marine and Fisheries, U.S. House of Representatives, June 30, 1983, p. 7.

79. Ibid.

80. "United States Arctic Policy and Arctic Policy Group," NSDM 144, December 22, 1971. "United States Arctic Policy," NSDD 90, April 14, 1983. Author interview with Al Chapman, polar affairs officer, U.S. Department of State.

81. Bouton, "A Reporter at Large," pp. 106–110.

82. Ibid.

83. Ibid., pp. 110, 112, 113, 120, and 121.

Chapter 8

1. James N. Barnes, Speech presented to the Antarctican Society, January 1984. Author's files (hereafter cited as Barnes's speech).

2. Barbara Mitchell, *Frozen Stakes: The Future of Antarctic Minerals* (London, England,

International Institute for Environment and Development, 1933) p. 42 (hereafter cited as Mitchell, *Frozen Stakes*).

3. Ibid.; "Antarctica: Memorandum by the Delegation of Malaysia on Antarctica as a Common Heritage of Mankind," NAC/CONF.7/INF.11, March 3, 1983, Seventh Conference of Heads of State or Government of Non-Aligned Countries, New Delhi, March 1983, p. 3 (hereafter cited as "Antarctica: Memorandum by the Delegation of Malaysia"); and Mahatir bin Mohomad, Speech reported in *The New York Times,* March 9, 1983.

4. "Text of paragraphs on Antarctica in the Non-Aligned Movement Meeting Communique" (adopted March 11, 1983, New Delhi), paragraph XVIII: 87. For Nehru's interest, see note 33 below.

5. For Malaysia, and Antigua and Barbuda difficulties, and the African nations' support, see Barnes's speech. The Peruvian diplomat was Alvaro de Soto, speaking at a 1979 International Institute for Environment and Development Symposium, as quoted in Barbara Mitchell and Richard Sandbrook, *The Management of the Southern Ocean* (London, England, International Institute for Environment and Development, 1980) p. 26 (hereafter cited as Mitchell and Sandbrook, *Management*). Other nonsignatory nations' statements on Antarctica prior to 1980 are quoted in ibid., pp. 25–30.

6. See "Antarctica: A Continent in Transition," prepared by the International Institute for Environment and Development, Washington, D.C., 1983, an information packet distributed at the United Nations Thirty-Eighth General Assembly (hereafter cited as IIED, Packet). For debate in the First Committee, see UN General Assembly, Thirty-Eighth Session, Official Records, Summary Record of the Forty-Second Meeting, A/C.1/38/PV.42, 29 December 1983. These documents contain records of national representatives' statements on the "Question of Antarctica" in the committee sessions.

7. UN General Assembly resolution A/RES/38/77, introduced November 28, 1983, and voted December 15, 1983.

8. Antarctic Treaty, Report of the Twelfth Consultative Meeting, Canberra, Australia, September 13–27, 1983; and Lee Kimball, "Unfreezing International Cooperation in Antarctica," *Christian Science Monitor,* Aug. 1, 1983.

9. Mitchell and Sandbrook, *Management,* p. 29.

10. Author interview with R. Tucker Scully, director of the Office of Oceans and Polar Affairs, U.S. Department of State. Also, Philip W. Quigg, *A Pole Apart,* A Twentieth Century Fund Report (New York, McGraw-Hill, 1983) p. 168 (hereafter referred to as Quigg, *Pole*); and IIED, Packet, "Introduction."

11. Author interview with the documents officer of the Sri Lanka Mission to the United Nations, New York City. For other spokesmen's views, see Mitchell and Sandbrook, *Management,* p. 25; and Quigg, *Pole,* pp. 167 and 168.

12. K. N. Menon, "The Unknown Antarctica," *India News,* April 11, 1983, p. 4; and K. S. Jayaraman, "Expedition to Antarctica," *Indian and Foreign Review,* Feb. 1, 1982, p. 20 (hereafter cited as Jayaraman, "Expedition").

13. Jayaraman, "Expedition," p. 20.

14. Ibid., pp. 20–22; and Deborah Shapley, "India in Antarctica: International Treaty Still on Ice," *Nature* vol. 301, Feb. 3, 1983, p. 362.

15. *Antarctic,* Journal of the Antarctic Section, Division of Scientific and Industrial Research of New Zealand (June) 1980, pp. 62 and 63. Information on Chinese accession is taken from author interview with R. Tucker Scully. Also see *National Geographic Atlas of the World* (5 ed., Washington, D.C., National Geographic Society, 1981) pp. 168 and 169.

16. IIED, Packet, "VI: National Activities, 1982–83 Antarctic Season"; and Deborah Shapley, "Arctic, Antarctic: A Polarization of Interests," *The Almanac of Sea Power, 1983* (Arlington, Va., Navy League of the United States, 1983) p. 105.

17. "Antarctica: Memorandum by the Delegation of Malaysia," note 4.

18. Gerald S. Schatz letter to Charles Warren, chairman of the Council on Environmental Quality, Jan. 10, 1979. Schatz letter to Cyrus R. Vance, secretary of state, Jan. 10,

1979, and M. D. Busby, acting deputy assistant secretary for oceans and fisheries affairs, U.S. Department of State, Letter to Schatz, Feb. 14, 1979. Author's files.

19. David B. H. Denoon, ed., *The New International Economic Order, A U.S. Response* (New York, New York University Press, 1979) pp. 3, 4, 13, 14, and 29.

20. See testimony of Leigh Ratiner in *Law of the Sea Conference,* Hearings before the Subcommittee on Domestic and International Scientific Planning, Analysis and Cooperation, Committee on Science and Technology, U.S. House of Representatives, 95 Cong., 1 sess. (April 26–28, 1977) pp. 42–51. Also see, "Draft Convention on the Law of the Sea (informal text)," Third United Nations Conference on the Law of the Sea, Resumed Ninth Session, Geneva, July 28–Aug. 29, 1980 (Washington, D.C., Department of State, 1980) pt. XI, "The Area," Articles 156–191, pp. 61–80 (hereafter cited as Draft LOS Convention).

21. Draft LOS Convention.

22. Edward E. Honnold, "Thaw in International Law? Rights in Antarctica Under the Law of Common Spaces," *The Yale Law Journal* vol. 87, no. 4 (March) 1978, pp. 804–859; and Edward Honnold, "Draft Provisions of a New International Convention on Antarctica," *Yale Studies in World Public Order* vol. 4 (1977).

23. Honnold, "Thaw," pp. 851–858. See also Ved P. Nanda and William K. Ris, Jr., "The Public Trust Doctrine: A Viable Approach to International Environmental Protection," *Ecology Law Quarterly* vol. 5, 1976, pp. 291–319.

24. See, for example, J. Kish, *The Law of International Spaces* (Leiden, The Netherlands, A. W. Sitjhoff, 1973).

25. "Charter of the United Nations," *Yearbook of the United Nations 1980* (New York, Department of Public Information, United Nations, 1983) vol. 34, App. II, pp. 1353 and 1354.

26. Robert D. Hayton, "The Antarctic Settlement of 1959," *American Journal of International Law* vol. 54, no. 2 (1960) pp. 366, 367, and p. 367 (note 59).

27. Ibid.

28. Alvaro de Soto, quoted in Mitchell and Sandbrook, *Management,* p. 26.

29. Full text of Article XII is found in Appendix A, The Antarctic Treaty.

30. Robert D. Hayton, "The Antarctic Settlement of 1959," *American Journal of International Law* vol. 54, no. 2 (1960).

31. For example, contrast Kish and Hambro with David A. Colson, "The Antarctic Treaty System: The Mineral Issue," *Law and Policy in International Business* vol. 12, no. 4, 1980, pp. 841–902.

32. Roberto E. Guyer, "The Antarctic System," 139 *Recueil de Cours* 150, vol. II, 1976, pp. 154–155.

33. Nehru was quoted in *The Hindustan Times,* May 16, 1958, as cited in Guyer, "The Antarctic System," p. 175 (note 7); see also, K. Ahlowalia, "The Antarctic Treaty: Should India Become a Party to It?," *Indian Journal of International Law* vol. 1, p. 483, as cited in Guyer, "The Antarctic System," p. 175 (note 8).

34. Guyer, "The Antarctic System," p. 224.

35. F. Zegers Santa Cruz, "El Sistema Antarctico y la Utilizacion de los Recursos" (Spanish and English), *University of Miami Law Review* vol. 33, no. 2 (December) 1978, p. 441.

36. Ibid., pp. 471 and 473.

37. R. Tucker Scully, "Alternatives for Cooperation and Institutionalization in Antarctica: Outlook for the 1990's." Paper presented at a conference sponsored by Universidad de Chile, Instituto de Estudies Internacionales, October 1982, Teniente Marsh Station, Chile.

38. "257 Believed Killed as a DC-10 Crashes on Antarctic Peak," *The New York Times,* Nov. 29, 1979; and author interview with Joseph Bennett of the National Science Foundation.

39. F. M. Auburn, *Antarctic Law and Politics* (London, C. Hurst, 1982) pp. 101–104 (hereafter cited as Auburn, *Antarctic Law*).

40. Author interview with John Behrendt of the U.S. Geologic Survey.

41. "Antarctic Minerals Regime, Beeby's Slick Solution," *Eco* vol. 23, no. 1, 1983.

42. Quigg, *Pole*, p. 199.

43. Finn Sollie, "The Political Experiment in Antarctica," in Richard S. Lewis and Philip M. Smith, eds., *Frozen Future* (New York, Quadrangle Books, 1973) p. 61.

44. Draft LOS Convention, "Continental Shelf," Articles 76–82, pp. 31–34.

45. Quigg, *Pole*, pp. 210–213.

46. James N. Barnes, "The Emerging Convention on the Conservation of Antarctic Marine Living Resources: An Attempt to Meet the New Realities of Resource Exploitation in the Southern Ocean," in Jonathan I. Charney, ed., *The New Nationalism and the Use of Common Spaces* (Totowa, N.J., Allanheld Osmun, 1982) p. 269.

47. Ibid., p. 270.

48. Ibid., p. 270, and note on pp. 112 and 113.

49. Gunnar Skagestad, "The Frozen Frontier: Models for International Cooperation," *Cooperation and Conflict* vol. 10, 1975, pp. 167–187. Auburn, *Antarctic Law*, pp. 199 and 200. Also, John Maddox, "Antarctic Mining Regime at Risk," *Nature* vol. 307 (January 12) 1984, pp. 105 and 106. See also "What To Do About Antarctica," *Nature* vol. 301 (February 17) 1983, pp. 551 and 552.

50. For Latin claims, see Robert D. Hayton, *National Interests in Antarctica: An Annotated Bibliography*, Item 1079, p. 76, and Item 558, p. 41.

51. Mitchell and Sandbrook, *Management*, pp. 77, 113, and 114.

52. Ibid., pp. 110 and 111.

53. James H. Zumberge, ed., *Possible Environmental Effects of Mineral Exploration and Exploitation in Antarctica* (Cambridge, England, Scientific Committee on Antarctic Research, 1979) pp. 47–50.

54. Creina Bond and Roy Siegfried, *Antarctica: No Single Country, No Single Sea* (New York, Mayflower Books, 1979) pp. 89, 96, and figs. 54, 55, and 56.

55. Walter Sullivan, *White Land of Adventure* (New York, McGraw-Hill, 1957) p. 71.

56. Bond and Siegfried, *Antarctica*, p. 96.

Sources of Illustrations

From the Collections of the Library of Congress
Page 4, W. Hodges (1777), "The Ice Islands, Seen the 9th of January 1773."
Page 4, H. G. Ponting (1914), "Ramparts of Mount Erebus."
Page 4, Shackleton's ship *Endurance*, 1915.
Page 10, Underwood & Underwood (1922), Roald Amundsen.
Page 10, Amundsen expedition, 1911–1912.
Page 10, Scott's Party at the South Pole.
Page 13, H. G. Ponting (1914), Scott expedition, 1914.
Page 13, H. G. Ponting, Amundsen expedition, 1911–1912.
Page 33, Byrd expedition, 1929.
Page 33, Roald Amundsen, Richard E. Byrd, and Floyd Bennett.
Page 37, Byrd expedition, 1929.
Page 45, Admiral Byrd on Broadway.

From the Collections of the National Archives
Page 13, Byrd's Second expedition.
Page 26, Palmer and von Bellinghausen meet off the South Shetland Islands. From Edmund Fanning, *Voyages Round the World* (New York, Collins and Hannay, 1833).
Page 29, Charles Wilkes. From Charles Wilkes, *Narrative of the U.S. Exploring Expedition*, 1845.
Page 29, Ko-Towatowa. From Charles Wilkes, *Narrative of the U.S. Exploring Expedition*, 1845.
Page 56, Richard E. Byrd.

From the National Science Foundation.
Page 134, Thompson, U.S. Navy, Wright Valley.
Page 174, Russ Kinne, U.S. South Pole station.
Page 193, William R. Curtsinger, The *Hero*.
Page 193, The *Eltanin*.

From the U.S. Coast Guard.
Page 193, The *Northwind*.

Photographs by A. W. Erickson
Page 109, Adélie penguin rookery.
Page 109, Antarctic night.
Page 109, Emperor penguins.
Page 116, Krill.
Page 116, Crabeater seals.
Page 116, Fur seals.
Page 118, Chinstrap penguins.
Page 175, McMurdo station.
Page 221, Drifting ice.
Page 247, Antarctic landscape.

Photographs, courtesy of Panarctic Oils Ltd.
Page 142, Diver.
Page 143, Offshore gas wells.

Photograph, courtesy of Mrs. Finn Ronne
Page 45, Finn Ronne.

Photograph, courtesy of Dome Petroleum Ltd.
Page 143, Offshore oil rig.

Sources of Maps

From the U.S. Central Intelligence Agency, *Polar Regions Atlas* (Washington, D.C., May 1978).

Endpapers, Antarctica
Page 70, Selected historic British expeditions (p. 40).
Page 80, Historic French expeditions (p. 41).
Page 80, Historic Australian expeditions (p. 41).
Page 80, Historic Norwegian expeditions (p. 41).
Page 80, Bellingshausen expedition, 1819–21 (p. 41).
Page 87, IGY Antarctic stations (p. 42).
Page 135, Comparison of Antarctica and the United States (p. 35).
Page 199, U.S. logistics at McMurdo (p. 49).

From Kenneth J. Bertrand, *Americans in Antarctica, 1775–1948,* Special Publication No. 39 (New York, American Geographical Society, 1971).
Page 24, Cruise of the *Hero* (p. 62).
Page 30, Wilkes' voyage (p. 182).
Page 36, First Byrd expedition (p. 294).
Page 39, Second Byrd expedition (p. 238).

From the Polar Information Service of the National Science Foundation.
Page 68, Seven announced claimants to Antarctica.

From the International Institute for Environment and Development, *Southern Ocean Workshop Conference Report* (April 1980).
Page 119, Krill harvests (p. 19, fig. 1).

From A. M. Ziegler, "Paleogeography," in M. W. McElhinny and D. A. Valencio, eds., *Paleoreconstruction of the Continents* (Washington, D.C., American Geophysical Union, 1981).
Page 126, Reconstruction of continental drift (pp. 33–36).

Adapted from I. O. Norton and J. G. Sclater, "A Model for the Evolution of the Indian Ocean and the Breakup of Gondwanaland," *Journal of Geophysical Research* vol. 84, no. B12 (Nov. 10, 1979).
Page 127, Subsequent breakup of Gondwana (pp. 6816, 6819, and 6821–6823).

Adapted from Compbell Craddock, "Tectonic Evolution of the Pacific Margin of Gondwanaland," in K. S. W. Campbell, ed., *Gondwana Geology* (Canberra, Australian National University Press, 1975).
Page 132, Geologic provinces of Antarctica (pp. 609–618).

Adapted from I. W. D. Dalziel and D. H. Elliot, "West Antarctica: Problem Child of Gondwanaland," *Tectonics* vol. 1, p. 14.
Page 138, West Antarctic microplates (p. 14, fig. 5).

From the National Academy of Sciences, "An Evaluation of Antarctic Marine Ecosystem Research" (Steele report) (Washington, D.C., 1981)
Page 152, Krill convention boundaries (p. 14).

From John Bengston, unpublished report to the Marine Mammal Commission, 1978.
Page 157, Effect of krill fishing on Antarctic ecosystem (p. 29, fig. 9).

Index

Numbers in *italics* denote figures; numbers followed by a "t" denote tables.

This book was set in Baskerville text and Benguiat display type by Circle Graphics, Washington, D.C. It was printed on 60 lb. white Mountie opaque stock by French/Bray, Inc., Glen Burnie, Maryland.

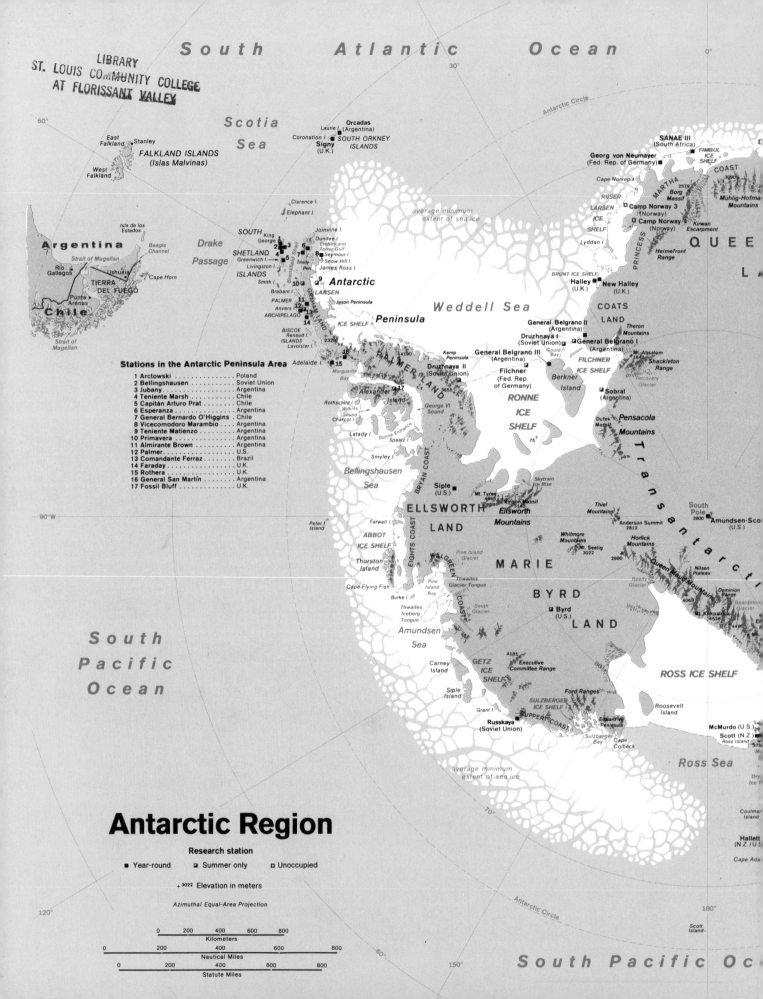

Antarctic Region